GENERAL ZOOLOGY
LABORATORY MANUAL

Fifth Edition

Stephen A. Miller
College of the Ozarks

Boston   Burr Ridge, IL   Dubuque, IA   Madison, WI   New York   San Francisco   St. Louis
Bangkok   Bogotá   Caracas   Kuala Lumpur   Lisbon   London   Madrid   Mexico City
Milan   Montreal   New Delhi   Santiago   Seoul   Singapore   Sydney   Taipei   Toronto

McGraw-Hill Higher Education

A Division of The **McGraw-Hill** *Companies*

GENERAL ZOOLOGY LABORATORY MANUAL, FIFTH EDITION

 This book is printed on recycled, acid-free paper containing 10% postconsumer waste.

9 0 QPD/QPD 0 9 8

ISBN 978-0-07-252837-4
MHID 0-07-252837-0

Executive editor: *Margaret J. Kemp*
Developmental editor: *Donna Nemmers*
Marketing manager: *Heather K. Wagner*
Project manager: *Rose Koos*
Production supervisor: *Sherry L. Kane*
Coordinator of freelance design: *David W. Hash*
Cover designer: *Nathan Bahls*
Senior photo research coordinator: *Lori Hancock*
Photo research: *LouAnn K. Wilson*
Executive producer: *Linda Meehan Avenarius*
Compositor: *GTS Graphics, Inc.*
Typeface: *10/12 Goudy*
Printer: *Quebecor World Dubuque, IA*

The credits section for this book begins on page 335 and is considered an extension of the copyright page.

Contents

UNIT I

Introductory Concepts and Skills

Exercise 1

The Microscope and Scientific Method   3

Exercise 2

Cells and Tissues   21

Exercise 3

Aspects of Cell Function   37

Exercise 4

Genetics   55

Exercise 5

Embryology   71

Exercise 6

Adaptations of Stream Invertebrates—A Scavenger Hunt   81

UNIT II

An Evolutionary Approach to the Animal Phyla

Exercise 7   𝒪K

The Classification of Organisms   99

Exercise 8

Animal-like Protists   109

UNIT III

An Evolutionary Approach to Vertebrate Structure and Function

Laboratory Times

Preface

This *General Zoology Laboratory Manual* is intended for students taking their first course in zoology. Exercises are provided that will help students: (1) understand the general principles that unite zoology with other fields of biology, (2) appreciate the diversity found in the animal kingdom and understand the evolutionary relationships that explain this diversity, (3) become familiar with the structure and function of vertebrate organ systems and appreciate some of the evolutionary changes that took place in the development of those organ systems, and (4) develop problem-solving skills.

The fifth edition of *General Zoology Laboratory Manual* has been carefully edited for clarity and conciseness, and changes have been made that make the laboratory manual more interactive. Relatively minor content changes have been made throughout the laboratory manual to keep the manual up to date. Other changes have been more extensive. For example, Exercise 2, Cells and Tissues, has been updated to reflect recent advances in our knowledge of membrane structure and function. Excercise 17, Chordata, has been updated to reflect current interpretations of higher taxonomy that have resulted from modern cladistic analysis. Other changes reflect a shift in emphasis in response to user feedback. For example, Exercise 15, Arthropoda, has undergone substantial revision. The dissection of the crayfish has been revised, especially coverage of the crayfish circulatory system, to make the dissection more manageable by students. Additional detail has been added to the observations of chelicerate and insect structure.

Pedagogy in the fifth edition has retained popular features from the fourth edition, including the review questions in the form of "Stop and Ask Yourself" boxes, and Student Worksheets, which appear at the end of each exercise. Worksheets are in a tear-out format so they can be used as graded homework assignments. The prelaboratory quiz at the beginning of each exercise helps students evaluate their readiness for the laboratory exercise and can be used by instructors to assess student preparation. Boldfaced type is used in the manual to emphasize important terms for each exercise. These include names of structures and concepts students will encounter on exams. A list of key terms is lo-cated at the end of each exercise and highlights important new terms and concepts introduced in each excercise. Activities to be carried out by students are highlighted by icons at the beginning (▼) and end (▲) of a set of instructions. In addition, the nineteen black-and-white plates of dissected specimens included in the fourth edition have been retained in this edition.

ORGANIZATION

Unit I contains exercises that focus on general biological principles, from cell structure and function to ecological and evolutionary principles. Exercise 1 includes an introduction to microscopes and scientific method. Problem-solving activities have been incorporated into exercises 1, 3, 4, and 6.

Unit II is a survey of the animal phyla. Exercise 7 is an introduction to taxonomy and cladistics. A brief "evolutionary perspective" begins other exercises in this section and describes the position of the phylum in question within the animal kingdom. Each exercise includes coverage of class representatives, and some activities that illustrate processes unique to the particular animal group. Most of these exercises conclude with a discussion of evolutionary relationships within the phylum studied and an activity involving cladistic analysis of these relationships.

Unit III is a systematic approach to the study of vertebrate structure, function, and evolution. The evolutionary orientation of unit III helps the student understand the changes that occurred in the evolution of vertebrate organ systems. Many activities ask the student to think critically about the adaptive significance of what is being observed. The shark is used in demonstration dissections, and the rat is the primary specimen for student dissection. Other vertebrate representatives including *Necturus*, the frog, and a bird are used to illustrate adaptations in other vertebrate classes.

This manual includes more than can be covered in a one-semester course, which allows some flexibility in course planning based on the preferences of individual instructors.

For example, some instructors have a very brief survey of the animal phyla in their labs and focus on units I and III. These instructors select a few activities in each of the exercises in unit II to survey the animal phyla in a few laboratory periods. Other instructors may omit most of unit I if the students had general biological principles in another course.

An Instructor's Resource Guide is available for all adopters of the fifth edition. It has been placed on the McGraw-Hill website and is available at www.mhhe.com/zoology (click on the cover of Miller and Harley's Zoology text). This guide lists all materials required for preparing each laboratory exercise. It also gives formulas for solutions and media, and includes suggestions for incorporating living material into the laboratory. Answers to the "Prelaboratory Quiz," "Stop and Ask Yourself" questions, and "Student Worksheet" questions are also included in the resource guide.

ACKNOWLEDGMENTS

I would like to thank those individuals who helped to make this laboratory manual possible. Many comments from users of past editions were most helpful in preparation of each new edition. The comments and suggestions of the following reviewers were most helpful:

Barbara J. Abraham   *Hampton University*
Sarah Cooper   *Beaver College*
Tom Dale   *Kirtland Community College*
Inez Devlin-Kelly   *Bakersfield College*
Peter K. Ducey   *State University of New York at Cortland*
DuWayne Englert   *Southern Illinois University*
Dennis Englin   *The Masters College*
Nels H. Granholm   *South Dakota State University*
W. E. Hamilton   *Penn State University*
Michael Hartmann   *Haywood Community College*
Kenneth D. Hoover   *Jacksonville University*
Barbara Hunnicutt   *Seminole Community College*
Karen E. McCracken   *Defiance College*
Thomas C. Moon   *California University of Pennsylvania*
John H. Roese   *Lake Superior State University*
Richard G. Rose   *West Valley College*
Fred H. Schindler   *Indian Hills Community College*

The writing of this laboratory manual would have been impossible without the help of the McGraw-Hill editors and their staff. The advice of Marge Kemp and Donna Nemmers has proven extremely valuable. The Project Manager for the fifth edition of *General Zoology Laboratory Manual* was Rose Koos. Her ability to keep the manuscript moving through all phases of production is greatly appreciated. I am also grateful for the support of Carol, Trey, Jeremy, Eric, and Bryan during this revision and for their understanding when this project encroached upon family time.

Stephen A. Miller, Ph.D.

Getting Started

The following section contains terms and concepts that will be used throughout this laboratory manual. Although most of these terms will be redefined within the exercises that follow, it is helpful to have some knowledge of them before beginning. The metric system is used throughout this laboratory manual. You must be familiar with metric units and know how to carry out conversions between larger and smaller units of measurement. The Terms of Direction and Body Planes and Sections are used frequently in units II and III for locating the parts of an animal and indicating how an animal or a structure is to be cut in a dissection. This information is located at the beginning of the laboratory manual for convenient, later reference.

The Metric System

Scientific measurements are usually carried out using the metric system. Metric units are used exclusively in this laboratory manual. The main advantage of the metric system is the convenience that results from its decimal nature. With the exception of units of time, units for measuring a given quantity are derived from a basic unit by multiplying or dividing by multiples of ten. The fundamental units of the metric system are

liter, a unit of capacity;
meter, a unit of length;
gram, a unit of mass;
second, a unit of time.

Units larger or smaller than these are formed by adding prefixes such as kilo (1,000), centi (1/100), and milli (1/1,000). Thus, kilometer = 1,000 meters, centimeter = 1/100 of a meter, and millimeter = 1/1,000 of a meter.

Other units used in measuring microscopic distances are the following:

The micrometer (μm) = 1/1,000,000 of a meter or
1×10^{-6} m.
The nanometer (nm) = 1/1,000,000,000 of a meter or
1×10^{-9} m.
The angstrom (Å) = 1/10,000,000,000 of a meter or
1×10^{-10} m.

Some commonly used units of conversion between metric and the English systems are given below.

English	Metric
1 mile	1.6 kilometers
1 inch	2.54 centimeters
0.39 inch	1 centimeter
1 ounce	28 grams
0.035 ounce	1 gram
1 fluid ounce	30 milliliters
0.033 fluid ounce	1 milliliter
1 pound	0.45 kilogram
2.2 pounds	1 kilogram

Microscope and Seat Assignment

Seat number _____
Microscope number _____

Microscope calibrations using an ocular micrometer (from exercise 1):

Low power (10 × objective). 1 ocular unit = _____ μm.
High power (40 × objective). 1 ocular unit = _____ μm.

Estimation of the diameter of the field of view (from exercise 1):

Low power (10 × objective). Diameter = approximately
_____ μm.
High power (40 × objective). Diameter = approximately
_____ μm.

Terms of Direction, Symmetry, and Body Planes and Sections

Terms of Direction
Terms of direction are used for locating body parts relative to a point of reference.

Aboral   The end opposite the mouth. Opposite—oral.
Anterior   The head end. Usually the end of a bilateral animal that meets its environment. Opposite—posterior.

Caudal   Toward the tail. Opposite—cephalic.

Cephalic   Toward the head. Opposite—caudal.

Distal   Away from the point of attachment of a structure on the body. (e.g., The toes are distal to the knee.) Opposite—proximal.

Dorsal   The back of an animal. Usually the upper surface. For animals that walk upright, dorsal and posterior are synonymous. Opposite—ventral.

Inferior   Below a point of reference. Opposite—superior.

Lateral   Away from the midsagittal plane of the body. Opposite—medial.

Medial (median)   On or near the plane of the body that divides a bilaterally symmetrical animal into mirror images. Opposite—lateral.

Oral   The end containing the mouth. Opposite—aboral.

Posterior   The tail end. Opposite—anterior.

Proximal   Toward the point of attachment of a structure on the body. (e.g., The hip is proximal to the knee.) Opposite—distal.

Superior   Above a point of reference. Opposite—inferior.

Ventral   The belly of an animal, usually the lower surface. For animals that walk upright, ventral and anterior are synonymous. Opposite—dorsal.

Symmetry

Symmetry describes how the parts of an organism are arranged around an axis or a point (fig. .001). The following three categories are most common.

Asymmetry   The arrangement of body parts without a central axis or point (e.g., the sponges).

Bilateral symmetry   The arrangement of body parts such that a single plane passing dorsoventrally through the longitudinal axis divides the animal into right and left mirror images (e.g., vertebrates).

Radial symmetry   The arrangement of body parts such that any plane passing through the oral—aboral axis divides the animal into mirror images (e.g., the cnidarians). Radial symmetry can be modified by the arrangement of some structures in pairs, or in other combinations, around the central axis (e.g., biradial symmetry in the ctenophorans and pentaradial symmetry in the echinoderms).

Body Planes and Sections

References to body planes are used to indicate the position of some structure relative to an imaginary plane passing through the body. References to body, organ, and tissue sections indicate how something is cut to observe internal structures.

Cross section   A section cut perpendicular to the longitudinal axis of a structure.

Frontal   A plane or section perpendicular to both sagittal and transverse planes. Divides an animal into dorsal and ventral regions.

Longitudinal section   A section cut parallel to the longitudinal axis of a structure.

Median   A plane or section passing through the longitudinal axis of a bilateral animal. Divides the animal into halves that are mirror images of each other.

Sagittal   A plane or section passing dorsoventrally through the body of a bilateral animal parallel to the longitudinal axis.

Transverse   A plane or section perpendicular to both sagittal and frontal planes. Divides an animal into anterior and posterior regions.

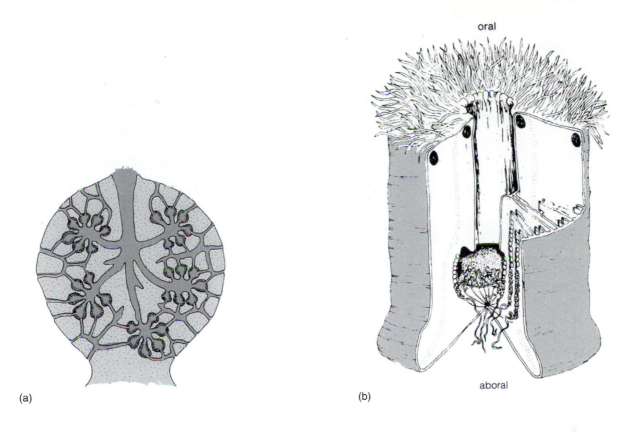

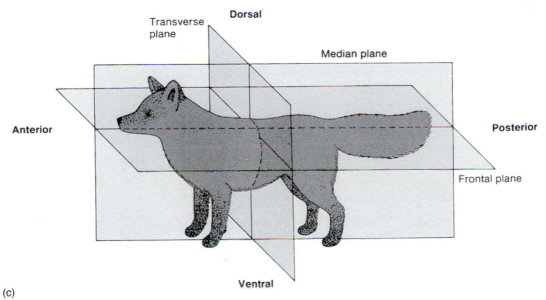

Figure .001 Symmetry, body planes, and terms of direction: (*a*) asymmetry, leucon sponge; (*b*) radial symmetry, a sea anemone; (*c*) bilateral symmetry, a vertebrate.

Introductory Concepts and Skills

Certain basic concepts unite the diverse subdisciplines within biology. All biologists must have a working knowledge of these important concepts. For example, outside of evolution, modern biology loses much of its meaning and excitement. For this reason, units II and III emphasize evolutionary relationships. Similarly, the modern biologist cannot function without some background in cell and tissue structure and function, genetics, embryology, and ecology. Unit I stresses these unifying concepts. All of these topics set the stage for material that is to come in units II and III. As you proceed through this course, you will find that what you learned in unit I is built upon and emphasized in different ways.

Exercise 1

The Microscope and Scientific Method

Learning Objectives

After completing this exercise, you should be able to do the following:

1. Identify and describe the functions of parts of the compound and dissecting microscopes.
2. Focus on objects under low and high power of the compound microscope.
3. Care for the compound and dissecting microscopes.
4. Make a wet mount.
5. Estimate the size of objects viewed under the compound microscope.
6. Describe the scientific method and define the following terms: hypothesis, scientific theory, and control group.
7. Identify three kinds of variables found in scientific experiments.
8. Investigate a question using scientific method.

Prelaboratory Quiz

Study this week's laboratory exercise and then complete the following quiz to assess your preparation for the laboratory.

The Microscope

1. During this week's laboratory, you will be learning to use compound and dissecting microscopes. Which of these two microscopes would you use to examine the whole-body structure of a honeybee?
2. The ability of a microscope to discern small objects that are very close together or to discern detail in an object is called
   a. total magnification.
   b. parfocal.
   c. resolution.
   d. discernment.
3. Which of the following parts of a compound microscope is adjusted to regulate the amount of light coming through the specimen?
   a. objective lens
   b. diaphragm
   c. condenser lens
   d. tube
4. Which of the following parts of a compound microscope is nearest the observer's eye when using the microscope?
   a. objective lens
   b. revolving nosepiece
   c. base
   d. ocular lens
5. The low-power objective of a compound microscope has a magnification of
   a. 4×.
   b. 10×.
   c. 40×.
   d. 100×.

6. Which of the following objective lenses should be used when beginning to focus a compound microscope?
   a. 4×
   b. 10×
   c. 40×
   d. the lowest power objective present on the microscope

7. A _____ is a small, nonunit scale in the eyepiece of a microscope that can be calibrated and used for measuring the size of microscopic objects.
   a. ocular micrometer
   b. stage micrometer
   c. objective micrometer
   d. focal micrometer

8. In this week's laboratory you will be making a temporary preparation that is useful in viewing fresh material. This temporary preparation is called a
   a. dry mount.
   b. wet mount.
   c. hanging mount.
   d. bubble mount.

Scientific Method

9. Define "science."

   _____

   _____

   _____

10. A generalized statement that has predictive value and that is supported by years of testing is called a(n)
    a. hypothesis.
    b. scientific theory.
    c. educated guess.
    d. postulate.

11. Parameters that could influence the results of an experiment and that must be held constant during the experiment are called
    a. dependent variables.
    b. controlled variables.
    c. independent variables.
    d. contingent variables.

12. The parameters that are monitored during an experiment while gathering data are called
    a. dependent variables.
    b. controlled variables.
    c. independent variables.
    d. contingent variables.

13. A test of a hypothesis should include all of the following EXCEPT
    a. replicates.
    b. a single controlled variable.
    c. a single independent variable.
    d. a control group.

14. "If/then statements" are used to
    a. reformulate a hypothesis that was proven wrong.
    b. predict the outcome of an experiment based on the hypothesis and the experimental methods.
    c. design a test of the hypothesis.
    d. formulate an initial hypothesis.

15. Replication of a scientific experiment is important because

    _____

    _____

    _____

    _____

    _____

The microscope is a tool that all biologists must be able to use correctly. The microscope enables biologists to extend their vision to a resolution of fractions of a micrometer. This degree of resolution requires sophisticated electron microscopes, not generally used by undergraduate students. Lower magnifications, however, are just as useful for most aspects of animal biology. Magnifications ranging from 5× to 1,000× will be commonly used in the studies that follow. All good-quality microscopes are expensive. Proper care and use of the microscope is, therefore, essential. The purpose of this exercise is to familiarize the student with the use and care of the two microscopes most frequently used in this course: the compound microscope and the dissecting microscope.

THE COMPOUND MICROSCOPE

The compound light microscope is used to examine details of the cellular and tissue structure of animals. The usefulness of a light microscope is limited not by its ability to magnify objects, but by its ability to discern small objects that are very close together. This is called **resolution** and is related to the wavelength of visible light used to illuminate an object. Shorter wavelengths of light (toward the blue end of the spectrum of visible light) are scattered less than longer wavelengths as they pass through material being viewed and, thus, distort the image less. Using short wavelengths of visible light, compound microscopes can resolve two points that are about 0.2 μm apart. Two points closer together than this appear as one point regardless of the magnification used. Resolution refers to the sharpness of the image. Electron microscopes are used when greater resolution is required.

As you work through this exercise, keep in mind that, although basically similar, microscopes vary depending upon make and quality. Expect that your microscope will differ in some details from the description that follows. Your instructor will point out these differences.

The compound microscope that is used in most zoology laboratories contains at least two lenses. These lenses bend (refract) light rays into one focal point—the eye(s) (fig. 1.1a). The **objective lens system** is nearest the object; the

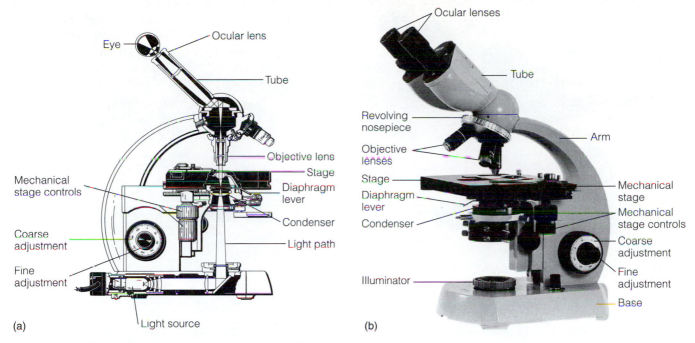

Figure 1.1 The compound microscope: (*a*) focusing of light rays; (*b*) parts of a microscope.

ocular lens system is nearest the eye. The total magnification of an object is the product of the two lens systems.

Ocular lens system	10×	10×
Objective lens system	10×	40×
Total magnification	100×	400×

▼ A microscope must be handled very carefully at all times. Always carry it upright with one hand under the base and the other hand around the arm. Obtain a microscope as directed by your instructor and place it on the bench in front of you. Unwind just enough of the cord to reach the outlet. Do not let extra cord hang over the edge of the bench. Plug in the microscope. Find and learn the functions of the microscope parts described in figure 1.1b and table 1.1. Compare your microscope to the variations shown in figure 1.2. ▲

TABLE 1.1   The Compound Microscope Parts and Their Function

Ocular Lens	Eyepiece containing a lens normally with a 10× magnification (usually stamped on the rim of the eyepiece). Often contains a pointer to aid in locating objects within the field in view.
Tube	Section between ocular lens and objective lenses. Contains system of lenses and mirrors.
Revolving Nosepiece	Holds objective lenses.
Objective Lenses	Lenses closest to the specimen. Usually stamped as follows: Scanning—4× Low power—10× High power—40× Oil immersion—100×*
Arm	Support running from base to tube.
Stage	Platform on which slide is positioned. Note the manner in which the slide is held in place.
Condenser	Lens just under the stage and the specimen. Concentrates light under the specimen.
Diaphragm	Controls the amount of light passing through the specimen and controls contrast and resolution of the image. Opens and closes by sliding a lever or turning a disc.
Base	Platform on which microscope rests. Contains either built-in light source or a mirror.
Coarse Adjustment	Larger knob on both sides of the arm. Objective lenses or stage move perceptibly when knobs are turned.
Fine Adjustment	Smaller knobs on both sides of the arm. Movement of the lenses or stage is nearly imperceptible unless looking through ocular.

*This lens, if present, should be used only with instructions from your instructor. When in use, the lens is immersed in a drop of oil that prevents the refraction of light between the microscope slide and lens. Refraction of light is a problem because of the small working distance (distance between the focal point of the objective lens and the specimen). Some immersion oil may be carcinogenic and should be handled with care.

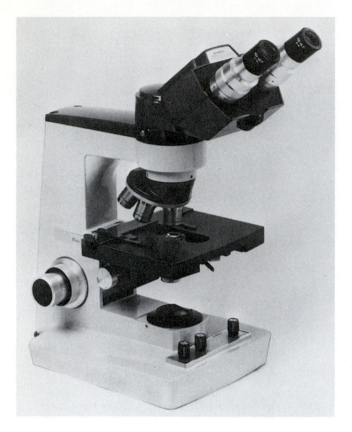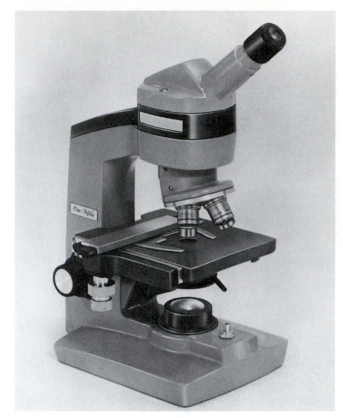

Figure 1.2 Variations in the compound microscope: *(a)* binocular; *(b)* turret nosepiece monocular with mechanical stage.

Care of the Microscope

Following is a list of precautions to observe when using the microscope. Become thoroughly familiar with them. They apply to both compound and dissecting microscopes.

1. Report any damage or malfunction immediately.
2. Carry the microscope carefully with one hand under the base and the other around the arm.
3. Do not remove the lenses from the microscope. If they are dirty inside, advise the instructor.
4. Clean the lenses with lens paper only. If gentle rubbing with lens paper does not adequately clean the lens, notify your instructor.
5. If anything is spilled on the microscope, clean it off immediately; if it is anything other than water, inform the instructor.
6. Unwind just enough cord to reach the outlet.
7. Before putting the microscope away, check for the following:
   a. Are the lenses clean?
   b. Is the low-power lens in position?
   c. Is the coarse focus knob turned down to its lower limit?
   d. Is the stage bare?
   e. Is the dustcover in place?

Use of the Compound Microscope

▼

1. Obtain a slide (letter slide or similar preparation) from your instructor. Clean and examine it visually to locate the position of the object on the slide (holding it up to a light may help).
2. Position the slide on the stage of the microscope so the object to be viewed is centered over the condenser.
3. Check to be sure the light is on, the low-power lens is in position, and the iris diaphragm is set at a medium opening. If your microscope has a mirror rather than a built-in light source, adjust the mirror to get a bright, evenly illuminated field of view. Use the concave surface if no condenser is present, the flat surface if a condenser is present. If your microscope has a voltage control knob, adjust this knob to a medium setting.
4. Looking from the side of the microscope, lower the objective (or raise the stage) until the objective reaches its lower (or upper in the case of the stage) limit, or until the objective is just about to touch the slide.
5. Now look through the microscope and focus by turning the coarse adjustment so that the slide and the lens move away from each other. Stop when the object comes into focus. Note: always focus initially using low power. If a voltage control knob is present, adjust the

light to a comfortable intensity. Adjust the iris diaphragm to achieve best contrast and resolution. If the iris diaphragm is closed too much, contrast will be high but resolution will be poor. Remember that thick specimens require more light than thin specimens.

6. Move the slide first to the left, then the right. Move the slide toward you, then away from you. What do you notice? _____

   Compare the orientation of the object on the stage with the orientation of the image. What do you notice? _____

7. With the object in focus and centered in the field of view, turn the nosepiece until the high-power objective clicks into position. Note that the object is still in focus or nearly so. Minor adjustments may be necessary. A microscope is said to be **parfocal** if it can be switched from low to high power and remain in focus. **Never focus with the coarse adjustment when the high-power objective is in position.** In switching from low power to high power, what do you notice about the brightness of the field of view? _____

   _____

   Why is it important to center the object in the field of view before switching to high power? _____

   _____

   _____ ▲

Estimating the Size of Microscopic Objects

It is useful to know the size of objects viewed in the microscope. A common technique to estimate size is to measure the diameter of the field of view and then to estimate the proportion of the field that the object occupies. From those figures, an approximate size can be obtained. The diameter of the low-power field can be measured directly using a millimeter rule or a stage micrometer.

▼ Place a millimeter ruler on the stage of your microscope and focus on the scale. Position the edge of the ruler across the diameter of the field of view with the 0 end of the millimeter scale at one side of the field of view. What is the diameter of the field of view of your microscope in millimeters?

_____

This value usually ranges between 1.5 and 1.8 mm. If an object takes up one half of a 1.6 mm field of view, what would be its approximate size? _____ ▲

The diameter of the high-power field cannot be measured directly because of its small size. It can be calculated as follows:

$$\frac{\text{Magnification of low-power objective}}{\text{Magnification of high-power objective}}$$

$$\times \text{ Diameter of low-power field}$$

▼ What is the diameter of the high-power field of view of your microscope? _____
This value usually ranges between 0.35 and 0.42 mm (350 and 420 μm). ▲

The Ocular Micrometer

More precise measurements can be made using an **ocular micrometer.** An ocular micrometer is a thin circle of glass or plastic etched with a nonunit scale usually ranging from 0 to 100. When placed in the ocular lens of the compound microscope, the scale image is superimposed over the image of the specimen. A stage micrometer is a glass slide etched with a scale of known units, usually divided into 0.01 mm and 0.1 mm graduations.

▼ The ocular micrometer is calibrated using the stage micrometer by aligning the images at the left edge of the scales (fig. 1.3). Calculate the distance between a given number of divisions on the ocular micrometer and substitute those values into the following.

1 ocular micrometer space = Distance between selected graduations on the ocular micrometer (from the stage micrometer)
                          ─────────────────────
                          Number of spaces measured on the ocular micrometer

= _____ mm

= _____ micrometers
  (1 μm = 1/1,000 mm)

This should be done for each objective lens and recorded under "Microscope and Seat Assignment" in the "Getting Started" section at the beginning of this manual. Assuming that one uses the same microscope in subsequent labs, these values can be used for measuring microscopic objects. Individual microscopes (even those of the same make and model) will vary with regard to these values and must be calibrated individually. ▲

Making a Wet Mount

The **wet mount** is a temporary preparation that is useful in viewing fresh material. A specimen is placed in a drop of water on a slide, then covered with a thin piece of glass or plastic called a coverslip (fig. 1.4). This technique keeps fresh material moist during observation.

▼ After adding the specimen to a drop of water on a clean slide, place one edge of a coverslip on the slide next to the drop. Incline the coverslip at a 45° angle and lower it slowly until the coverslip touches the water drop and rests on the slide. The water now forms a thin sheet under the "floating" coverslip. This procedure should eliminate most air bubbles that might obscure the specimen.

   Make a wet mount of a small length of hair. Estimate its diameter under low and high power.

Low power _____
High power _____

Which figure is more accurate? _____
Why? _____

_____

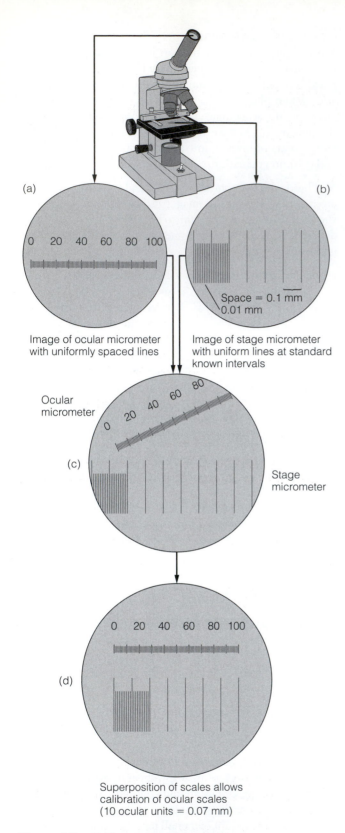

0 20 40 60 80 100

(a)

(b)

Space = 0.1 mm
0.01 mm

Image of ocular micrometer with uniformly spaced lines

Image of stage micrometer with uniform lines at standard known intervals

Ocular micrometer

0 20 40 60 80

0 20 40 60

(c)

Stage micrometer

0 20 40 60 80 100

(d)

Superposition of scales allows calibration of ocular scales (10 ocular units = 0.07 mm)

Figure 1.3 Calibrating an ocular micrometer.

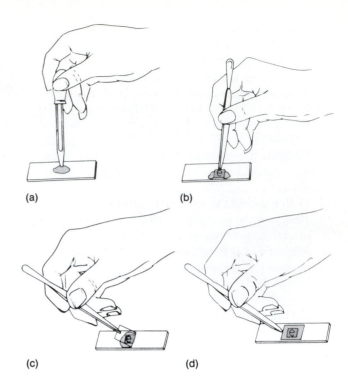

(a)

(b)

(c)

(d)

Figure 1.4 Making a wet mount.

Look for any air bubbles. What do they look like under the microscope? _____ ▲

Other Exercises

▼

1. Practice making wet mounts and estimating sizes using drops of water from a sample of pond water.
2. Make a wet mount of two crossed hairs or threads of different colors. Using high power, focus on the regions where the hairs or threads cross. When focusing down with the fine adjustment, notice that the hair or thread on top will come into focus first. Can you tell which is on top? This exercise demonstrates that focusing up and down through a relatively thick specimen can give one an appreciation of the three-dimensional structure of a specimen. ▲

STOP AND ASK YOURSELF

1. Locate each of the following on your compound microscope and briefly describe its function.

   a. ocular lens _____

   _____

   b. stage _____

   _____

   c. condenser lens _____

   _____

d. diaphragm _____

_____

2. What is the total magnification of a microscope that has a 10× ocular lens and a 20× objective lens? _____

3. What is the correct procedure for carrying the microscope? _____

_____

4. What is an ocular micrometer? _____

_____

_____

_____

THE DISSECTING MICROSCOPE

The dissecting microscope is useful when magnifications between 5× and 50× are desired. It is useful for viewing entire specimens of small animals or body parts of larger animals. It has the added advantage of giving a three-dimensional view of the object.

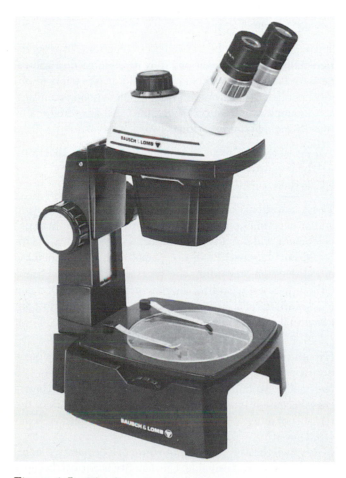

Figure 1.5 The dissecting microscope.

Use of the Dissecting Microscope

All precautions and rules that applied to the compound microscope also apply to the dissecting microscope. Some dissecting microscopes have built-in light sources; others do not. Obtain a dissecting microscope and light source. Note that the microscope is binocular and has a single focusing knob and a lens system that allows you to change magnifications by rotating a dial or revolving a nosepiece (fig. 1.5).

▼ Your instructor has provided some materials for viewing under the dissection microscope. Adjust the eyepieces to fit the width of your eyes. (Some eyepieces can be adjusted for slight differences in each eye. Your instructor will describe those procedures if they apply to your microscope.) Examine your specimen under low and high magnifications. Note the three-dimensional image. Compare the orientation of the image to the orientation of the specimen on the stage. Move the specimen to the right and left. What do you notice?

_____

The stage of the microscope is usually equipped with a plate that can be reversed to provide either a white or black background. Use the background that will provide better contrast. Try switching back and forth.

Living and preserved specimens are best viewed while completely immersed in water. This prevents a distorted image due to the refraction of light off moist surfaces. ▲

STOP AND ASK YOURSELF

5. How is an image seen in the compound microscope different from that seen in the dissecting microscope?

_____

_____

6. How does one improve contrast when using a dissecting microscope? _____

_____

_____

AN INTRODUCTION TO SCIENTIFIC METHOD

The purpose of this exercise is to help you to understand that science is a process, not a body of facts to be memorized in a classroom. After completing the activities described in this section, you should understand that scientists use their senses, or technological extensions of their senses, in investigating natural occurrences. You should also become familiar with scientific method and use that method to carry out an experiment that requires critical thinking, prediction of results, collection and organization of data, and analysis of data.

Science is an objective process used to investigate natural occurrences. It is knowledge gained by observation. The goal of science is often said to be the description of theories. A theory in the scientific context is different from the everyday use of the term to mean a guess. Theories are not developed as a result of any one experiment or, as commonly portrayed, a hypothesis that has been supported. Instead, **scientific theories** are generalized statements that have predictive value. These statements are supported by many experiments, each testing one small aspect of a larger concept. In other words, a theory is a concept that has usually undergone years of testing and has been supported again and again. It is a concept that is as certain as anything can be in the realm of science. The uncertainty implied in the word "theory" reflects the fact that scientists should never imply that they know everything there is to know about any concept. New technologies, ideas, and experiments may bring new information to light that modifies a scientist's understanding of the concept. Therefore, details of a theory may change; however, the essence of a theory is unlikely to change. A theory is thought to accurately reflect how things happened in the past and how they will occur in the future.

In order for our concept of science to be valid, scientists must make three assumptions that are impossible to prove but seem to be true based upon experience. (1) Nature is understandable. Scientists are not perfect in their observations, but the work of science involves testing and retesting by individuals working independently or together. This retesting eventually illuminates errors and helps to improve our understanding of natural events. (2) Nature is real. What we observe with our senses in the laboratory or in field studies is reality. (3) Cause-and-effect relationships are consistent in nature. This concept is sometimes called uniformitarianism. It means that the processes that shaped our natural world in the past are the same as those occurring now and the same as those that will occur in the future. Stated another way, an experiment that we performed yesterday and repeated today should have produced the same results both days. The results should also be the same if the experiment is performed a year from now. It is this latter concept that allows scientists and our students to make reliable predictions regarding the outcome of an experiment.

Science is not without its limits. It is important that students realize that science is a process that can be used to investigate natural events. In order to investigate anything scientifically, scientists use their senses to weigh, measure, or otherwise observe an occurrence. Scientists often use microscopes and electronic devices in their work; however, these instruments are extensions of scientists' senses. Science cannot be used to investigate the supernatural. Many problems arise when individuals try to use scientific methods to study matters of the spirit. Similarly, science is ethically neutral. Science is not good or bad, moral or immoral. Scientists have no special authority when it comes to making ethical decisions. We can, however, all benefit from the information presented by scientists when we make ethical decisions.

Scientific Methods

Scientists use a variety of methods when investigating natural events. Even though experimentation and the scientific method are fundamental scientific approaches to solving problems, many questions are answered by careful, detailed observation. Detailed observations are fundamental to the work of a biologist studying the mating behavior of a bird or the structure of the genetic material (DNA). Without careful observations, we would know far less about the world around us. The following exercise stresses both careful observation and the use of the scientific method.

The Scientific Method

The **scientific method** is a series of steps that reflects a logical approach to solving problems in our physical world. It is a series of steps that we often use without realizing it.

Questions Scientists usually begin with a question or a set of questions based on observations, previous research, reading, or discussions with other scientists. They are questions that can be answered using our senses or extensions of our senses. Examples of questions that can be answered using scientific method include the following:

1. How many eggs does the average robin lay during its reproductive season?
2. What is the effect of acid deposition on a pond ecosystem?

These questions can be answered by some form of observation and/or experimentation. It is possible to find robins' nests and count the number of eggs laid. It is possible to observe ponds and experimentally manipulate pond organisms.

Examples of questions that cannot be answered scientifically include the following:

1. At what stage of development, if any, should abortions be permitted?
2. Is there a god?

These questions cannot be answered with observation and/or experimentation. Scientists can describe human development and provide information on the viability of embryos, but scientists have no special authority in making moral decisions regarding the practice of abortion. Similarly, a question regarding the presence or absence of a god is a supernatural query. Science cannot be used to test supernatural events.

Hypothesis Scientists begin the process of answering questions by posing plausible explanations. A **hypothesis** is a reasonable explanation for a question. It is more than a guess because it is usually supported by previous work of the scientist or others. A hypothesis is sometimes called an "educated guess."

A hypothesis is stated in the form of a declarative statement and must be falsifiable. Falsifiability is critical, because it is possible to prove a hypothesis false with a single experiment. On the other hand, one cannot prove a hypothesis to be true. A hypothesis may be supported by the results of a particular experiment, but new evidence or another experiment with slightly different conditions might prove it to be false. For example, suppose an experimenter formulated the hypothesis that nutrient A enhanced the regeneration of *Dugesia* (planaria). She carried out an experiment under conditions specified by her experimental protocol and found results consistent with her hypothesis. Did she prove her hypothesis to be true? What if she repeats the experiment under slightly different conditions of light, temperature, and moisture? It is possible that she could prove the hypothesis false. Her hypothesis was supported by the first experiment but may be disproved by the second experiment. Now the scientist can return to the hypothesis and modify it based upon the new information provided by the second experiment. The new hypothesis is then tested. The support for a hypothesis grows increasingly strong with repeated experimentation (often over many years), but the hypothesis is never proven true.

If a scientist is not careful, it is easy to feel ownership of, and attachment to, a particular hypothesis. A hypothesis should not be viewed as something that one wants to confirm. A hypothesis that is proven false provides as much (sometimes more) information as one that is supported. Scientists must avoid becoming personally attached to a hypothesis—a mind-set that leads to blinding oneself to contradictory data.

A Scientific Experiment—The Test of the Hypothesis

All scientific experiments have sets of variables tested, manipulated, or controlled. The **dependent variable** is the parameter that the scientist is measuring. It is what will be affected during the experiment. In many experiments, scientists must monitor more than one dependent variable. The **independent variable** is the parameter being tested. It is what the scientist varies. In any experiment, there must be only one independent variable. When there is only one independent variable, any changes in the dependent variable(s) must be the result of the scientist's manipulation of the independent variable. All other variables that might influence the outcome of an experiment are called **controlled variables.** Controlled variables are held constant in an experiment so that they do not affect the outcome. For example, a student plans an experiment to test the hypothesis that *Dugesia* are negatively geotactic, that is, *Dugesia* move away from the direction of the pull of gravity. In this experiment, gravitation is the independent variable, and the response of *Dugesia* (movement relative to the direction of the pull of gravity) is the dependent variable. Controlled variables would include light intensity, water temperature, and others.

After the hypothesis is formulated and the variables are identified, an experimental procedure is outlined. An experimental procedure is often based upon the work of others,

consultation with peers, and a scientist's own creativity. The level of treatment is based on previous experience, background reading, or intuition followed by trial and error experimentation. For example, the procedure used in the *Dugesia* experiment could be determined by background reading or by observation of the animal in its natural habitat. Experimental conditions that closely mimic the animal's natural environment usually produce the most meaningful results.

While outlining an experimental procedure, two final considerations are critical. One consideration is replication. A **replicate** is the duplication of an experiment. When a scientist performs a well thought-out experiment more than once, the results from all replicates should be consistent. Consistency of results is one source of reassurance that an experiment was well conceived and carried out. (Recall the uniformitarianism concept.) A second consideration is the inclusion of a control group. A **control group** is an experimental setup in which the independent variable has been eliminated or set to some standard value. In the *Dugesia* experiment, a control group might consist of the experiment being carried out in a container oriented in another direction relative to the pull of gravity.

Finally, after an experiment is designed, a scientist should be able to make **predictions** regarding the outcome of the experiment. These predictions are in the form of "if/then statements." If/then statements reinforce the concept of cause and effect relationships in nature. Predictions provide a useful reference point for drawing conclusions from an experiment. The "if" portion of the statement is the condition assumed by the hypothesis. The "then" portion of the statement is the outcome expected given the hypothesis and the conditions of the experiment. An if/then statement that predicted the outcome of the *Dugesia* experiment could be worded as follows: "If *Dugesia* respond negatively to the pull of gravity, then they will crawl to the highest point in a dish containing water and a rock tower."

PHOTOTAXIS OF *DUGESIA*

Dugesia (planaria) is a flatworm (phylum Platyhelminthes) that lives on the bottom of freshwater lakes, ponds, and rivers. It feeds on decaying plant and animal materials. Like all animals, *Dugesia* must respond to changes in its environment. A response to an environmental condition is called a *taxis*. For example, a response to light is called phototaxis. If an animal responds by orienting toward or moving toward light, it is positively phototactic. If an animal responds by orienting or moving away from light, it is negatively phototactic.

▼ The following activities should be completed individually or in groups prior to coming to laboratory. The laboratory time should be used for carrying out your experiment. Examine figure 11.1 to familiarize yourself with the structure of *Dugesia*. Note the general shape of the body. Notice the eyes and the nervous system (see fig. 11.1d). Consult your textbook for additional information on *Dugesia's* ecology and behavior. Record these background observations in worksheet 1.1.

Hypothesis

Based on the observations that you have made on the structure, habitat, and feeding habits of *Dugesia*, formulate a hypothesis regarding phototaxis of *Dugesia*. In formulating your hypothesis, consider the following:

1. Do you think *Dugesia* will display phototaxis? Why or why not? _____
   _____
   _____
   _____

2. If it does display phototaxis, do you think it will be positive or negative? Why? _____
   _____
   _____
   _____

3. Formulate your hypothesis and record it in worksheet 1.1.

4. Look at your hypothesis again. A hypothesis is always in the form of a declarative statement, never in the form of a question. Is your hypothesis in the form of a declarative statement? _____ If not, reword it.

5. Look at your hypothesis again. A hypothesis must be falsifiable. What set of results would prove your hypothesis incorrect? _____
   _____
   _____

Materials and Methods

Your instructor will have equipped the laboratory with pond water, rectangular glass dishes, beakers, white and black paper, tape, aluminum foil, light sources, and other equipment that you may find useful in setting up your experiment. Design a test of your hypothesis by considering the following:

1. What is the single independent variable that you will test in this experiment? Be as specific as possible to ensure that you have the single variable clearly in mind. In other words, think about whether you are concerned with the direction of light, intensity of light, color of light, or some other aspect of the quality or quantity of light. _____

2. Design a simple experimental setup using the rectangular glass dish, paper, light source, and other materials. Describe or draw an illustration showing that setup in the "Materials and Methods" section of worksheet 1.1. Explain where animals will be placed at the beginning of the experiment.

3. What dependent variable(s) will you be monitoring during the experiment? These could include the orientation of the *Dugesia* in the dish relative to the light source, the position of the *Dugesia* relative to some fixed point or line in the dish, or some other measuring criterion. _____
   _____

4. What controlled variables will be held constant during this experiment? Think about the effects of the following variables on your experiment. In the space next to each, describe how the variable can be controlled. Add other controlled variables as you think of them.
   a. overhead lighting _____
   _____
   b. window lighting _____
   _____
   c. heat from the lamp _____
   d. _____
   _____
   e. _____
   _____
   f. _____
   _____

5. What should be the nature of your control group?
   _____
   _____
   _____

6. How many *Dugesia* should be in your control and experimental groups? (Your instructor will probably impose some limits on this aspect of your experiment.)

7. How many replicates of this experiment will you carry out?
   _____

8. Write a complete description of your experiment in the "Materials and Methods" section of worksheet 1.1.

9. How will the data from the experiment be recorded? Will you be recording data in timed increments? If so, what time increments will you use? Remember that this exercise must be designed to be carried out within the time allotted for the laboratory and that you may want to perform one or more replicates of your experiment. _____
   Sketch a blank table in the "Results" section of worksheet 1.1 for recording data during and at the conclusion of the experiment.

10. Remember that the results of an experiment must be communicated to others. The results of an experiment consist of raw data from replicates of an experiment. These data must be combined, averages or percentages calculated, or otherwise presented for interpretation.

What is the most effective form for the presentation of your results to others in your class? Will your data be presented in a table, a graph, or in some other format (describe)? _____

_____

Sketch the table or graph in the "Results" section of worksheet 1.1. Be sure to write in column headings in your table and/or label the axes of your graph.

Results

When the laboratory begins, you will carry out your experiment and fill in the data table that you constructed in worksheet 1.1. Perform any calculations required and transfer the data for presentation.

Discussion

Look at your data. Do they support or refute your hypothesis? If they refute your hypothesis, suggest an alternative hypothesis that could be tested in another experiment. If your hypothesis is supported, describe how the results of your experiment help explain where *Dugesia* live in ponds and streams. Turn in your results to your laboratory instructor. ▲

7. What is scientific method? _____

_____

_____

8. What is a hypothesis? _____

_____

_____

9. What is a control group in an experiment?

_____

_____

_____

10. Why is an experiment that tests more than one variable at a time a poorly designed experiment?

_____

_____

_____

KEY TERMS

control group   11
controlled variable(s)   11
dependent variable(s)   11
hypothesis   10
independent variable(s)   11
objective lens system   4

ocular lens system   5
ocular micrometer   7
parfocal   7
predictions   11
replicate   11

resolution   4
science   10
scientific method   10
scientific theory   10
wet mount   7

WORKSHEET 1.1   Phototaxis of *Dugesia*

Question
    What is the response of *Dugesia* to light?

Observations from Background Reading and Observations of *Dugesia*
    Presence and location of the eyes

    Organization of the nervous system

    Habitat

    Food

    Other observations

Hypothesis

Materials and Methods

The Experimental Setup

Complete Description of the Experiment

Results

Data Table for Collection of Raw Data

Data Analyzed and in Presentation Form (Use the accompanying graph paper if necessary.)

Discussion

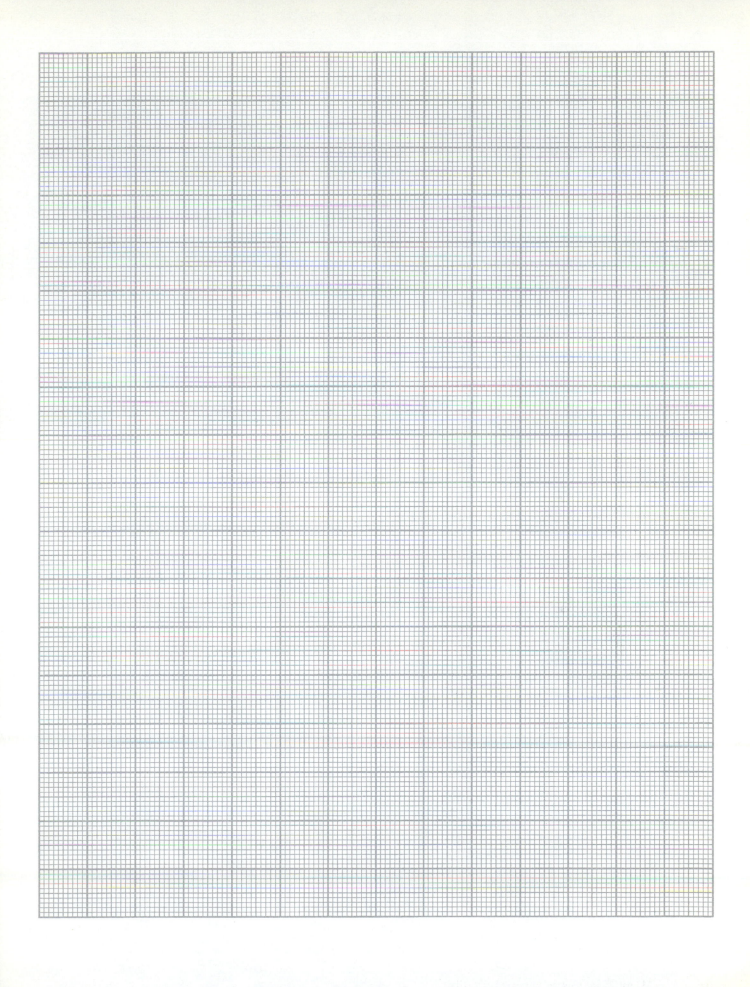

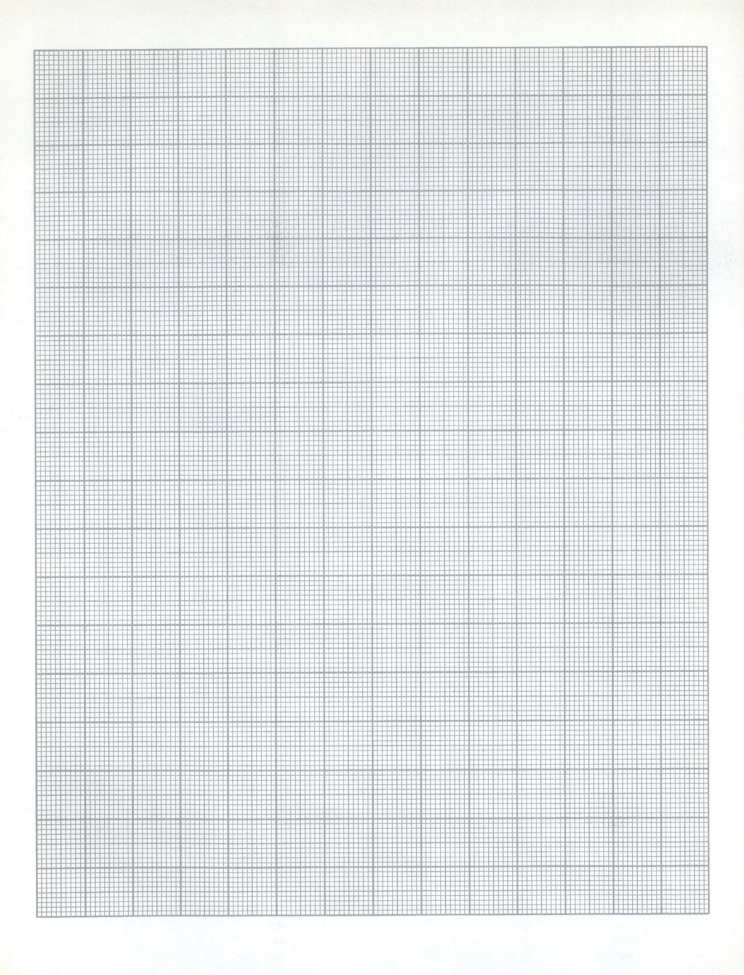

WORKSHEET 1.2   Microscope and Scientific Method

The Compound Microscope

1. What is meant by the resolving power of a microscope? Why is magnification without resolution useless for zoological studies?

2. What are three things that should be checked before putting a microscope away?

3. What is the procedure used in focusing a compound microscope?

4. An object's length takes up one-fourth of the low-power field of view of a compound microscope. What is its approximate length?

5. Describe the procedure used in making a wet mount. Why is it useful in studying fresh material?

6. The compound microscope provides a two-dimensional image of an object. In spite of this, the observer can appreciate its three-dimensional structure through proper use of the microscope. Explain.

The Dissecting Microscope

7. Compare and contrast the usefulness of compound and dissecting microscopes in zoological studies.

An Introduction to Scientific Method

8. What is science? What do you think most beginning students think science is after seeing their science textbooks for the first time?

9. How is the word "theory" similar to, or different from, the everyday use of the word?

10. Formulate two questions that could be answered using scientific methods.

11. Formulate two questions that could not be answered using scientific methods.

12. Why is it impossible to prove a hypothesis true?

13. A scientist wants to test the effects of nutrient A on the growth of mice.
    a. What questions might this scientist be asking?

    b. Write out a hypothesis that could be the basis for this scientist's experiment.

    c. What is (are) the dependent variable(s) in this experiment?

    d. What is the independent variable?

    e. What are the controlled variables that the scientist should keep constant during the experiment?

    f. Outline a test procedure that could be used in this experiment.

    g. What control group(s) could be used in this experiment?

Exercise 2

Cells and Tissues

Learning Objectives

After completing this exercise, you should be able to do the following:

Cell Structure

1. Describe the significance of the cell as the fundamental unit of life.

2. Describe four variations in the size and shape of animal cells.

3. Identify the cellular organelles characteristic of animal cells as seen in electron micrographs.

4. Describe the functions of the cellular organelles characteristic of animal cells.

Histology

5. Name and describe the functions of the four kinds of tissues found in animal systems.

6. Recognize and describe the function and location of the following types of epithelium:
   a. simple squamous epithelium,
   b. simple cuboidal epithelium,
   c. simple columnar epithelium, and
   d. stratified epithelium.

7. Describe the functions of general and special connective tissues.

8. Recognize and describe the function and location of the following types of connective tissue:
   a. dense connective tissue,
   b. loose connective tissue, and
   c. adipose tissue.

Prelaboratory Quiz

Study this week's laboratory exercise and then complete the following quiz to assess your preparation for the laboratory.

Cell Structure

1. Name two of the three concepts that make up cell theory. _____
   _____

2. Using a light microscope, one can clearly understand
   a. details of the organization of cellular organelles.
   b. the size, shape, and diversity of cells.
   c. how cells are organized to form tissues.
   d. All of the above are correct.
   e. Both b and c are correct.

3. The cellular organelle responsible for the synthesis of proteins in a cell is the
   a. mitochondrion.
   b. centriole.
   c. smooth endoplasmic reticulum.
   d. ribosome.

4. Ribosomes are synthesized at the
   a. centriole.
   b. mitochondrion.
   c. Golgi apparatus.
   d. nucleolus.

5. A _____ is a cylindrical arrangement of proteins that produces movement within cells and supports cell processes.
   a. microtubule
   b. microfilament
   c. flagellum
   d. rough endoplasmic reticulum

6. Vesicular structures that contain enzymes capable of digesting biologically important macromolecules are called
   a. vacuoles.
   b. lysosomes.
   c. secretory vesicles.
   d. ribosomes.

Histology

7. What is histology?

8–11. Name the four kinds of animal tissues and give one example of each type of tissue.

Name	Example
8. _____	_____
9. _____	_____
10. _____	_____
11. _____	_____

12. An epithelium that consists of a single layer of cells resting on the basement membrane is called a
    a. squamous epithelium.
    b. stratified epithelium.
    c. stratified squamous epithelium.
    d. simple epithelium.

13. A connective tissue that functions in binding skin to muscle is
    a. dense connective tissue.
    b. loose connective tissue.
    c. adipose tissue.
    d. cartilage.

14. All of the following are components of connective tissues EXCEPT
    a. cells.
    b. protein fibers.
    c. a basement membrane.
    d. a matrix.

15. A connective tissue specialized for storing fat is
    a. adipose tissue.
    b. dense connective tissue.
    c. lymph.
    d. cartilage.

CELL STRUCTURE

To understand cellular function is to understand much of life. According to **cell theory,** cells (1) are the basic units of life, (2) possess all characteristics of life, and (3) arise only from preexisting cells. These apparently simple statements are the result of years of work by early cell biologists, and they have important implications for all biologists. While working through this course, you will find that you must repeatedly return to the level of the cell to understand animal function. For example, while one can study the contractile properties of a muscle by observing an entire muscle, it is impossible to understand how a muscle contracts without studying the muscle cell. Similarly, it is impossible to understand the function of the nervous system without an appreciation of the nerve cell, or to understand the function of the endocrine system without familiarity with models of the plasma membrane. Just as animal function is tied to cell function, animal structure is dependent on the organization of cells into tissues. Today it is more important than ever for the biologist to understand cell structure and function.

ANIMAL CELLS—LIGHT MICROSCOPE

Detailed study of cell structure requires the use of the electron microscope. Few details of subcellular structure can be studied under the light microscope. The purpose of this study is not to examine details of structure, but to gain an appreciation of some of the diversity found in animal cells. As you work through the exercise, record your observations in table 2.1 in worksheet 2.1. Draw and label the cells observed.

▼ Using the flat end of a clean toothpick, a tongue depressor or a popsicle stick, gently scrape the inside of your cheek. Swirl the scrapings in a drop of water on a slide. Cover with a coverslip, place the slide on the stage of the microscope, and examine using reduced light and low power. Look for small, flattened, and transparent epithelial cells. Stain the preparation with methylene blue by placing a small drop of stain at one edge of the coverslip. Draw the stain under the coverslip by touching the edge of a paper towel to the opposite edge of the coverslip. As water is drawn from under the coverslip, it will be replaced by stain. Examine the preparation under low power, then high power.

The bulk of the cell's interior is **cytoplasm.** The cytoplasm contains organelles that carry out specific functions. These structures usually cannot be seen clearly under the light microscope. The outer boundary of the cell is defined by the **plasma membrane.** Note the **nucleus,** the genetic control center of the cell. If your microscope has an ocular micrometer, measure the diameter of the cell using the calibration factor you calculated in exercise 1. If your microscope does not have an ocular micrometer, estimate the cell's diameter using the method described in exercise 1. Record your observations in table 2.1 in worksheet 2.1.

Examine a spinal cord smear and look for star-shaped nerve cells. These cells show a distinct **nucleolus** within a relatively clear nucleus. The nucleolus is an area of active RNA synthesis. Estimate the size of these cells and record your observations in table 2.1.

Examine a blood smear, first under low power, then under high power. Biconcave red blood cells will be stained pink. Do you see any evidence of a nuclei in any of the red blood cells? Red blood cells are formed in the bone marrow and spleen. They lose their nuclei prior to being released into the bloodstream. Estimate their size. Using low power, look for cells containing darkly stained nuclei and relatively clear cytoplasm. Examine one of these white blood cells under high power. What is the shape of the nucleus? _____

Estimate the size of the white blood cell you are observing. Record your observations in table 2.1.

Examine a preparation of *Amphiuma* liver. Note the closely packed cells. Focus on one or two cells under high power. Note the nucleus and the very small, dark dots in the cytoplasm. These densely packed **mitochondria** are responsible for energy conversions in the cell. Estimate the size of a cell and record your observations in table 2.1.

Examine a preparation of small intestine epithelium stained with Golgi stain. Look for tall cells lining the cavity of the intestine. These cells are actively secreting enzymes into the intestine. What is the shape of the nucleus? _____

The **Golgi apparatus** is responsible for packaging proteins prior to secretion. Look for the black, layered membranes of the Golgi apparatus on the lumenal side (toward the cavity of the intestine) of the nucleus. The darkly stained particles at the lumenal surface of the cell are the secretory products that have been released at the plasma membrane. Estimate the longest dimension of this cell and record your observations in table 2.1 in worksheet 2.1. ▲

ANIMAL CELLS—ELECTRON MICROSCOPE

The structure of a cell as seen with the electron microscope is called the cell's ultrastructure. The previous activity was designed to give you an appreciation for a small part of the diversity found in animal cells. It was also designed to give you an appreciation for the minuteness of cell organelles. Electron micrographs are needed to study details of cell structure. In the following activity, you will study cell ultrastructure based on the diagrammatic representation in figure 2.1 and the electron micrographs and drawings of figures 2.2–2.12.

The Plasma Membrane

The cell has many membranes. The plasma membrane separates the contents of the cell from its surroundings, but it has many other functions as well. Proteins in the plasma membrane create channels for substances moving passively

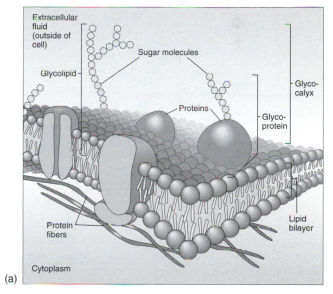

(a)

(b)

Figure 2.2 The plasma membrane: (*a*) the fluid mosaic model, showing the lipid bilayer and intrinsic proteins; (*b*) transmission electron micrograph of a portion of a plasma membrane. Note the organization into a double layer (×250,000).

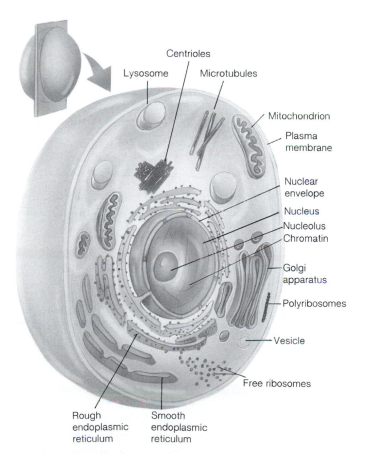

Figure 2.1 Diagrammatic representation of the ultrastructure of an animal cell.

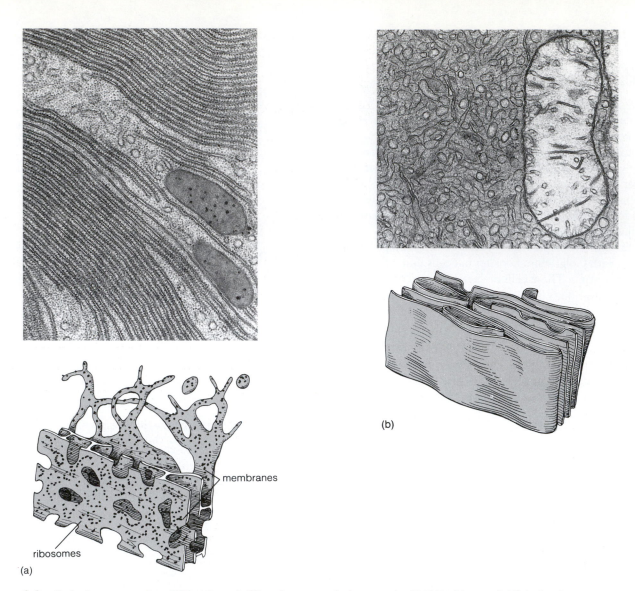

Figure 2.3 Endoplasmic reticulum (ER): (*a*) rough ER with associated ribosomes (×60,000); (*b*) smooth ER lacks ribosomes (×64,000). A mitochondrion is shown at the right side of the electron micrograph.

through the plasma membrane. Proteins also serve as carriers for substances that are actively transported across the membrane. In addition, proteins serve as receptor sites for hormones, antibodies, and other biologically important compounds. The plasma membrane provides a location for enzymatic reactions and is involved with cell adhesion. A current model of the plasma membrane that explains those functions is the **fluid mosaic model** (fig. 2.2). This model describes the plasma membrane as a bimolecular lipid layer. Hydrophilic ends of lipid molecules are oriented to the outside of the membrane, while hydrophobic ends are oriented to the inside of the membrane. Some proteins are loosely associated with the surface of the membrane (extrinsic proteins); other proteins are embedded in the membrane (intrinsic proteins). When carbohydrates unite with lipids, they form glycolipids, and when they unite with proteins they form glycoproteins. Surface carbohydrates, lipids, and proteins make up the **glycocalyx,** which is necessary for cell-to-cell recognition.

The membrane can be visualized as a very fluid structure that can invaginate to form and pinch off vesicles or unite with vesicles to release materials to the outside of the cell.

Endoplasmic Reticulum and Ribosomes

The **endoplasmic reticulum** (ER) is a complex network of membranes that runs through much of the animal cell (fig. 2.3). This membrane system is the site of many of the synthesis reactions, and it has important transport functions within the cell. Very small particles, called **ribosomes,** are frequently associated with the ER. In this state, the endoplasmic reticulum is called **rough ER.** When devoid of ribosomes, the endoplasmic reticulum is called **smooth ER.** Ribosomes and ER are functionally interrelated. Proteins are synthesized at the ribosome and then transported throughout the cell within the ER. Ribosomes are also found "floating" in the cytoplasm, probably interconnected by fine pro-

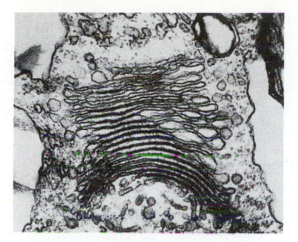

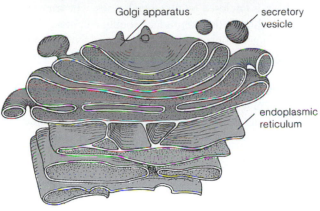

Figure 2.4   The Golgi apparatus. Note the vacuoles pinching free of the cisternae (×37,250).

tein strands. Smooth ER is the site of synthesis of steroids, detoxification of organic molecules, and the storage of ions such as Ca^{++}.

The Golgi Apparatus

The Golgi apparatus is a membranous structure that packages products of cellular metabolism (fig. 2.4). It comprises stacks of **cisternae** that are often associated with the endoplasmic reticulum. In figure 2.4, note the vesicles that have been released to the cytoplasm. Golgi apparatuses are particularly numerous in cells of the pancreas, in cells lining the cavity of the small intestine, in cells of endocrine glands, and in other secretory structures. Can you describe why the endoplasmic reticulum, ribosomes, and Golgi apparatuses are functionally interrelated? _____

_____

_____

_____

_____

The Nucleus and Nucleolus

The nucleus is the genetic control center of the cell and the most obvious structure in most cells. The nucleus contains DNA and protein, usually in a highly dispersed state called

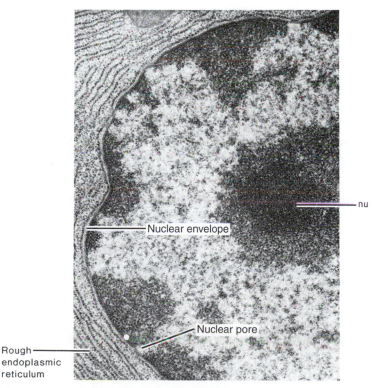

Figure 2.5   The nucleus. Note the nucleolus (nu), chromatin (containing DNA), the double-layered nuclear envelope, and nuclear membrane pores. Note the rough endoplasmic reticulum outside the nucleus (×31,500).

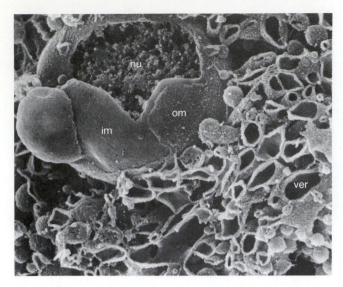

Figure 2.6 A scanning electron micrograph showing the nucleus (nu), with its inner (im) and outer (om) membranes, and the endoplasmic reticulum (ver) (×22,000).

chromatin. During cell division, chromatin condenses (winds) into chromosomes. The nucleus is surrounded by a double membrane called the **nuclear envelope.** The nuclear envelope has relatively large pores and is continuous with endoplasmic reticulum (figs. 2.5 and 2.6). Nuclear pores allow RNA to move between the nucleus and the cytoplasm. Inside the nucleus one frequently finds a nucleolus. This is where the RNA of ribosomes is synthesized.

Mitochondria

Most energy conversions in the cell take place within mitochondria (fig. 2.7). In these reactions, energy in carbohydrates, fats, and proteins is converted into a form usable by the cell called adenosine triphosphate (ATP). Each mitochondrion is enclosed by two membranes. The inner membrane is tightly folded to produce shelflike **cristae.** Between cristae is a gelatinous **matrix** containing DNA, ribosomes, and enzymes. Mitochondrial DNA and ribosomes are similar to bacterial DNA and ribosomes. This has led biologists to believe that mitochondria are self-replicating units and that they may have evolved from symbiotic, bacterialike cells. Enzymes located in the matrix and on the cristae catalyze cellular energy conversions.

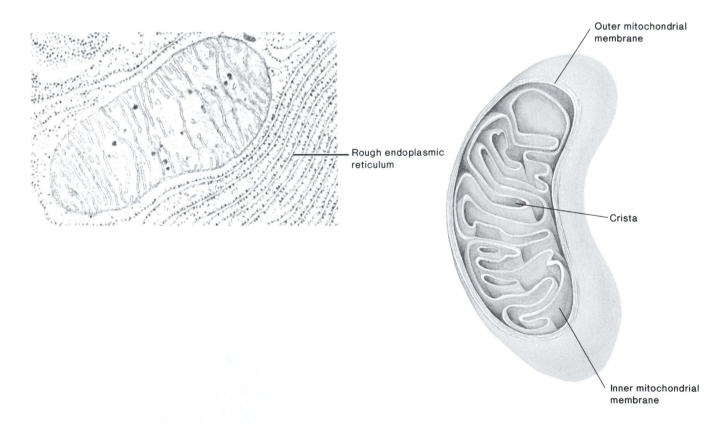

Figure 2.7 The mitochondrion. Note the inner and outer membranes, the cristae, and the matrix. Also note the rough endoplasmic reticulum outside of the mitochondrion (×36,300).

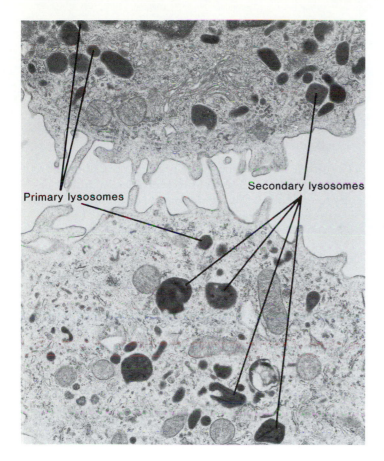

Primary lysosomes   **Secondary lysosomes**

Figure 2.8 Lysosomes in white blood cells (macrophages) of the lungs. After a lysosome (primary lysosome) fuses with a vacuole containing foreign or waste material, it is often called a secondary lysosome ($\times 10{,}400$).

Lysosomes

Lysosomes are vesicular structures that contain enzymes capable of digesting all biologically important macromolecules (fig. 2.8). They are responsible for digesting particles ingested by the cell as well as old, nonfunctional organelles. Lysosomes are formed at the Golgi apparatus when enzymes, produced at the ribosomes and transported to the Golgi apparatus by the endoplasmic reticulum, are enclosed within a vesicle.

Microtubules and Microfilaments

Microtubules and microfilaments are the basis of cellular locomotion and structure. **Microfilaments** are protein fibers that are arranged in linear arrays or networks. Microfilaments are organized similar to actin, one of the proteins present in muscle. Actin, along with other proteins, is involved with the shortening of muscle cells during contraction. Similar microfilaments are involved with amoeboid locomotion and other cytoplasmic movements, movement of vesicles and other structures within the cell, and the maintenance of cell shape. The array of microtubules and microfilaments within a cell is called the cytoskeleton.

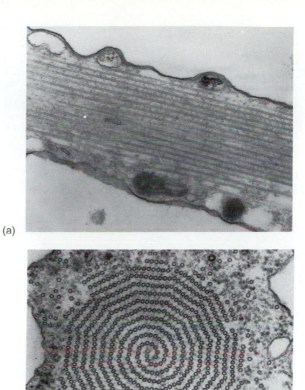

(a)

(b)

Figure 2.9 Ultrastructure of an axopodium (slender cell extension used in feeding and locomotion) of the protozoan *Echinosphaerium nucleofilum* showing microtubules in (*a*) longitudinal ($\times 30{,}000$), and (*b*) cross sections ($\times 4{,}000$).

Microtubules are cylindrical arrangements of proteins that, like microfilaments, are the basis of cellular locomotion and structure (fig. 2.9). Microtubules support elongate cell processes (axons of nerve cells), produce movement within cells (movement of chromosomes in cell division), and provide movement for entire cells or substances outside of cells (cilia and flagella).

Cilia and Flagella

Cilia and **flagella** are hairlike organelles that project from the surface of some cells. Cilia and flagella cause cells to move through their media or cause fluids or particles to move over the cell surface. Cilia and flagella (excluding bacterial flagella) are similar in structure. The primary difference is in length, flagella being longer than cilia. Just inside the plasma membrane of a cilium or flagellum are nine double microtubules. These microtubules originate at the base of the cilium or flagellum at a structure called the **basal body.** These microtubules run the length of the cilium or flagellum and surround a core of two central microtubules (fig. 2.10). This is often referred to as a 9 + 2 organization. Movement of the cilium or flagellum is believed to occur as adjacent double tubules slide past one another. The interaction between

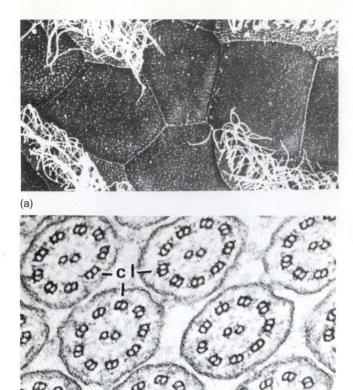

(a)

(b)

Figure 2.10 The structure of cilia: *(a)* scanning electron micrograph of cilia from the gills of an amphibian embryo (*Ambystoma mexicanum*) (×600); *(b)* cross sections through the cilia (cl) of the protozoan *Euplotes eurystomes*. Note the 9 + 2 pattern of microtubules.

microtubules is believed to result from small "arms" on one member of a pair of microtubules interacting with an adjacent pair of microtubules. The basal body (or kinetosome) organizes the formation of the microtubules. It is a short cylinder in which microtubules are arranged in nine triplets around a core that lacks central tubules (9 + 0).

The Centrosome and Centrioles

The **centrosome** is a structure that occurs outside the nucleus of animal cells. It consists of two **centrioles** set at right angles to each other. The centrosome plays a critical role in organizing the tubule system (mitotic spindle) that is involved with chromosome movements during cell division. Centrosomes also organize the array of microtubules and microfilaments that make up the cytoskeleton of a cell. Centrioles are similar in structure to basal bodies (fig. 2.11).

Identification of Subcellular Structures

▼ Identify and describe the function of the subcellular structures shown in figure 2.12. Record your answers in table 2.2 in worksheet 2.1. ▲

STOP AND ASK YOURSELF

1. What is the function of the following organelles of the cell?

   a. nucleus _____

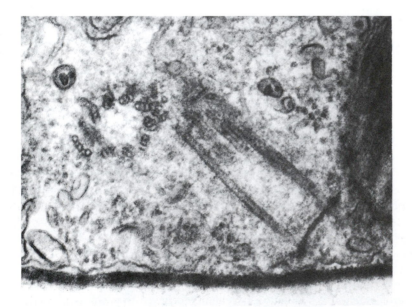

Figure 2.11 Cross section and longitudinal section of a pair of centrioles. Note the 9 + 0 pattern (×100,000).

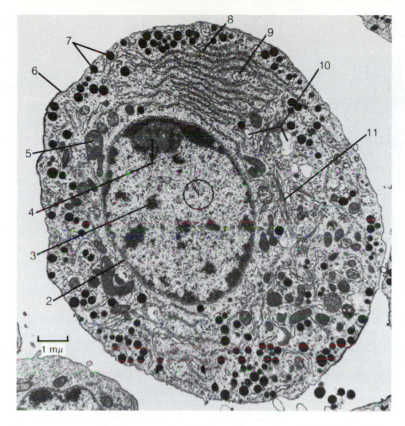

Figure 2.12 Transmission electron micrograph of an animal cell (photomicrograph ×8,000). Identify the organelles indicated and enter their names and functions in table 2.2 in worksheet 2.1.

b. plasma membrane _____

_____

_____

c. Golgi apparatus _____

_____

d. mitochondrion _____

_____

2. How would you describe the structure of the mitochondrion? _____

_____

_____

3. What is the difference between smooth and rough ER? _____

_____

4. How would you describe the fluid mosaic model of membrane structure? _____

_____

_____

_____

_____

HISTOLOGY: EPITHELIAL AND CONNECTIVE TISSUES

Tissues are groups of cells of similar structure and origin that function together. The study of the structure of tissues is called **histology.** There are four kinds of tissues: **epithelial, connective, muscular,** and **nervous.** In the following section, you will study epithelial and connective tissues as they occur in vertebrates. (Although invertebrate tissues are organized in a similar manner, there are striking differences that cannot be covered here.) A study of most of the special connective tissues, muscular tissue, and nervous tissue will be delayed until unit III. The structure of these tissues is closely tied to their physiology. One can best understand these tissues by studying structure and function together.

▼ When examining each tissue, read the accompanying description first and then examine the slide as directed. Keep in mind that you will be looking at sections through organs (assemblages of tissues of various kinds) and that you want to look in specific regions for the desired tissue. Also, thin sections for mounting specimens on slides are often made at angles such that the structure of a cell may not look as you expect. For example, a columnar cell cut in cross section may look cuboidal. You must look at more than one cell, or possibly more than one slide, to appreciate the actual structure of the tissue. Compare what you see to the appropriate figure. ▲

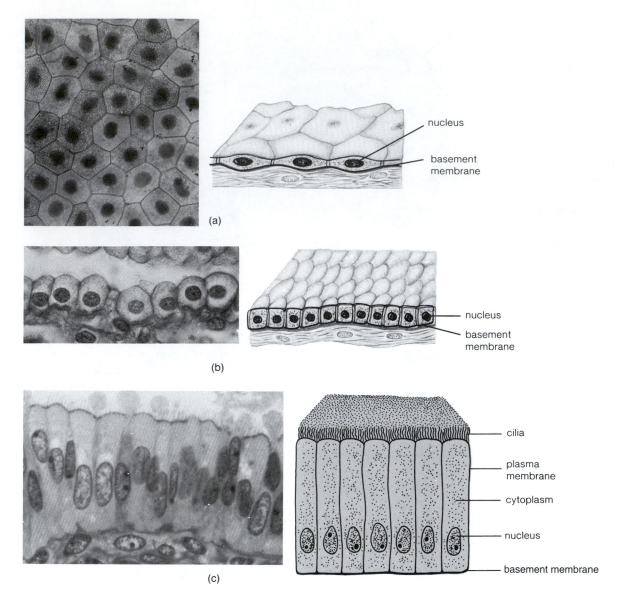

Figure 2.13 Epithelial tissues: (*a*) simple squamous epithelium (×750); (*b*) simple cuboidal epithelium (photomicrograph ×250); (*c*) simple columnar epithelium (photomicrograph ×1,200).

EPITHELIAL TISSUE

Epithelial tissue consists of a sheet of cells that covers the surface of the body or of one of its cavities. One side of a layer of epithelial tissue is, therefore, exposed to air or body fluids and the other side rests against a noncellular layer produced by the epithelial tissue and the underlying connective tissue called the **basement membrane.** In most preparations, the basement membrane cannot be directly observed, but its position is indicated by the lower limit of the inner cell layer. The chief functions of epithelium are protection, secretion, absorption, and providing surfaces for diffusion. Epithelial tissues are classified according to shape and layering.

Simple Epithelium

Simple epithelium consists of a single layer of cells, all of which reach the basement membrane. Simple epithelium is found in areas of little wear and tear or where diffusion or absorption occurs.

Simple Squamous Epithelium

▼ In a diagram, this type of epithelium appears to consist of thin, flat cells. Nuclei often appear to bulge from the middle of the cell and are flattened dorsoventrally (fig. 2.13a). In the lungs, simple squamous cells line the alveoli (tiny air sacs) and make up the thin layer across which gases diffuse.

Examine a prepared slide of lung tissue. Under low power, note the loose, airy consistency. The sectioned larger ducts and vessels represent branches of the pulmonary circulation and respiratory passageways. The alveoli are the smallest, blind-ending sacs. Focus on a group of these under high power. Note the thin layer of cells bordering the alveoli. You will not be able to see cell boundaries; however, individual cells are represented by the presence of darkly stained nuclei.

Simple squamous epithelium also lines all vessels in the circulatory system. Capillaries are entirely formed from this tissue and represent the single cell layer for diffusion of substances between blood and tissue spaces. Examine a cross section through a vein or artery. These vessels consist of layers of epithelial, connective, and muscular tissues. On the lumenal (inner) surface is a layer of simple squamous epithelium. Focus on this surface with low and then high power. At intervals along the surface, you should see darkly stained nuclei protruding from the inner vessel lining. Cell boundaries of these simple squamous cells will not be obvious. ▲

Simple Cuboidal Epithelium

Simple cuboidal epithelium consists of cells that are approximately as wide as they are high and deep. The nuclei usually appear round (fig. 2.13b). It is again a relatively delicate tissue that occurs in areas where transport is important. This tissue performs very important transport functions in the tubule system of the kidney.

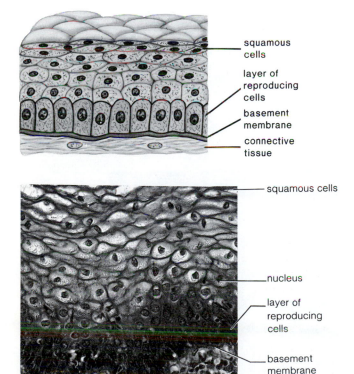

squamous cells

layer of reproducing cells

basement membrane

connective tissue

— squamous cells

— nucleus

— layer of reproducing cells

— basement membrane

Figure 2.14 Stratified squamous epithelium (photomicrograph ×100).

▼ Examine a prepared slide of a kidney section. The kidney is divided into two regions. The outer region is the cortex (containing glomerular capsules for the filtration of blood), and the inner region is the medulla (containing the bulk of the tubule system). Under low power, distinguish between these two regions. Within the medulla, note sectioned tubules. In the outer medulla and inner cortex, you are likely to see tubules in cross section. In the inner medulla, more tubules are likely to be longitudinally sectioned. In tubule cross sections, cells will be arranged circularly around the tubule lumen. Under high power, note the spherical nuclei and cuboidal cells. (Cuboidal cells often appear square when sectioned, but they are really hexagonal.) Deeper in the medulla, the tubules will appear as linear rows of cells on either side of the tubule lumen. Note again the cuboidal cells.

The thyroid gland (endocrine in function) consists of follicles containing precursors of hormones produced by the simple cuboidal cells lining the follicles. Examine a thyroid section under low power. The follicles can be seen as relatively large, uniformly stained structures. The uniformly stained interior of the follicle represents the stored secretory product. Under high power, note the cuboidal cells lining the follicle. ▲

Other cuboidal cells can be observed in the epithelial lining of the ovary. These cells function in gametogenesis in the developing ovary. In the salivary glands, cuboidal cells function in the production and secretion of saliva.

Simple Columnar Epithelium

Simple columnar epithelium has cells taller than they are wide. Nuclei are flattened laterally (fig. 2.13c). Simple columnar epithelium is found where wear and tear is relatively pronounced, but where absorption and secretion still occur. It is most commonly found lining the digestive tract.

▼ Examine a section of the small intestine under low power. The inner lining is folded into many fingerlike villi that increase the surface area for secretion and absorption. Focus on the surface of one of these villi and switch to high power. Under high power, note the single layer of tall, thin cells. Note that the darkly stained nuclei are flattened laterally. With some stains, the surface of the cells will appear more intensely stained with faint vertical striations. These striations are microvilli. They comprise the **brush border,** which greatly increases the intestinal surface area for cellular transport processes (see fig. 22.4). ▲

Some simple columnar cells in the digestive tract, called **goblet cells,** are modified for the production and secretion of mucus. These appear as translucent cells interspersed with the cells just studied. Mucus lubricates, buffers, and facilitates the adherence of food and fecal particles.

Stratified Epithelium

Stratified epithelium is found where there are two or more layers of cells (i.e., the outermost layer of cells does not reach the basement membrane). Only the bottommost layer of cells is reproducing. As new cells are produced, older cells are pushed toward the surface.

Stratified epithelium is classified on the basis of the shape of the outermost layer of cells. Stratified epithelium occurs in areas where wear and tear are common. It serves as a protective barrier between the outside and the tissues below. Little transport occurs across stratified epithelium. Stratified squamous epithelium occurs in the skin, vagina, and cornea. Stratified cuboidal epithelium occurs in sweat glands, and stratified columnar epithelium occurs in the male urethra.

▼ Examine a section of human skin. Under low power, find the outer cell layers. Note the rows of darkly stained cells. The innermost row of cells sits on the basement membrane. This wavy boundary separates dermal and epidermal layers of the skin (fig. 2.14).

Switch to high power. Compare the shape of the inner and outer cell layers. Note the transition between the cuboidal cells near the basement membrane and the squamous cells at the outside. Recall that stratified epithelium is named on the basis of the shape of the outer layer of cells (hence, stratified squamous epithelium). In this tissue, the outer cells are largely dead or dying; they are so far from the blood supply in the dermis that they can no longer be nourished by diffusion. In addition, those outer cells contain a tough protein (keratin) that makes this layer relatively impermeable to water, bacteria, and other external agents. Can you appreciate why this tissue is largely protective? ▲

CONNECTIVE TISSUE

Connective tissues bind, anchor, and support the body and its organs. They consist of general connective tissues and special connective tissues.

General Connective Tissues

General connective tissues bind body structures together and give soft body parts structural support. They are made of a gel-like matrix (ground substance) in which various kinds of cells and protein fibers are embedded. We will study two examples of general connective tissues. Dense connective tissue and loose connective tissue are characterized by the composition and arrangement of fibers present in the tissue.

1. **Dense Connective Tissue** (fig. 2.15)
   a. Fibers are primarily collagenous and parallel to one another.
   b. Fibers are oriented along lines of stress.
   c. Dense connective tissue is resistant to stretching.
   d. Example: tendon.

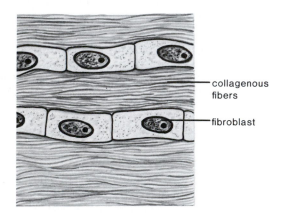

collagenous fibers

fibroblast

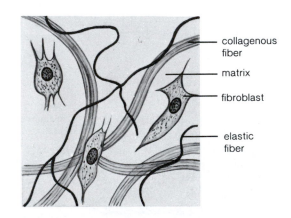

collagenous fiber

matrix

fibroblast

elastic fiber

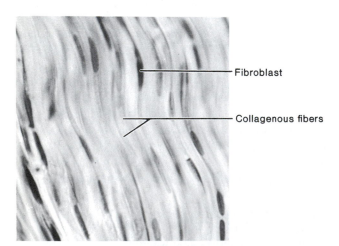

Fibroblast

Collagenous fibers

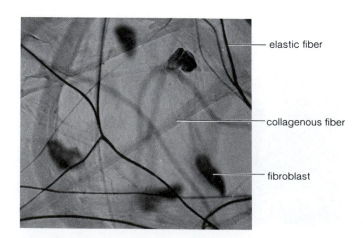

elastic fiber

collagenous fiber

fibroblast

Figure 2.15 Dense (fibrous) connective tissue (photomicrograph ×400).

Figure 2.16 Loose connective tissue (photomicrograph ×250).

▼ Examine a slide of longitudinally sectioned dense (white fibrous) connective tissue. (These slides are usually prepared from a piece of tendon.) Note the parallel strands of collagenous fibers separated by rows of tendon cells (nuclei darkly stained). The bulk of this tissue is made of bundles of collagenous fibers. The matrix and cells are sandwiched between the fibers. ▲

2. **Loose Connective Tissue** (fig. 2.16)
   a. There are three fiber types: collagenous, reticular, and elastic.
   b. Fibers are oriented randomly.
   c. Very common—fastens skin to muscle, muscle to muscle, blood vessels and nerves to other body parts, etc.
   d. Example: subcutaneous layer of the skin.

▼ Examine a slide of loose (areolar) connective tissue under low and high power. This slide will most likely be a section from the subcutaneous layer of the skin, which fastens the skin to muscle. Note the irregularly arranged fibers and scattered cells. All cells and fibers are embedded in a clear matrix. ▲

Special Connective Tissues

Special connective tissues consist of cartilage, bone, blood, lymph, and adipose tissue. They are connective tissues with specialized structures and functions. We will postpone our study of cartilage and bone until the laboratory on vertebrate musculoskeletal systems and our study of blood until the laboratory on vertebrate circulation.

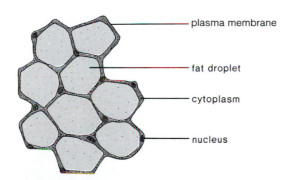

plasma membrane

fat droplet

cytoplasm

nucleus

Figure 2.17 Adipose tissue (photomicrograph ×250).

Adipose Tissue

Adipose tissue is derived from loose connective tissue (fig. 2.17). Adipose accumulates in the subcutaneous layer of the skin, around the heart, kidneys, etc. As adipose cells mature, fat accumulates in the cell and pushes cytoplasm to the periphery.

▼ Examine a slide showing adipose tissue. Use a slide especially prepared to show adipose tissue or use the skin section previously studied. If the latter, look deep in the section within the subcutaneous layer. This area contains loose connective tissue as well as adipose tissue. Observe under low and high power. Note the large, clear, fat storage area. The fat has been dissolved by the clearing agent used in preparing the slide. Identify the nucleus and cytoplasm of the fat cells. ▲

MUSCULAR TISSUE

There are three types of muscular tissue: skeletal, cardiac, and smooth (visceral). We will postpone the study of muscle histology until the laboratory exercise on the muscular system.

NERVOUS TISSUE

The basic unit of the nervous system is the neuron. The histology of the nervous system involves the study of the different kinds of neurons and how various neurons are associated with each other. We will postpone the study of nervous tissue until the laboratory exercise on the nervous system.

S T O P   A N D   A S K   Y O U R S E L F

5. What is histology? _____
_____

6. What are the functions of simple epithelium, and where does it occur? _____
_____
_____

7. What are the functions of stratified epithelium, and where does it occur? _____
_____
_____

8. How would you distinguish loose connective tissue from dense connective tissue when viewing them in a microscope? _____
_____
_____
_____

9. Where could you find deposits of adipose tissue? ____
_____
_____

KEY TERMS

WORKSHEET 2.1   Cells and Tissues

Cell Structure

1. What is the usefulness of the light microscope in the study of cell structure?

2. What is the usefulness of the electron microscope in the study of cell structure?

3. What cellular structure would you associate with
   a. energy conversions within the cell?

   b. protein synthesis?

   c. packaging products of cellular metabolism?

   d. cytoplasmic movements within a cell?

   e. enzymatic digestion?

TABLE 2.1	Observations of Selected Animal Cells as Seen through the Compound Microscope (Recall that 1 mm = 1,000 μm.)			
Cell Type	Size (μm)	Cell Shape	Organelles Visible	Special Characteristics
Cheek epithelial cell				
Nerve cell				
Red blood cell				
White blood cell				
Amphiuma liver cell				
Intestinal epithelial cell				

TABLE 2.2   The Identification and Description of Subcellular Structures in Figure 2.12

Number	Name	Function
1		
2		
3		
4		
5		
6		
7		
8		
9		
10		
11		

Histology: Epithelial and Connective Tissues

4. Define the term "epithelium."

5. What kinds of functions do connective tissues perform?

6. Give an example of where one could find
   a. dense connective tissue.

   b. loose connective tissue.

   c. simple cuboidal epithelium.

   d. stratified epithelium.

7. Describe the appearance of adipose tissue when it is sectioned, stained, and viewed through a compound microscope.

Exercise 3

Aspects of Cell Function

Learning Objectives

After completing this exercise, you should be able to do the following:

Transport Into and Out of Cells

1. Explain why cells must exchange materials with their environment.
2. Describe processes involved in
   a. diffusion,
   b. osmosis,
   c. pinocytosis,
   d. phagocytosis, and
   e. exocytosis.
3. Describe the role of the plasma membrane in transport processes.
4. Give examples of transport processes as they relate to specific organ systems.

Aerobic Cellular Respiration

5. Describe the significance of cellular respiration in animals.
6. Describe the relationship between organismal and cellular function.
7. Apply scientific method by designing and carrying out an experiment relating to cellular respiration.
8. Analyze data and prepare it for presentation in a scientific report.

Mitosis

9. Describe the significance of mitosis and meiosis in animal systems.
10. Describe the events of mitosis in animal cells.
11. Recognize stages of mitosis in the whitefish blastula.

Prelaboratory Quiz

Study this week's laboratory exercise and then complete the following quiz to assess your preparation for the laboratory.

Transport Into and Out of Cells

1. In this week's laboratory, you will be studying cellular transport processes. Which of the following processes involve passive transport?
   a. diffusion only
   b. osmosis only
   c. bulk transport
   d. diffusion and osmosis
2. Which of the following is necessary in order for materials to move by diffusion?
   a. energy expenditure by a cell
   b. a selectively permeable membrane
   c. a concentration gradient
   d. a membrane carrier
3. In this week's laboratory, you will be observing osmosis across the membrane(s) of
   a. a thistle tube apparatus.
   b. cheek cells.
   c. white blood cells.
   d. red blood cells.
4. If a cell is placed in a hypotonic solution it will
   a. swell.
   b. shrivel.
   c. remain unchanged.
5. When large molecules are transported across the plasma membrane of a cell by invagination of the plasma membrane, forming a vacuole around the molecule, the process is called
   a. passive transport.
   b. active transport.
   c. bulk transport.

Aerobic Cellular Respiration

6. In the experiment on aerobic cellular respiration, you will be measuring the rate of cellular respiration by monitoring the rate at which
   a. oxygen is produced.
   b. oxygen is used.
   c. carbon dioxide is used.
   d. carbon dioxide is produced.

7. The purpose of the potassium hydroxide in the cellular respiration apparatus is to
   a. absorb carbon dioxide.
   b. absorb oxygen.
   c. absorb water.
   d. provide oxygen.

8. Which one would NOT be an act of plagiarism?
   a. using a direct quotation without quotation marks and a citation
   b. using a direct quotation with a citation but without quotation marks or some other form of setting the quotation apart from the rest of the text
   c. rewording an idea of another person without a citation
   d. rewording an unsubstantiated opinion of another person without a citation
   e. All of the above would be acts of plagiarism.

9. A(n) _____ is a portion of a scientific paper that presents a concise summary of information presented in the paper.
   a. abstract
   b. introduction
   c. materials and methods section
   d. discussion section

10. Tables, graphs, and other data presentations are usually found in the _____ section of a scientific paper.
    a. abstract
    b. introduction
    c. materials and methods
    d. results
    e. discussion

Cell Division

11. The division of the cytoplasm during cell division is
    a. mitosis.
    b. prophase.
    c. anaphase.
    d. telophase.
    e. cytokinesis.

12. When chromosomes are lined up along the equator of a dividing cell, the cell is in _____ of mitosis.
    a. prophase
    b. metaphase
    c. anaphase
    d. telophase

13. Chromatids reach the poles of a cell and begin to "unwind" into chromatin, the nuclear envelope and nucleolus reappear, and spindle fibers disappear during _____ of mitosis.
    a. prophase
    b. metaphase
    c. anaphase
    d. telophase

14. In comparison to the cells that enter into the process of mitotic cell division, the daughter cells
    a. have exactly the same genetic composition.
    b. have one-half the number of chromosomes.
    c. have chromosomes consisting of two chromatids.
    d. have double the number of chromosomes.

15. All of the following are valid comparisons of mitosis and meiosis EXCEPT one. Select the one exception.
    a. Meiosis results in the production of gametes, and mitosis results in the production of other body cells.
    b. A single mitotic division results in the formation of four daughter cells, and a single meiotic division results in the formation of two daughter cells.
    c. Meiosis results in the formation of haploid cells, and mitosis results in the formation of diploid cells.
    d. Mitosis occurs in all tissues of an animal, and meiosis occurs only in gonads.

TRANSPORT INTO AND OUT OF CELLS

A cell is not independent of its environment. It is continuously exchanging gases, nutrients, and wastes between intracellular and extracellular compartments. What can be exchanged and how exchange occurs is largely determined by the properties of the plasma membrane and the substance involved. Exchange occurs by **passive transport, active transport,** and **bulk transport.** Passive transport involves the movement of substances without the expenditure of energy by the cell. In active transport, protein carriers in the plasma membrane use energy in transporting substances across the plasma membrane. Bulk transport occurs when large molecules must be transported; it involves invagination or evagination of broad areas of the plasma membrane.

PASSIVE TRANSPORT— DIFFUSION AND OSMOSIS

Diffusion is the movement of a substance from an area of high concentration to an area of low concentration and is the result of the constant random motion of the molecules involved. As molecules randomly vibrate, they collide with one another and rebound. The result is a net movement of molecules toward areas of lower concentration. The difference in concentration of a substance between two points of reference is a **concentration gradient.** Diffusion is important in virtually all animal systems. It is the means by which gases are exchanged between cells and their environment, many nutrients are taken into cells, waste products are lost from cells, and ions are exchanged in muscle, nerve, and kidney functions.

Brownian Movement

▼ Use a teasing needle to get a small amount of powdered carmine dye into a drop of water on a microscope slide. Cover with a coverslip. Focus on some of the smallest particles using high power. Notice that the dye particles are vibrating slightly. This vibration is Brownian movement. It is the result of water molecules bumping into dye particles. Trace the course of your dye particle with a pencil in the space below. ▲

Is there any net movement, or does the particle just move around the same average point? _____

Diffusion

▼ Diffusion can occur through gases, liquids, and gels. If there are no air currents in your room, open a container of butyric acid (or some other volatile liquid) in one corner of the room. What do you observe over the course of the period?

Drop a small crystal of potassium permanganate into a test tube of water. Similarly, drop a crystal of potassium permanganate onto a plate of agar. Observe both during the remainder of the period. Do not disturb the test tube. Explain your observations. _____

_____

_____ ▲

Osmosis

Osmosis is the diffusion of water through a selectively permeable membrane. In living systems, it is the plasma membrane that serves as the selectively permeable membrane. A substance's permeability in the plasma membrane is determined by that substance's solubility in lipid, size, and/or charge. Substances soluble in lipid diffuse readily through the lipid portion of the plasma membrane (e.g., oxygen). Substances smaller than eight angstroms (e.g., water) can diffuse rapidly through channels of the plasma membrane. Substances carrying negative charges pass through membrane channels more readily than positively charged substances of similar size. In osmosis, the selectively permeable

membrane separates two solutions containing different concentrations of a nondiffusible solute. Water diffuses along its concentration gradient until concentrations are equal or osmotic pressure balances water's diffusion gradient. **Osmotic pressure** is a measure of the pressure that would be needed to prevent the net diffusion of water from one solution to another. Envision a red blood cell placed in a solution containing a lower solute concentration than the cytoplasm of the red blood cell. Water will diffuse into the red blood cell until the force of water pushing out on the inside of the cell (the hydrostatic pressure) balances osmosis. At equilibrium, as much water is forced out of the cell by hydrostatic pressure as moves into the cell by osmosis.

▼ Osmosis is important in kidney function, blood pressure regulation, exchanges at capillary beds, and many other water-related processes. Osmosis can be demonstrated by a variety of experimental procedures. Your instructor will have prepared the following solutions of sodium chloride: 0%, 0.6%, 0.9%, and 1.5%. Fill each of four test tubes about one-fourth full with one of these solutions. Sterilize the tip of your finger with alcohol, allow to dry, and puncture with a sterile lancet. **Be sure to work only with your own blood and never reuse a lancet.** Add a drop of blood to each test tube. Swirl to mix. Make a wet mount of each of the solutions, examining the cells in the 0.9% solution first. Compare all other cells to these. If the solution outside the cell has a lower concentration of nondiffusible solute than inside the cell, the solution is **hypotonic** to the cell. If the solution outside the cell has a higher concentration of solute, it is **hypertonic** to the cell. If the concentrations are equal, the solution and the cell are **isotonic** to one another (fig. 3.1). Use these terms in filling in table 3.1. ▲

ACTIVE TRANSPORT

Substances that need to be accumulated within cells against a concentration gradient must move by active transport. Active transport involves the expenditure of energy and the use of protein carriers in the plasma membrane. Can you think of situations where substances must move by active transport? _____

_____

As you proceed through this course, you will become aware of the importance of this process in nerve cells, muscle cells, and the endocrine and excretory systems.

The laboratory exercise on the excretory system (exercise 23) has a demonstration of active transport of chlorophenol red across the goldfish kidney tubule. Your instructor may want you to turn to that section at this time.

BULK TRANSPORT

Bulk transport occurs when large molecules must be transported across the plasma membrane without being broken down. These molecules are too large to diffuse through

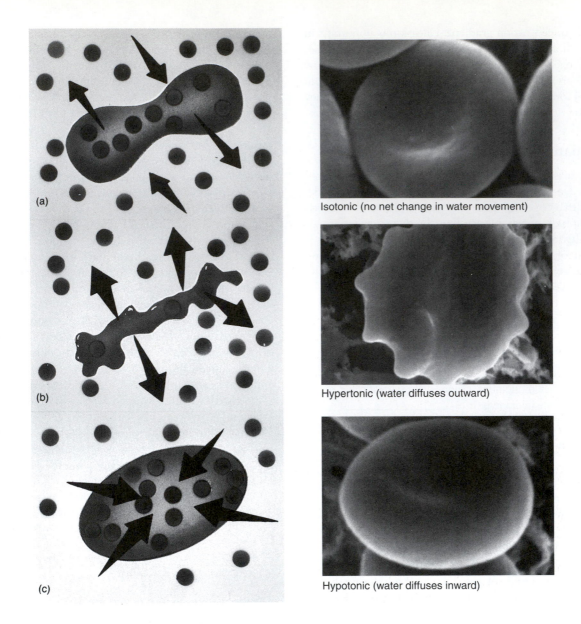

Figure 3.1 Osmosis across red blood cell plasma membranes. *(a)* An isotonic solution has the same concentration of salts as intracellular fluids, and the normal cell shape is retained (×6,900). *(b)* A hypertonic solution causes the cell to lose water and results in a shrunken appearance (×9,000). *(c)* A hypotonic solution causes the cell to gain water and swell. The cells in this scanning electron micrograph are in a slightly hypotonic solution and are just starting to swell (×6,900).

Isotonic (no net change in water movement)

Hypertonic (water diffuses outward)

Hypotonic (water diffuses inward)

TABLE 3.1   Results of the Osmosis Exercise

Solution	Cell Shape	Explanation	The solution is ? to the cell.
0%			
0.6%			
0.9%			
1.5%			

membrane pores or to be carried by membrane carriers. For example, nursing mothers provide their infants with maternal antibodies (large proteins) that must be transported across the gut epithelium of the infant without being broken down. Breaking antibodies down would destroy their function. This transport is accomplished by the plasma membrane invaginating to form a vacuole around the antibody, thus incorporating the antibody into the cell. Bulk transport depends on a very fluid plasma membrane and occurs in a variety of ways. None of the mechanisms will be demonstrated, but two common forms it can take are described in the following two paragraphs.

Endocytosis is a form of bulk transport in which materials are taken into a cell. It occurs by **pinocytosis** (cell drinking) or **phagocytosis** (cell eating). Pinocytosis is the nonspecific uptake of small droplets of extracellular fluid and the substances dissolved in the fluid. Invagination of the plasma membrane and vacuole formation incorporate the fluid into the cell. In phagocytosis, cell processes extend around a solid, engulfing it as an amoeba engulfs a food particle or a white blood cell engulfs a foreign substance. Exercise 8 (Animal-like Protists) has a demonstration of endocytosis by *Paramecium*.

Exocytosis occurs when large molecules are moved out of the cell. A vacuole fuses with the plasma membrane and an opening in the plasma membrane develops to release the contents of the vacuole to the outside. The vacuolar membrane is then simply incorporated into the plasma membrane.

STOP AND ASK YOURSELF

1. What is Brownian movement? _____

   _____

   _____

2. What is the concentration of dissolved salts in isotonic intravenous injections?

   _____

   _____

3. Red blood cells are placed in a 1.2% solution of sodium chloride. This solution is _____ (hypotonic, isotonic, hypertonic) to the cells.

4. In living systems, what serves as the selectively permeable membrane across which osmosis occurs?

   _____

   _____

5. What is active transport? _____

   _____

   _____

AEROBIC CELLULAR RESPIRATION: AN EXERCISE IN SCIENTIFIC METHOD

Measuring the Oxygen Consumed in Cellular Respiration

Aerobic cellular respiration is the process by which most animals make energy available for cellular functions. It consists of a series of reactions, occurring in the cytoplasm and mitochondria, that can be represented by the following equation:

$$C_6H_{12}O_6 + 6O_2 \xrightarrow[\text{mitochondria}]{\text{enzymes}} 6CO_2 + 6H_2O + ATP$$

In these reactions, energy in the form of glucose is converted into a form used by the cell, adenosine triphosphate (ATP). It is assumed that students have covered these reactions in lecture sections.

In detecting aerobic cellular respiration, one can look at the quantity of oxygen used by an organism. When an animal breathes air into its lungs, oxygen is transported by the blood to the cells where it is used in cellular respiration. The rate at which oxygen is used by cells largely determines the rate at which oxygen is delivered to the tissues. This regulation is a complex process involving an interaction of the oxygen levels in body fluids with respiratory centers in the nervous system. Details of that regulation cannot be covered here. It is important to realize, however, that the oxygen needs of an animal are directly correlated to the metabolic demands of its cells. Therefore, it is possible to evaluate cellular respiratory rates by measuring rates of oxygen consumption. Metabolic demands of cells vary depending upon a variety of internal and external conditions (temperature, time of year, body size, hormonal states, etc.).

In this exercise, you will be designing an experiment to test the influence of one variable on the rate of oxygen consumption in an experimental animal. It will be followed by a formal written report in which you will present and discuss your results. (Exercise 1 has a discussion of scientific method.)

The Experimental Apparatus

The experimental setup will be similar to that shown in figure 3.2. A widemouthed flask or similar container is fitted with an airtight stopper. Glass tubing connected to a short length of plastic tubing is fitted into the stopper. The suction end of a 1-ml pipette (0.1-ml gradations) is connected to the tubing. A 5-ml syringe is forced through the stopper as shown. You will be measuring oxygen consumption by measuring the change in gas volume inside the chamber. As you can see in the equation for aerobic cellular respiration, a molecule of CO_2 is released with every molecule of O_2 used. To see a change in the volume of gas in the system, the CO_2 given off must be removed. This is accomplished by enclosing potassium hydroxide (KOH) or soda lime in the chamber.

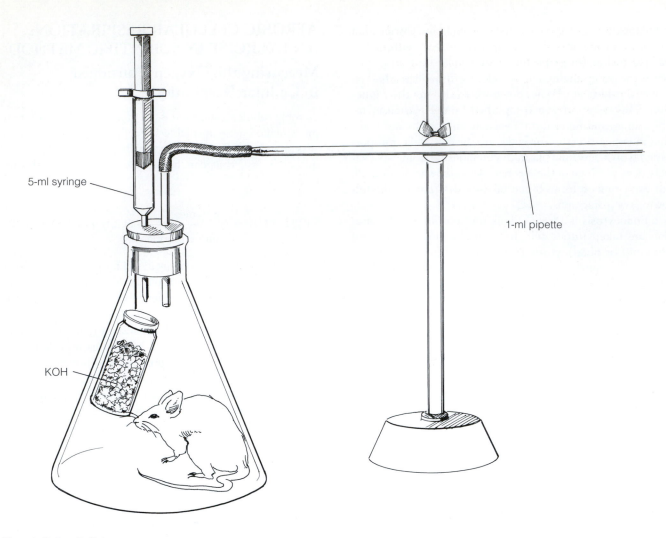

Figure 3.2 Cellular respiration apparatus.

Now, as the animal uses oxygen, the volume of gas in the system will decrease. Decreases in gas volume are measured by introducing a colored liquid into the delivery end of the pipette. By timing the movement of the colored liquid, one can calculate rate of oxygen consumption. When the liquid reaches the end of the graduations of the pipette, the syringe is used to force air into the system, thus returning the colored liquid to the delivery end of the pipette.

Procedure

1. A little water must be added to the flask for aquatic or semiaquatic animals.
2. Keep KOH away from the experimental animal by suspending a vial of KOH from the stopper. KOH is a strong base and can cause burns.
3. Before beginning, weigh your animal(s) so that oxygen consumption can be expressed on a per-gram basis.
4. After placing the test animal in the container, let the animal calm down for a few minutes before starting. Do not disturb the animal during the experiment.
5. Allow the temperature in the container to stabilize before beginning and do not handle the container during the experiment. Temperature changes in the container will cause pressure changes that will result in inaccurate readings. For small mammals, a three-holed stopper may be used to accommodate a thermometer. Temperature changes can be minimized by immersing the bottom of the container in a water bath.
6. In preparing to seal the container, pull the plunger on the syringe to the 5-ml mark before inserting the stopper. Be sure the stopper fits into the container tightly to prevent air leaks. The first run begins on the introduction of the colored liquid into the pipette. For subsequent runs, carefully introduce air into the chamber using the syringe. Be sure to time the movement of the fluid so that oxygen consumption per unit time can be calculated.
7. One control must be in the form of a thermobarometer. This consists of a second setup

excluding the experimental animal. Adjust the colored liquid to the 0.5-ml level. Changes in atmospheric pressure and environmental temperature can be expected to change the volume of gas in all systems. In analyzing data, adjust changes in the other system(s) by adding or subtracting values obtained from the thermobarometer.

8. To make values comparable to published data, they must be corrected to standard temperature and pressure. From your data, calculate the rate of oxygen consumption in ml O_2 consumed/minute, then substitute into the following formula:

$$V_2 = \frac{P_1\,(273)}{T_1\,(760)} \times V_1$$

P_1 = Atmospheric pressure (mmHg) during the experiment.

T_1 = Temperature (°K) during the experiment (°K = °C + 273).

V_1 = Uncorrected volume of oxygen used per minute.

V_2 = Corrected volume of oxygen used per minute.

Presenting Your Results

Writing well is one of the most important skills you can develop. In scientific writing, the format of the paper may be slightly different, but the overall emphasis is the same as in writing for most other disciplines. Be conscious of

• thoroughness and accuracy
• conciseness
• grammar
• plagiarism (the lack of it is explained in the following discussion)
• originality

Preciseness

Most student papers are generally longer than needed. Preciseness in writing means presenting information thoroughly and accurately, yet doing it simply and concisely. Do not use excess words and phrases like "it has been shown." Too many "whiches," "howevers," and "therefores" detract from a paper. By eliminating excess words, most students could shorten their papers by 25% without loss of information.

Science (like any other discipline) has its own technical terminology. It is difficult enough to understand these terms without compounding the difficulty by adding other obscure words. The simpler the better, as long as the meaning doesn't change. One recent student wrote that an insect ". . . commences its feeding activity." Why not say ". . . begins to feed"? Wordiness and obscure writing result in a confused reader. Sentence-by-sentence proofreading is the only way to prevent such problems.

Plagiarism

Plagiarism is the presentation of another's words, ideas, or opinions as one's own. Anything that is not originally yours or that is not common knowledge (e.g., snakes are reptiles, or birds and bats are the only vertebrates that fly) must be cited. Direct quotations must be set off by quotation marks and cited. All other nonoriginal information must be reworded and cited.

Originality

Scientific research and writing is more than putting together others' ideas and reporting on them. While that is an important part of the initial phases of research design, it should stimulate the scientist to think beyond that printed material. Original research and writing means that, even though others may have performed similar research and even though the writer has certainly been influenced by previous work, the essence of what is presented is based on research and ideas that are the writer's own.

Format

The format for a scientific paper varies slightly depending upon the journal for which it is prepared. A typical format, however, would include the following elements:

• title
• abstract
• introduction
• materials and methods
• results
• discussion
• literature cited

Title A short descriptive title should include three elements: (1) the independent variable being tested (e.g., the effect of altered photoperiod), (2) the process being investigated (i.e., aerobic cellular respiration), and (3) the species name of the experimental animal (e.g., a crayfish, *Cambarus setosus*).

Abstract The abstract is a concise summary of information presented in the paper. It should include a statement of purpose, a brief summary of the results, and a statement of the conclusions drawn. The purpose of an abstract is to familiarize the reader with the content of a paper. The reader can then decide whether to read the entire paper.

Introduction The introduction gives a statement of the hypothesis and information on the questions and observations that led to the formulation of the hypothesis. It is not a complete literature review, but a description of other work that influenced the research being described.

Materials and Methods This section precisely describes all equipment and procedures used in carrying out the research. The purpose of this section is to present enough detail so that someone else could carry out the same procedures and confirm the results. It allows a critical analysis of the adequacy of the research methodology.

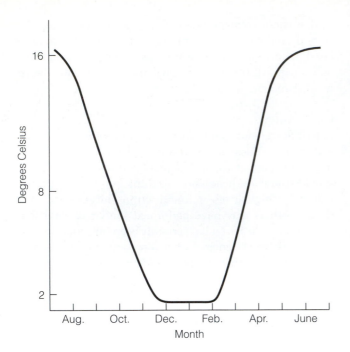

Figure 1 Temperature fluctuations in Parfrey's Glen Creek.

Figure 3.3 Format for a figure.

Table 1. The monthly weight gain in rats on a diet deficient in iodine.

month	mean weight gain
jan.	4.6 g
feb.	5.2 g
—	—
—	—
—	—
—	—
—	—
—	—

Figure 3.4 Format for a table.

Results The data are presented in the results section. Data should be organized into a clearly understandable form. Data might be presented on a graph or in a table. If so, all information for interpretation should be on the graph or in the table. Each graph or table should be on a separate page, should have a reference in the text (Figure 1 shows . . .), and should appear in the format shown in figures 3.3 and 3.4. Figures and tables should be used when they make results easier to understand. No interpretation of the results is appropriate in the results section.

Discussion In the discussion, results are interpreted and related to other studies. Ideas regarding how procedures might be improved and the need for future research can also be discussed. No new data are presented here.

Citations Citations within the body of the paper are made by using author and date. They may be placed at the end of a sentence or incorporated within the sentence. If there are more than two authors, *et al.* may be used. Following are illustrations of citations.

Pennak (1989) states _____

_____ (statement) _____ (Pennak, 1989).

For more than two authors:

_____ (statement) _____ (Dhar et al., 1997).

Everything cited in the body of the paper must have a reference in the literature cited.

Literature Cited This is an alphabetical listing of all resources mentioned in the paper; nothing should appear in the literature cited that is not mentioned in the paper. Alphabetize by the first author's last name. Do not use *et al.* here. Use the format shown in the following examples. Note spaces to be left between the author's name, the date, and the other items.

Format for citing a journal article:
    Dhar, S. R., A. Logan, B. A. McDonald, and J. E. Ward. 1997. Endoscopic investigations of feeding structures and mechanisms in two plectolophous brachiopods. *Invertebrate Biology,* 116 (2): 142–150.

Format for citing a book:
    Pennak, R. W. 1989. *Freshwater Invertebrates of the United States.* New York. John Wiley & Sons, Inc. xvi + 628 pp.

Getting Started

▼ The nature of your experiment will be limited by the variety of organisms available for study and the equipment. Those limits will be defined by your instructor. You may be asked to work in groups, and your instructor may give you specific suggestions on your project. Read through the sections of your textbook covering aerobic cellular respiration and temperature regulation. As you do, consider the following: What environmental factors might affect the rate of aerobic cellular respiration? What is its role in generating metabolic heat for temperature regulation? Is this latter role similar in cold-blooded animals (ectotherms) and warm-blooded animals (endotherms)? Is there any reason that the rate of cellular respiration would vary because of size, age, or sex? Record your observations and questions in worksheet 3.1. Narrow your observations and questions down to a single question and formulate a hypothesis based on that question. Be sure that the hypothesis is in the form of a declarative statement involving a single independent variable. Record your hypothesis in worksheet 3.1. Design a test of your hypothesis and record the procedure in worksheet 3.1. Be sure to record your independent variable, your dependent variable(s), your controlled variables, and construct a data table for recording results. Consult with your instructor for further instructions and then carry out your experiment. You may be asked to present your results as a formal written report using the guidelines presented on pages 43–44. ▲

STOP AND ASK YOURSELF

6. What is plagiarism? _____

_____

7. What is the purpose of KOH in the cellular respiration apparatus? _____

_____

_____

8. What is a thermobarometer? _____

_____

9. List the typical format headings in a scientific paper.

_____

_____

CELL DIVISION

Mitosis is the process by which a nucleus divides, and **cytokinesis** is the process by which the cytoplasm divides. Together, they result in the division of the entire cell. From the single-celled zygote arise all of the billions of cells that make up the adult organism. Not only is it a division of one cell to form two, but it is an exact division that results in two cells genetically identical to the original. Each cell of an organism must have a complete set of genetic instructions to function properly.

The **cell cycle** is shown in figure 3.5. Note that the cell spends most of its time in **interphase.** During interphase, the cell is carrying on processes characteristic of that cell. With regard to events related to mitosis, interphase is divided into three stages: **gap 1 (G₁), synthesis (S),** and **gap 2 (G₂).** G_1 stage is a period of active RNA and protein synthesis. It immediately follows mitosis. The synthesis phase is a period in which DNA **(chromatin)** is replicated (duplicated) so that each **chromosome** will consist of two chromatids when mitosis begins. The two **chromatids** of a chromosome are joined by a differentiated region of DNA, the **centromere.** In different chromosomes, the position of the centromere may vary from the middle to either end of the chromosome. The G_2 stage is a period of protein syn-

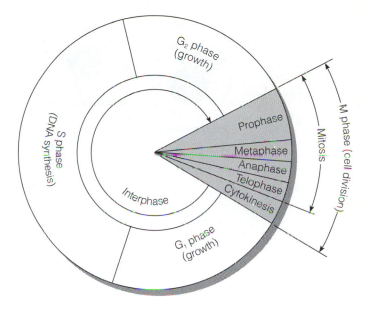

Figure 3.5 The cell cycle.

thesis prior to mitosis. The duration of interphase varies, depending on species and cell type, from hours (embryonic cells) to years (adult bone cells).

Mitosis is divided into four phases that represent artificial points of reference for study of the process: prophase, metaphase, anaphase, and telophase. In reality, mitosis is a smooth, continuous series of changes with no sharp breaks between stages. This can be emphasized by applying the terms "early" or "late" to cells that seem to be between two stages.

MITOSIS IN THE WHITEFISH BLASTULA

▼ Obtain a slide of a whitefish blastula. The blastula is an early stage in animal development. It is a hollow ball of rapidly dividing cells. Your slide will show one or more thin sections through that ball of cells. Read through the following descriptions and then examine your slide under low power. Switch to high power, looking for cells representative of each of the stages of mitosis (figs. 3.6 and 3.7). You may need to look at other sections or other slides to see all stages. Remember that mitosis is a continuous process and that you are likely to see intermediate stages. Make drawings of what you see. ▲

Prophase

The beginning of prophase is signaled by the migration of pairs of centrioles to opposite ends (poles) of the cell and the breakdown of the nuclear envelope and nucleolus. Chromatin

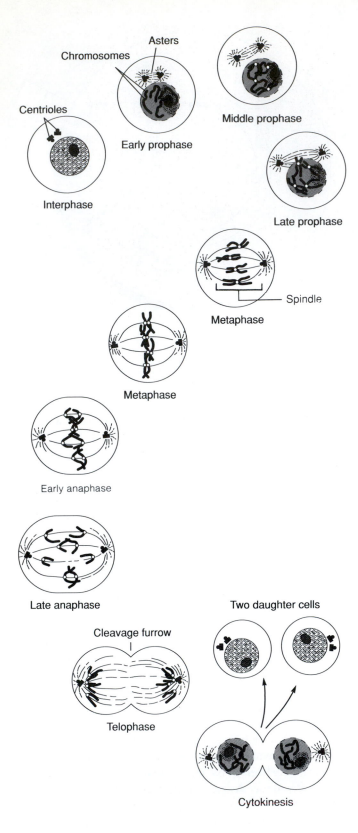

Figure 3.6 Animal mitosis.

begins to condense (winding and folding of DNA strands) into chromosomes, and microtubules begin to form from proteins of the cytoplasm. In prophase, chromosomes consist of two chromatids (called sister chromatids) joined at the centromere. Centrioles organize microtubule assembly. The role of centrioles in cell division is still not well understood and is under investigation. Microtubules are involved with the movement of chromosomes in the stages that follow. When viewed under the light microscope, it seems that the microtubules push and pull chromosomes.

Metaphase

In metaphase, chromosomes move to the equator of the cell. Some microtubules (**spindle fibers**) originate at the poles and attach at the centromere of each chromosome at a disk of protein called the kinetochore. Other microtubules run from pole to pole. Microtubules radiating from the poles of the cell form the **aster.** The centrioles and the microtubules radiating between them form the **mitotic spindle.** Late in metaphase, chromosomes are in position along the equator of the cell and the centromeres divide. This latter event frees the sister chromatids from each other (see fig. 3.7a).

Anaphase

Anaphase begins as chromatids of each chromosome separate and move toward opposite poles of the cell. At this point, the sister chromatids are considered full-fledged chromosomes. Depending on the position of the centromere, chromosomes will appear straight, J-shaped, or V-shaped, with the centromere at the apex of each leading the chromosome to the pole. Activities of the pole-to-pole spindle fibers result in an elongation of the entire spindle. Anaphase ends as chromosomes approach the poles of the cell (see fig. 3.7b).

Telophase

Telophase is essentially the reverse of prophase. Chromosomes have reached the poles of the cells, furrowing of the plasma membrane is initiated, spindle fibers disappear, chromosomes "unwind" into chromatin (dispersed DNA and protein strands), and the nuclear envelope and nucleolus reappear. Upon completion of furrowing, cytokinesis is considered to be ended and two new daughter cells result. Each daughter cell enters G_1 interphase with chromosomes consisting of one chromatid each and with exactly the same quality of DNA as the parent cell. DNA is duplicated in S phase, so each chromosome will have two chromatids before the next division. A second pair of centrioles is also synthesized during S phase.

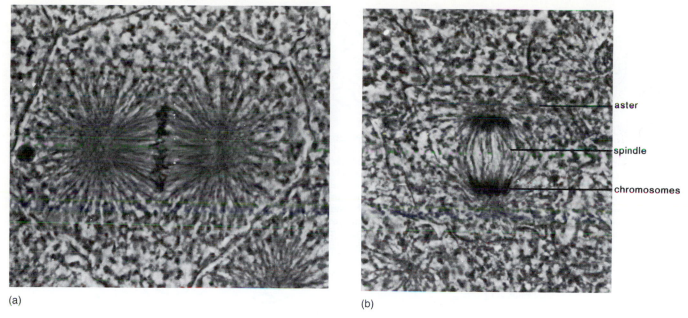

(a)   (b)

Figure 3.7 Mitosis in the whitefish blastula: (*a*) metaphase; (*b*) late anaphase.

CYTOKINESIS IN THE WHITEFISH BLASTULA

Look again at your slide of a whitefish blastula and find a cell in late anaphase of mitosis. As chromosomes approach the poles of a dividing animal cell, cytokinesis begins. Cytokinesis is the division of the cytoplasm by cleavage of the cell at the equator, a process called **furrowing.** Microfilaments are responsible for this gradual constriction of the plasma membrane.

A COMPARISON OF MITOSIS AND MEIOSIS

Meiosis is a form of cell division that produces the cells involved with sexual reproduction. The products of this form of cell division are usually called eggs and sperm in animals and are referred to as **gametes.** During sexual reproduction, gametes fuse to form a single cell called a **zygote.** The zygote then divides by mitosis to produce the various stages characteristic of an animal's embryonic development.

To maintain a constant number of chromosomes in the next generation, gametes must be produced that have one-half the chromosome number of ordinary body cells. Chromosomes exist in pairs called homologous pairs. Each member of a pair of chromosomes is similar in size and shape and carries genes that code for the same traits of the animal. When a cell contains both members of all homologous pairs, it is diploid (2N). When a cell contains only one member of all pairs of chromosomes, it is haploid (1N). All body cells of most animals, except sperm or eggs, possess a diploid (2N) number of chromosomes. Half of these chromosomes came to the animal through the sperm cell of the father, and half of the chromosomes came in the egg cell of the mother. Meiosis is the form of cell division that produces haploid gametes. Fertilization results in the union of sperm and egg cells and the restoration of the diploid number of chromosomes.

Compare the processes of mitosis and meiosis as shown in figure 3.8. Unlike mitosis, meiosis consists of two nuclear divisions, called meiosis I and meiosis II. During meiosis I, homologous pairs of chromosomes pair in a process called synapsis. The pairs then separate and move toward opposite poles of the cell. The two cells produced by this first division contain half of the number of chromosomes present in the parental cell. The chromosomes of these cells consist of two chromatids. Meiosis II is just like mitosis, and the cells produced have chromosomes consisting of one chromatid.

The overall result of mitosis is the production of two cells that are genetically identical to the parent cell. It is the form of cell division that results in the production of all of the cells of an animal's body from a single fertilized egg. In meiosis, chromosome numbers are halved. Members of a pair of chromosomes sort independently during the first division, and parts of homologous chromosomes can be exchanged. The result is that the cells produced are always genetically different from one another.

Exercise 24 has an activity that helps one understand the processes of spermatogenesis and oogenesis. Your instructor may want you to turn to that exercise at this time.

STOP   AND   ASK   YOURSELF

10. What occurs during the S phase of the cell cycle?

_____

_____

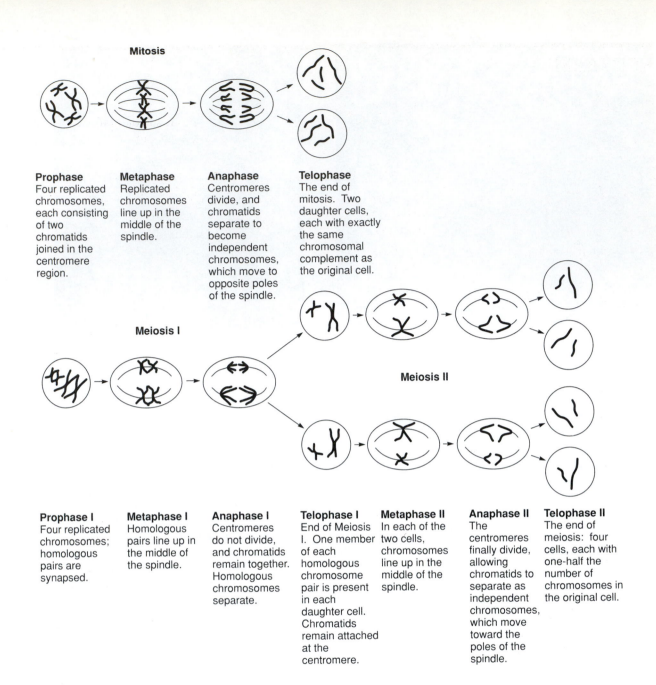

Mitosis

Prophase
Four replicated chromosomes, each consisting of two chromatids joined in the centromere region.

Metaphase
Replicated chromosomes line up in the middle of the spindle.

Anaphase
Centromeres divide, and chromatids separate to become independent chromosomes, which move to opposite poles of the spindle.

Telophase
The end of mitosis. Two daughter cells, each with exactly the same chromosomal complement as the original cell.

Meiosis I

Meiosis II

Prophase I
Four replicated chromosomes; homologous pairs are synapsed.

Metaphase I
Homologous pairs line up in the middle of the spindle.

Anaphase I
Centromeres do not divide, and chromatids remain together. Homologous chromosomes separate.

Telophase I
End of Meiosis I. One member of each homologous chromosome pair is present in each daughter cell. Chromatids remain attached at the centromere.

Metaphase II
In each of the two cells, chromosomes line up in the middle of the spindle.

Anaphase II
The centromeres finally divide, allowing chromatids to separate as independent chromosomes, which move toward the poles of the spindle.

Telophase II
The end of meiosis: four cells, each with one-half the number of chromosomes in the original cell.

Figure 3.8 A comparison of mitosis and meiosis.

11. At what stage of mitosis are chromosomes lined up along the equator of the cell? _____

_____

12. At what stage of mitosis do chromatids of a chromosome separate and begin moving toward opposite poles of the cell?

_____

13. At what stage of mitosis is furrowing completed?

_____

14. Describe the events of prophase of mitosis. _____

_____

_____

_____

_____

KEY TERMS

active transport 38

aster 46

bulk transport 38

cell cycle 45

centromere 45

chromatids 45

chromatin 45

chromosome 45

concentration gradient 39

cytokinesis 45

diffusion 38

endocytosis 41

exocytosis 41

furrowing 47

gametes 47

gap 1 (G_1) 45

gap 2 (G_2) 45

hypertonic 39

hypotonic 39

interphase 45

isotonic 39

meiosis 47

mitosis 45

mitotic spindle 46

osmosis 39

osmotic pressure 39

passive transport 38

phagocytosis 41

pinocytosis 41

spindle fibers 46

synthesis (S) 45

zygote 47

WORKSHEET 3.1   Aerobic Cellular Respiration

Observations and Questions from Background Reading

Hypothesis

Materials and Methods
   Identify the variables:
      Independent variable

      Dependent variable(s)

      Controlled variables

   Describe your experiment

Use an "if/then statement" to predict the results of your experiment.

Results
Construct a data table for recording data during your experiment.

WORKSHEET 3.2   Aspects of Cell Function

Transport Into and Out of Cells

1. Define osmosis. Give an example of where it occurs in animal systems.

2. What will happen to a blood cell placed in a hypertonic solution?

3. What is a concentration gradient? What is the importance of a concentration gradient in active and passive transport mechanisms?

Aerobic Cellular Respiration

4. Use the formula for cellular respiration to explain why oxygen consumption during gas exchange can be used in measuring rates of cellular respiration.

5. Why is it necessary to standardize the data derived in this exercise to conditions of standard temperature and pressure?

Cell Division

6. Compare and contrast the quantity and composition of genetic material of a cell entering mitosis to that of the daughter cells that result from mitosis.

7. Compare and contrast the composition of genetic material of a cell entering meiosis to that of the daughter cells that result from meiosis.

8. Why is S phase of interphase critically important to the process of mitosis?

9. Draw an animal cell in metaphase of mitosis, and then in anaphase of mitosis.

Exercise 4

Genetics

Learning Objectives

After completing this exercise, you should be able to do the following:

1. Predict the results of crosses involving one or two autosomally linked traits.

2. Predict the results of crosses involving single X-linked traits.

3. Perform crosses using the classical genetic organism, *Drosophila melanogaster*.

4. Analyze data using a simple statistical test.

5. Describe the principles of segregation and independent assortment and use these principles in solving problems involving monohybrid, dihybrid, and X-linked crosses.

Prelaboratory Quiz

Study this week's laboratory exercise and then complete the following quiz to assess your preparation for the laboratory.

1. Two chromosomes containing genes coding for the same group of traits are called
   a. allelic chromosomes.
   b. homologous chromosomes.
   c. homozygous chromosomes.
   d. heterozygous chromosomes.

2. A gene that, when present, is always expressed is said to be
   a. dominant.
   b. recessive.
   c. homozygous.
   d. heterozygous.

3. The genes that code for alternate expressions of a trait are called
   a. alleles.
   b. homologous genes.
   c. homozygous genes.
   d. recessive genes.

4. Use the word "alleles" or "genes" to fill in the following blanks.
   a. The two _____ for body color studied in this week's laboratory are ebony and wild.
   b. Diploid organisms possess two sets of chromosomes containing _____ that code for the same traits, but not necessarily the same expression of those traits.
   c. When a fly has the recessive ebony phenotype, both _____ coding for the body color trait must be the same.

5. True/False   A fruit fly that has sex combs on its prothoracic legs; a genital plate; and a dark, wide band at the tip of the abdomen is a female.

6. True/False   Fruit flies have four larval instar stages and a pupal stage.

7. True/False   In fruit flies, the genetic symbol used to represent a trait is usually taken from the mutant expression of the trait.

8. True/False   The expression "e^+" is indicating a recessive allele derived by mutation from the wild allele, e.

9. True/False   The expression "B" is indicating a dominant mutant derived from a recessive wild allele indicated by B^+.

10. True/False   The expression "dp^+" is indicating the dominant wild allele of dp.

11. True/False   The chi-square test is used in this laboratory exercise to determine whether or not the results of your crosses approximate the results expected based on Mendelian inheritance patterns.

12. True/False   The time from egg-laying to emergence of fruit fly adults spans approximately two weeks.

13. True/False   This week's laboratory exercise involves two sets of fruit fly crosses.

14. True/False   The sex-linked trait used in this week's laboratory exercise is a wing-shape trait.

15. True/False   The second fruit fly cross in this week's laboratory exercise involves a demonstration of Mendel's principle of independent assortment and genes carried on a single pair of homologous chromosomes.

Genetics is the study of the mechanisms of transmission of genes from parents to offspring. These mechanisms are dependent on the behavior of chromosomes during the formation of gametes and the way in which gametes are brought together in fertilization. The processes occurring at the cellular level are those that take place during meiosis. In this exercise, it is assumed that the student is thoroughly familiar with meiosis. If not, review the process of meiosis. Be able to explain the results of these exercises based on what occurs in meiosis.

Diploid organisms possess two sets of chromosomes containing genes that code for the same traits. Chromosomes containing genes that code for the same traits are referred to as homologous chromosomes, and the genes that code for alternate expressions of a trait are referred to as alleles. Genes are often either dominant or recessive. A dominant gene is one that, when it is present, is always expressed. A recessive gene is one that is expressed only when a similar recessive gene is on the homologous chromosome. When the two genes that determine a trait are alike, the organism is homozygous for that trait. When the two genes are different, the organism is heterozygous for that trait. Thus, a dominant trait is expressed in either the homozygous or heterozygous state; a recessive trait is expressed only in the homozygous state.

INHERITANCE IN *DROSOPHILA*— MATERIALS AND METHODS

In this exercise you will be making crosses involving the fruit fly, *Drosophila melanogaster*. The development from egg to adult spans approximately two weeks, as shown in figure 4.1. Because of this, the following exercise will span the next few laboratory periods. You should be prepared to come to the laboratory outside of the normal class period to analyze your results.

Anesthetizing Fruit Flies

▼ Obtain a vial of mixed fruit flies. Charge an etherizer with ether. Gently tap the flies to the bottom of the vial. Remove the plug from the vial and tap the flies into the etherizer. Allow the etherizer to sit until the activity has ceased. At this time, charge a reetherizer (funnel with cotton forced into the stem) with ether. Set the reetherizer open-end-down on the table to prevent evaporation of ether. If the flies start to wake while working with them, place the reetherizer over them. If done properly, any counting or sorting done to that point will not be disrupted. An alternate anesthetizing procedure may be described by your instructor using a commercially prepared fruit fly anesthetic. ▲

Determination of Sex

▼ When flies are anesthetized, empty them onto a white 3 × 5 card. Examine them under a dissecting microscope. Table 4.1 and figure 4.2 show five characteristics that can be used in distinguishing sexes. All characteristics are easily visible under the dissecting microscope. To see clearly the sex combs on the prothoracic legs (front legs), use the high power of the dissecting microscope. In manipulating flies, use a fine-bristled brush. Separate flies into males and females. You will find the banding pattern on the abdomen and the appearance of the genitalia easiest and quickest to use. Have the instructor confirm your decisions. ▲

Distinguishing Phenotypes

▼ Notice that the flies are of a number of different body forms. Wild-type flies have long wings, gray-brown bodies, and red eyes. A mutation called "ebony" is expressed as a

TABLE 4.1	Distinguishing Sexes in *Drosophila melanogaster*	
	Male	**Female**
Overall size	Smaller	Larger
Sex combs on prothoracic legs	Present	Absent
Abdomen shape	Rounded	Tapering
Abdomen color pattern	Large terminal dark band	More, narrower bands
Genitalia	Sclerotized	Nonsclerotized

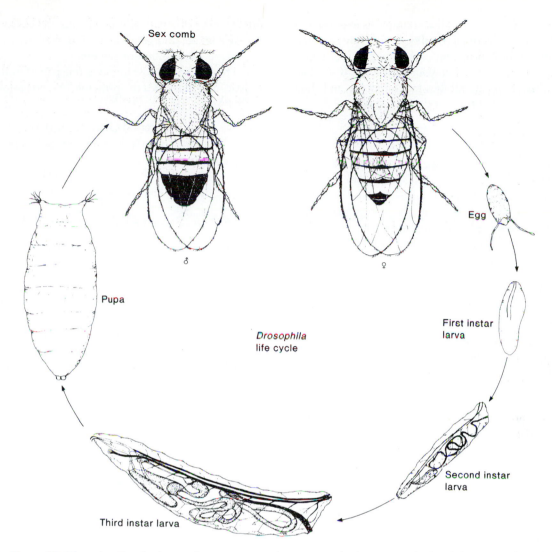

Figure 4.1 *Drosophila* life cycle. The third instar larva is present about one week after eggs are laid. Adults emerge from the pupal case two weeks after eggs are laid.

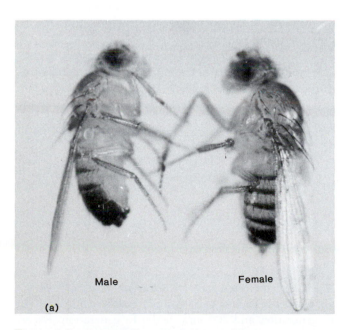

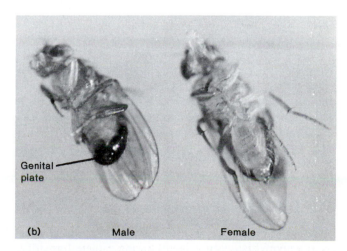

Figure 4.2 Distinguishing sexes of *Drosophila*.

black-bodied fly. A mutation called "dumpy" is expressed in wings that just reach the tip of the abdomen or have a notch along the inside terminal-margin of the wing. Other mutations may have been selected by your instructor. If so, they will be described. Separate all flies by phenotype and then have your instructor confirm your decisions. ▲

GENETIC SYMBOLS

The conventional method of representing alleles for a particular locus is based on the name for the dominant expression of the trait. The dominant allele is represented by an uppercase letter from the name of the dominant expression. The recessive allele is represented by the lowercase of the same letter. In the fruit fly, all traits are derived from mutations of wild-type genes. The wild-type expressions of traits are usually dominant. Using an uppercase W to represent many different wild-type traits would be impossible. The alternative convention with the fruit fly is to use a symbol derived from the mutant trait to represent the mutant allele. If the mutant allele is recessive, it is represented by a lowercase letter; if it is dominant, the mutant allele is represented by an uppercase letter. The wild-type allele is represented by the same upper- or lowercase letter followed by a superscript $^+$.

STOP AND ASK YOURSELF

1. Genes coding for different expressions of a particular trait are called _____ .

2. When two genes that determine a trait are alike, the organism is _____ for the trait.

3. What characteristics can be used to distinguish male and female fruit flies? _____
_____

4. Sepia is a recessive mutant that results in a brown eye color when it is in the homozygous state in a fruit fly. Propose genetic symbols to represent the sepia and wild alleles. Sepia _____ ; Wild
_____

EXPERIMENTAL PROCEDURE

Depending on the amount of time available, you will be provided either with parental flies or progeny of the parental flies (first filial or F_1 generation). If the former, your crosses will span four or five weeks; if the latter, two or three weeks. The timing of the manipulations involved is indicated in table 4.2 in worksheet 4.1. Fill in the dates for later reference.

Cross Involving One Pair of Characters— The Principle of Segregation

Mendel's principle of segregation states that pairs of genes are distributed between gametes during gamete formation. Segregation occurs during meiosis when genes carried on

one member of a pair of homologous chromosomes end up in one gamete, and genes carried on the other member are segregated into a different gamete.

The following cross involves one pair of fruit fly characteristics and illustrates the principle of segregation. Cross I is as follows:

Parental Generation (P)
Homozygous Wild × Homozygous Ebony
↓
First Filial Generation (F_1)
F_1 × F_1 (Brother—Sister mating)
↓
Second Filial Generation (F_2)

Determine the genotypes of all flies involved in this cross, and complete the Punnett squares for Cross I in worksheet 4.1.

▼ Place two etherized males and two etherized virgin females of appropriate genotypes in each of two vials and then plug the vials. Keep the vials on their sides until the flies become active again. This prevents flies from sticking in the media. Label the vials with the cross (Cross I), the generation (P or F_1), the date, and your initials. You should score at least 50 flies in the F_1 generation and 100 flies in the F_2 generation. Record your data in table 4.3 in worksheet 4.1. ▲

Cross Involving Two Pairs of Characters— The Principle of Independent Assortment

Mendel's principle of independent assortment states that during gamete formation, pairs of genes carried on different pairs of homologous chromosomes segregate independently of one another. Independent assortment means that during meiosis, when homologous pairs of chromosomes line up at metaphase I and then segregate, the behavior of one pair of chromosomes does not influence the behavior of any other pair. Offspring of a cross between individuals carrying genes for two traits should show random combinations of the expressions of the two traits.

The following cross involves two pairs of fruit fly characteristics and illustrates the principle of independent assortment. Cross II is as follows:

Parental Generation (P)
Homozygous Wild × Homozygous Dumpy Ebony
↓
F_1 Generation
F_1 × F_1 (Brother—Sister mating)
↓
F_2 Generation

Determine the genotypes of all flies involved in this cross, and complete the Punnett squares for Cross II in worksheet 4.1.

▼ Make your crosses as before using two males and two virgin females in each of two vials. Score at least 50 flies in the F_1 generation and 200 flies in the F_2 generation. Record your data in table 4.4 in worksheet 4.1. ▲

Cross Involving X (Sex) Linkage

Genes carried by sex (X) chromosomes show different inheritance patterns depending on whether the male or female introduces a particular expression into a cross. When a trait is determined by genes carried by the X chromosome, it is said to be X-linked, and the inheritance pattern is called sex-linked inheritance. The symbols used to represent X-linked traits are an X—to indicate that the trait is X-linked—followed by a superscript—to indicate the particular allele. Females have two X chromosomes, and both Xs are indicated in the genotype. In the male, a single X is present, and the Y chromosome replaces the second X in the genotype.

You should work in teams in Cross III. Each team will receive flies for the following crosses.

Cross A		Cross B	
White-Eyed × Wild Male	(P)	Wild × White-Eyed	
Female		Female	Male
↓		↓	
F₁ Generation	Brother—	F₁ Generation	
	Sister		
F₁ × F₁	matings	F₁ × F₁	
↓		↓	
F₂		F₂	

Determine the genotypes of all flies involved in these crosses and complete the Punnett squares for Cross III in worksheet 4.1.

▼ Each half of a team should be responsible for the care and scoring of cultures for either Cross IIIA or IIIB. At the end of the exercise, share results. Members of each team should make crosses as before using two males and two females in each of two vials. Score 200 flies in both the F_1 and F_2 generations. Be sure to record both sex and eye color (tables 4.5 and 4.6 in worksheet 4.1). ▲

Written Report

▼ An informal written report is due one week after the F_2 generations are scored. Summarize the results of your crosses by answering the following for each cross.

1. Which character is dominant and which is recessive?
2. How do you know?
3. What is the ratio of dominant to recessive in the F_2 generation?

4. What ratio would you expect (see worksheet 4.1)? How many flies of each phenotype would you expect assuming a total equal to the total number of flies you counted?
5. To test whether your results conform to what you expected (from question 4), perform chi-square tests. The chi-square test is a statistical tool that can be used to determine the probability of obtaining a set of results, assuming some theoretical expectation. You will be testing your results based on the predictions of Mendelian genetics. If the probability of obtaining a set of results (given the theoretical expectation) is less than 0.05 (5%), the results are said to be significantly different than expected. If the probability of obtaining a set of results is greater than 0.05, the results are said to conform to the theoretical expectation. Rather than calculating a probability, you will calculate a unitless number, the chi-square value. The chi-square formula is as follows:

$$X^2 = \Sigma \frac{(O - E)^2}{E}$$

X^2 = Chi-square value

O = Observed number of offspring in one phenotypic class

E = Expected number of offspring in the same phenotypic class

Σ = The summation of $\frac{(O - E)^2}{E}$ for all phenotypic classes

The chi-square value becomes greater as the difference between observed and expected increases. If this chi-square value is greater than the chi-square value associated with a 0.05 probability, the results are interpreted as being different than expected. If the calculated chi-square value is less than that associated with a 0.05 probability, the results conform to the expectation. Normally, the chi-square value associated with 0.05 probability is taken from a table of chi-square values. Tables 4.7–4.9 in worksheet 4.1 give instructions for calculating the chi-square value and give the chi-square value associated with the 0.05 probability for each of your crosses.

6. Explain any results that deviate from what you expected.
7. In Cross III, explain the differences between crosses A and B. ▲

KEY TERMS

WORKSHEET 4.1 Genetics

	Approximate Elapsed Time	Cross I	Cross II	Cross III
TABLE 4.2 Dates of Fruit Fly Manipulations				
Parental cross	(Day 0)			
Parents removed	(Day 7)			
F₁ Scored	(Day 14)			
F₁ Cross performed	(Day 14)			
F₁ Parents removed	(Day 21)			
F₂ Scored	(Day 28)			

1. Expected Results from Cross I
   Symbols used to represent the alleles involved in this cross:

   ebony—e
   wild—e^+

   Genotypes and phenotypes of individuals and gametes:
   (a)   Parental generation

   homozygous ebony   homozygous wild

   (b)   Parental gametes

   (c)   F₁ generation genotype

   (d)   F₁ generation phenotype

(e) F₁ gametes

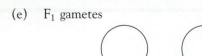

(f) Punnett square for F₂ progeny.

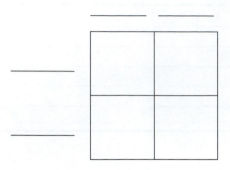

(g) Expected phenotypic ratio for the F₂ generation

2. Expected Results from Cross II
Symbols used to represent the alleles involved in this cross:

dumpy—dp ebony—e
wild—dp^+ wild—e^+

Genotypes and phenotypes of individuals and gametes:
(a) Parental generation

homozygous dumpy ebony   homozygous wild

(b) Parental gametes

(c)   F₁ generation genotype

(d)   F₁ generation phenotype

(e)   F₁ gametes

(f)   Punnett square for F₂ progeny

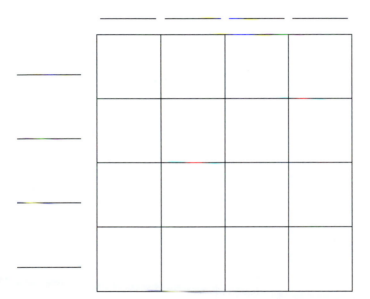

(g)   Expected phenotypic ratio for the F₂ generation

3. Expected Results from Cross III
   Symbols used to represent the alleles involved in this cross:

   wild eye—X^{w+}
   white eye—X^{w}

Genotypes and phenotypes of individuals and gametes:

(a) Parental generation
     Cross A

wild male   white-eyed female

     Cross B

white-eyed male   wild female

(b) Parental gametes
     Cross A

male   female

     Cross B

male   female

(c) F₁ generation genotype
     Cross A

male   female

     Cross B

male   female

(d) F_1 generation phenotype
      Cross A

       male   female
      Cross B

       male   female

(e) F_1 gametes
      Cross A

 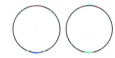

         male   female

      Cross B

         male   female

(f)   Punnett square for cross A F_2 progeny

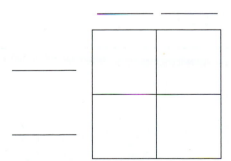

(g)   Phenotypic ratio for cross A F_2 progeny

(h)   Punnett square for cross B F_2 progeny

___ ___		
___		
___		

(i)   Phenotypic ratio for cross B F_2 progeny

TABLE 4.3 Data from Cross I	
Wild Flies	Ebony Flies
F_1	
F_2	

TABLE 4.4 Data from Cross II			
Wild Wild	Wild Ebony	Dumpy Wild	Dumpy Ebony
F_1			
F_2			

TABLE 4.5 Data from Cross IIIA			
Cross A White-Eyed Female $\times$ Wild Male			
White-Eyed		Wild	
Male	Female	Male	Female
F_1			
F_2			

TABLE 4.6 Data from Cross IIIB			
Cross B Wild Female $\times$ White-Eyed Male			
White-Eyed		Wild	
Male	Female	Male	Female
F_1			
F_2			

TABLE 4.7   Chi-square Calculations for Cross I, F_2 Generation

	Ebony Flies	Wild Flies	Total
Observed number (O)			
Expected number (E)			
$(O - E)^2$			*
$\dfrac{(O - E)^2}{E}$			

$$X^2 = \Sigma \frac{(O - E)^2}{E}$$

$$X^2 = \underline{\hspace{3cm}}$$

X^2 at 0.05 probability $= 3.84$
*Total is not needed here.

TABLE 4.8   Chi-square Calculations for Cross II, F_2 Generation

	Wild Wild	Wild Ebony	Dumpy Wild	Dumpy Ebony	Total
Observed number (O)					
Expected number (E)					
$(O - E)^2$					*
$\dfrac{(O - E)^2}{E}$					

$$X^2 = \Sigma \frac{(O - E)^2}{E}$$

$$X^2 = \underline{\hspace{3cm}}$$

X^2 at 0.05 probability $= 7.81$
*Total is not needed here.

TABLE 4.9 Chi-square Calculations for Cross III, F_2 Generations (Test the cross you scored.)

	White-Eyed Males	Wild Males	White-Eyed Females	Wild Females	Total
Observed number (O)					
Expected number (E)					
$(O - E)^2$					*
$\dfrac{(O - E)^2}{E}$					

$$X^2 = \Sigma \frac{(O - E)^2}{E}$$

$$X^2 = \underline{\hspace{2cm}}$$

X^2 at 0.05 probability $= 7.81$
*Total is not needed here.

WORKSHEET 4.2   Genetics Problems

1. A geneticist crossed a homozygous ebony-bodied fruit fly with one that is heterozygous for the ebony body color. What are the genotypes of both flies and phenotypic and genotypic ratios of flies in the next generation?

2. Indicate the genotypes of the parents in the following crosses.

   Parents   Progeny

   a. _____ ebony × _____ wild   82 ebony   78 wild
   b. _____ ebony × _____ wild   0 ebony   74 wild
   c. _____ wild × _____ wild   39 ebony   118 wild

3. A geneticist crossed a wild-bodied, ebony fruit fly with one that is dumpy-bodied and wild in color. Assuming that both flies are heterozygous for the wild traits, what are the phenotypes and ratios expected for the next generation of fruit flies?

4. A geneticist crossed a dumpy ebony fruit fly to a wild fly that is heterozygous for both traits. What phenotypic ratio of offspring would you expect to see in the next generation?

5. Supply the genotypes of the fruit flies described below.
   a. A wild fly that had one dumpy ebony parent:
      Dumpy ebony parent   Wild fly

   b. A dumpy-bodied, wild-colored fruit fly that had a wild-bodied, ebony offspring:
      Dumpy wild fly   Wild ebony offspring

   c. A fly that produced all of the following kinds of gametes: dpe^+, dp^+e^+, dpe, dp^+e.

6. Describe the results of a cross between a white-eyed female fruit fly and a wild-eyed male fruit fly.

7. Describe the results of a cross between a female that is heterozygous for the white-eye trait and a white-eyed male.

8. Explain why all of the male offspring of a white-eyed female will have white eyes, while none of the male offspring of a white-eyed male will have white eyes unless the mother also carries a white-eye allele.

Exercise 5

Embryology

Learning Objectives

After completing this exercise, you should be able to do the following:

1. Describe the goals of embryological studies.

2. Describe the events of fertilization and early embryology for the sea urchin and the frog.

3. Explain how the yolk influences patterns of cleavage.

4. Recognize stages in the embryology of the sea urchin and the frog.

5. Describe the significance of the three embryonic germ layers—ectoderm, endoderm, and mesoderm—including the variety of structures that each forms in the adult animal.

Prelaboratory Quiz

Study this week's laboratory exercise and then complete the following quiz to assess your preparation for the laboratory.

1. Cleavages of a fertilized egg usually begin at the
   a. animal pole.
   b. vegetal pole.
   c. yolky end of the zygote.
   d. point of fertilization.

2. The embryonic germ layer that gives rise to the embryonic gut is the
   a. ectoderm.
   b. endoderm.
   c. mesoderm.

3. The embryonic germ layer that gives rise to muscles, blood, and blood vessels is the
   a. ectoderm.
   b. endoderm.
   c. mesoderm.

4. In this week's laboratory, we will be studying the development of
   a. an echinoderm only.
   b. an amphibian only.
   c. an echinoderm and an amphibian.
   d. an echinoderm and a bird.

5. This structure blocks the entry of additional sperm into the egg after penetration by the first sperm cell.
   a. gray crescent
   b. blastopore
   c. yolk plug
   d. fertilization membrane

6. Gastrulation results in the formation of the embryonic gut or
   a. coelom.
   b. archenteron.
   c. mesoderm.
   d. chordamesoderm.

7. A hollow ball of cells that follows the morula stage is called the
    a. gastrula.
    b. pluteus.
    c. blastula.
    d. neurula.

8. Expansion of the ectoderm toward the vegetal pole of an amphibian embryo is called
    a. involution.
    b. epiboly.
    c. neurulation.
    d. differentiation.

9. The notochord of an amphibian embryo forms from
    a. somites.
    b. neural plate.
    c. chordamesoderm.
    d. ectoderm.

10. The neural tube of an amphibian embryo forms when _____ roll(s) up and over the dorsal midline of the embryo.
    a. notochord
    b. neural folds
    c. somites
    d. archenteron cells

11. True/False   The coelom of an echinoderm embryo forms from outpocketings of the gut tract.

12. True/False   The blastopore of an echinoderm embryo forms as a result of gastrulation.

13. True/False   A thickened and flattened surface on the dorsal aspect of an amphibian embryo is called the chordamesoderm.

14. True/False   The coelom of an amphibian embryo forms by hollowing out blocks of mesoderm sandwiched between ectoderm and endoderm.

15. True/False   The yolk plug of an amphibian embryo forms during epiboly.

Embryology is the study of the changes a fertilized egg goes through in developing to the point at which major organ systems are differentiated and the basic body plan is established.

Gaining an understanding of embryological processes is a basic part of any zoologist's training. Embryological studies are often used to gain an understanding of the evolution of a particular group of organisms. The embryological development of an organ or organ system may parallel the changes that occurred in evolution. This "ontogeny-recapitulates-phylogeny" concept has been misused in carrying it to the extreme, but when used within reasonable limits, it remains a useful concept in evolutionary studies.

Embryological studies are also performed by developmental biologists. They study the cellular processes that control various aspects of development. The goal in developmental biology is to describe the mechanisms that determine why certain cells develop into one set of structures and other cells into other structures. Developmental biologists use this information to try to determine what happens to cells when they malfunction (e.g., become malignant).

EGG TYPES AND CLEAVAGE PATTERNS

All of the cells of an adult animal are the products of the single cell, called the zygote, that resulted from fertilization. Mitotic divisions of the zygote are called **cleavages** and result in cells called **blastomeres.** The eggs of most animals have one end that is less dense, the **animal pole.** The initial cleavage frequently begins at the animal pole. The opposite end (usually where yolk accumulates) is the **vegetal pole.**

Differences in egg size and cleavage patterns among animal species are reflections of a species' evolutionary history and give clues to developmental processes. Closely related animals usually have similar eggs, cleavage patterns, and developmental stages. The quantity and distribution of the embryonic food reserve—the yolk—often reflect the length of independent development. In other words, embryos of species that survive for long periods before feeding begins (or nourishment is provided by a parent) usually have large yolk supplies. The size of the egg sometimes is a reflection of the size of the offspring.

The size of the egg and the amount of yolk influence the size of the blastomeres that result from cleavage and the rate of cleavage. Cleavages that completely divide an egg are holoblastic cleavages. In eggs with moderate to large quantities of unevenly distributed yolk, the cleavages slow, or even stop, when yolk is encountered. If cleavages cannot completely divide the yoke, the cleavages are meroblastic. The development of these embryos occurs around, or on top of, the yolk.

The early cleavage patterns and later development of animal embryos also vary depending on the orientation of the mitotic spindle, the source of mesodermal cells, and the manner in which the coelom is formed. These differences are used by zoologists to help us understand evolutionary relationships.

THE EMBRYONIC GERM LAYERS AND THEIR DERIVATIVES

Tissues and organs of animals arise from layers, or blocks of embryonic cells called embryonic germ layers. Their development from a nondescript form in the early embryo to their form in late embryonic through adult stages is called **differentiation.** One layer, **ectoderm,** gives rise to the outer body wall. **Endoderm** forms the inner lining of the digestive cavity. **Mesoderm** gives rise to muscles, blood and blood vessels, skeletal elements, and other connective tissues.

The following two exercises illustrate the early embryological stages of an echinoderm (sea urchins and their relatives) and an amphibian (frogs and their relatives). These

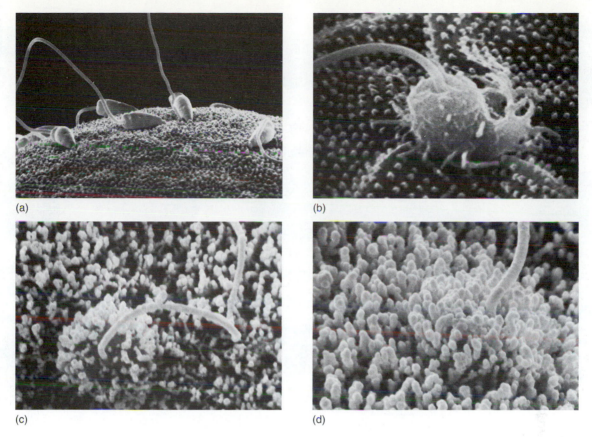

(a)

(b)

(c)

(d)

Figure 5.1 Scanning electron micrograph of fertilization of the sea urchin egg: (*a*) attachment of sperm to the egg surface; (*b–d*) sperm penetration. The fertilization membrane has been removed in (*c*) and (*d*) (*b*: ×7,500; *c*: ×4,800; *d*: ×7,125). (*c,d*) Reprinted with permission from M. Tegner, "Sea Urchin Sperm-Egg Interaction Studied with the Scanning Electron Microscope" in *Science*, 179:687, February 1973. Copyright © by American Association for the Advancement of Science.

animals have similar developmental stages that reflect common evolutionary ties. Their developmental patterns, however, are influenced by the fact that sea urchin eggs have a very small amount of evenly distributed yolk, and amphibian eggs have a moderate amount of unevenly distributed yolk. As you work through these exercises, note the similarities and differences in the embryology of these two organisms.

ECHINODERM EMBRYOLOGY

Studies of both living and prepared slide material are possible because mature sea urchins are readily available from biological suppliers. If facilities are available, students may have an opportunity to study both. Because sea urchins have little yolk, cleavages are holoblastic and equal. This means that cleavages pass completely through the egg and result in cells approximately equal in size. If living materials are not available, proceed to the slide study.

Fertilization and Development of Living Sea Urchin Eggs

If living eggs are available, their development should be initiated at the beginning of the laboratory. Two precautions are necessary throughout this exercise. In handling adults, eggs, sperm, and developing embryos, use only the seawater provided. Fresh water will kill these marine organisms. Also, seawater is extremely corrosive. Microscopes or other equipment that come into contact with seawater must be thoroughly cleaned.

▼ Your instructor will have induced the shedding of gametes by injecting the sea urchin with a potassium chloride solution. The sperm suspension is a milky white color, and the egg suspension is yellow. Place five drops of egg suspension into a petri dish half full of seawater. Using a separate pipette, place one or two drops of a dilute sperm solution into the petri dish. Swirl the suspension. This batch of fertilized eggs should be kept at a temperature not exceeding 22° C (70° F) and will serve as the stock from which subsequent samples will be drawn. Record the time when sperm was added to the egg suspension. _____ ▲

Next we will observe fertilization and the formation of the fertilization membrane. Remember that you will be looking at a two-dimensional image of a three-dimensional egg. It is unlikely that you will see the one sperm that actually enters the egg. By focusing up and down, however, you will be able to see sperm accumulating around the egg. You will not be able to see details of sperm structure or the microvilli that surround and draw a sperm into the egg (fig. 5.1).

Immediately after the penetration of the first sperm, ionic changes on the surface of the egg make the egg plasma

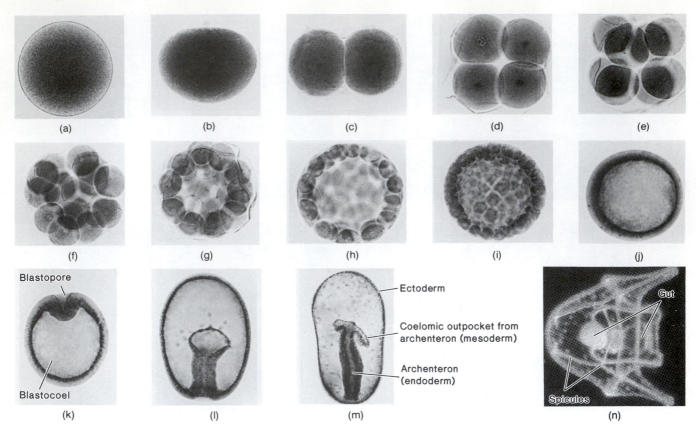

Figure 5.2 Stages in sea star development: *(a)* egg; *(b)* zygote; *(c–h)* 2-, 4-, 8-, 16-, 32-, and 64-celled stages; *(i)* nonmotile blastula; *(j)* ciliated blastula; *(k)* early gastrula; *(l)* midgastrula; *(m)* late gastrula; *(n)* pluteus. Note the skeletal spicules and fully formed gut tract of the pluteus.

membrane unresponsive to other sperm. Granules in the outer (cortical) region of the egg cytoplasm discharge their contents into the area between the egg plasma membrane and a layer of mucopolysaccharide, called the vitelline membrane, that covers the egg. An osmotic influx of water into the area between the plasma membrane and the vitelline membrane causes the latter to rise off the surface of the egg. The vitelline membrane is then called the **fertilization membrane** and blocks the entry of more sperm into the fertilized egg. Other changes in the cell are initiated in response to sperm entry, including the activation of mitochondria and ribosomes. Collectively these are called **cortical reactions,** and they are important in initiating developmental processes.

▼ Place a drop of fresh egg suspension on a slide. Cover with a coverslip. Observe the eggs, using a compound microscope. Focus on a group of eggs near the center of the coverslip. Take a drop of dilute sperm suspension and place it along one edge of the coverslip. As sperm cells move beneath the coverslip, they will fertilize the eggs they encounter. Watch the eggs closely in your field of view to see if you can observe fertilization. The halolike fertilization membrane should form within two to five minutes after fertilization.

The first cleavage occurs 60 to 70 minutes following fertilization. At that time, prepare a slide from your petri-dish culture and observe. While waiting for that cleavage, move on to the study of later developmental stages using prepared

slides and embryos from eggs fertilized by your instructor. After leaving the laboratory today, try to return as often as possible (or as your instructor directs) to observe developmental stages in eggs that you have fertilized. ▲

Slide Study of Sea Urchin Development

▼ Examine slides of developing sea urchin embryos. Prepared slides usually contain a variety of developmental stages from the egg to the pluteus stage. It is possible, however, that one or more of the following stages may be missing. If so, after studying your slide, trade with another student or obtain a second slide. Note the following stages.

Two-Cell Stage (60–70 Minutes After Fertilization)
The two-cell stage results from the first cleavage that proceeds from animal to vegetal pole. It consists of distinct blastomeres (fig. 5.2a–c).

Four- and Eight-Cell Stages (4–5 Hours After Fertilization)
The four- and eight-cell stages result from the second and third cleavages. The second cleavage proceeds from animal to vegetal pole. The third cleavage is equatorial (fig. 5.2d, e).

Morula (6 Hours)

The **morula** consists of 16 to 64 cells. It is in the form of a solid ball of cells (fig. 5.2f, g).

Blastula (9–11 Hours)

The **blastula** is a hollow ball of cells. Note the outer cells that surround the lighter inner cavity **(blastocoel).** Note also that up to this point, there has been no increase in the overall size of the embryo. (The blastula is approximately the same size as the egg.) Cells of the blastula develop hair-like cilia used in locomotion (fig. 5.2h–j).

Gastrula (32 Hours)

Gastrulation occurs by invagination of the blastula. In figure 5.2k, invagination is shown as an inward movement of cells at one point on the blastula, forming a small "pocket" on one side of the embryo. As gastrulation proceeds, the invagination becomes larger and the blastocoel smaller. The point of invagination is referred to as the **blastopore,** and the new cavity formed is the **archenteron.** In the fully formed **gastrula,** two germ layers are apparent. The lining of the archenteron will form the lining of the gut tract. The outer cell layer gives rise to the outer body wall, nervous system, and glands. Note the increased size of the gastrula (fig. 5.2k–m). Near the end of gastrulation, outpockets of the archenteron form the first mesoderm, and cavities of these outpockets are the embryonic coelom (fig.5.2m).

Pluteus (48–68 Hours)

The pluteus is a ciliated, free-swimming larval stage. The mesoderm, which initially formed late in the gastrula stage, proliferates and gives rise to supportive structures, muscle, and circulatory elements (fig. 5.2n). ▲

STOP AND ASK YOURSELF

1. What is embryology? _____

_____

2. Cleavage of an echinoderm zygote is holoblastic. What does that mean? _____

_____

3. What are cortical reactions? _____

_____

4. What event (visible under the light microscope) occurs within five minutes of the fertilization of a sea urchin egg? _____

5. The archenteron of a sea urchin embryo forms during the _____ stage. Describe how the archenteron forms. _____

_____

AMPHIBIAN EMBRYOLOGY

In this exercise, you should study living embryos and stained sections. If living embryos are available, they will likely be at the same stage of development. Your instructor may ask you to return to the laboratory outside of the regular class period to observe as many stages as possible. Most of your time today will be spent in the following slide study. Compare your specimens to figures 5.3 and 5.4.

▼ Using low power, examine a prepared slide showing a fertilized amphibian egg. This zygote demonstrates the influence a moderate amount of yolk has on embryological processes. Yolk is distributed throughout, but it is especially concentrated in the vegetal (less pigmented) half of the egg. This makes the vegetal pole of an amphibian egg heavier, orienting it downward when floating in water. The color pattern you are observing helps camouflage the developing embryo. The light color of the vegetal pole makes the egg difficult for a fish to see from below because the egg blends with the sky above. The dark color of the animal pole makes the egg blend with the substrate of the pond or lake when viewed from above. The dark color also absorbs heat and blocks ultraviolet radiation. Note the **gray crescent** on the margin of the pigmented area. It forms on the side opposite the point of sperm penetration. ▲

Early Cleavages

▼ Examine slides that show the sequence of cleavages from the fertilized egg through the blastula. The first cleavage begins at the animal pole and then slows upon entering the yolk-filled vegetal half of the egg (fig. 5.3a, b). The second division, which may begin before the first is completed, runs from pole to pole at right angles to the first (fig. 5.3c). The third division is meridional, that is, at right angles to the first two cleavages and in the animal half of the embryo (fig. 5.3d). The eight-celled embryo consists of four smaller cells in the animal region and four larger, yolk-filled cells in the vegetal region. Subsequent divisions result in numerous small cells at the animal pole and fewer, larger cells at the vegetal pole. These early cleavages are holoblastic, but the divisions are unequal because of the yolk in the vegetal pole. In the blastula, note the blastocoel. The blastocoel is in the animal half of the embryo because of the unequal distribution of yolk (fig. 5.3e,f). ▲

Gastrulation and Germ Layer Formation

Gastrulation of the amphibian embryo cannot occur by simple invagination because of the heavy, yolk-filled cells in the vegetal region. Gastrulation begins in the area of the gray crescent as cells from the region divide and migrate to the interior (fig. 5.3g). This process, called **involution,** results in a small, slitlike blastopore. As involution continues, the blastopore slit spreads transversely, and the archenteron begins to form internally. The lining of the archenteron is

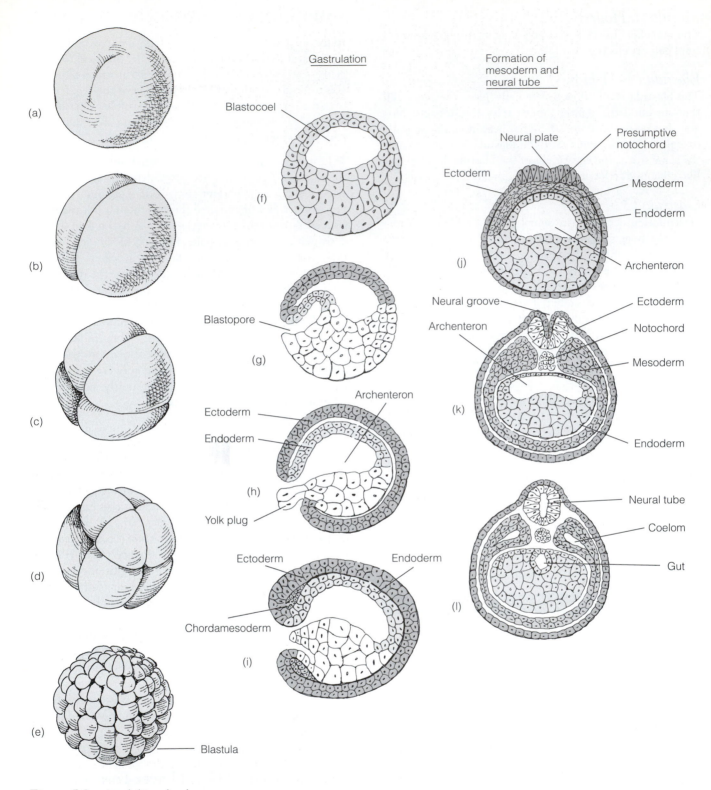

Gastrulation

Blastocoel

(f)

Blastopore

(g)

Ectoderm
Endoderm
Archenteron

(h)

Yolk plug

Ectoderm
Chordamesoderm
Endoderm

(i)

(a)

(b)

(c)

(d)

(e) —— Blastula

Formation of
mesoderm and
neural tube

Neural plate
Ectoderm
Presumptive
notochord
Mesoderm
Endoderm
Archenteron

(j)

Neural groove
Archenteron
Ectoderm
Notochord
Mesoderm
Endoderm

(k)

Neural tube
Coelom
Gut

(l)

Figure 5.3 Amphibian development.

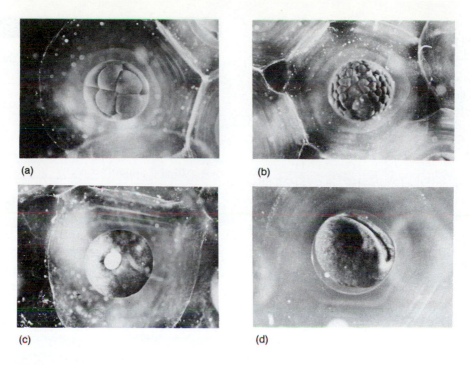

(a)

(b)

(c)

(d)

Figure 5.4 Amphibian development: (a) 8-cell stage; (b) early blastula; (c) late gastrula (yolk-plug stage); (d) neurula.

the endoderm. Cells rolling over the dorsal rim of the blastopore **(dorsal lip)** are replaced by cells moving over the surface from the animal pole toward the dorsal lip. The archenteron gradually increases in size, and the blastocoel decreases in size.

▼ Examine a sagittal section of an early gastrula under low power. Locate the animal pole, vegetal pole, blastocoel, blastopore, dorsal lip of the blastopore, and archenteron.

As gastrulation continues, the embryo begins to elongate, and ectodermal cells covering the surface of the embryo proliferate. As ectoderm proliferates, a spreading and thinning occurs. This expansion of the ectoderm toward the vegetal pole is called **epiboly.** During epiboly, the blastopore moves toward the vegetal pole and takes the form of a ring that surrounds yolk-filled vegetal cells. Vegetal cells protruding from this ring are called the **yolk plug** (fig. 5.3h). The ring of ectoderm around the yolk plug eventually closes to form a small blastopore. Later, the anus forms in the region of the blastopore. Internally, the archenteron increases in size until the blastocoel is obliterated. Examine a sagittal section of a late gastrula (yolk-plug stage) under low power. Locate the blastopore, yolk plug, dorsal lip of the blastopore, ectoderm, endoderm, archenteron, and the nearly obliterated blastocoel. ▲

Mesoderm Formation

Some of the last cells to move into the blastopore over the dorsal lip are presumptive (will develop into) mesoderm and presumptive notochord (a supportive element of vertebrates). Initially these cells make up the dorsal lining of the

archenteron near the blastopore. Later they detach from the endoderm and move between the endoderm and ectoderm in the region of the dorsal lip of the blastopore (fig. 5.3i). Some of this mesoderm, called **chordamesoderm,** spreads anteriorly between ectoderm and endoderm. The chordamesoderm differentiates into the **notochord** (fig. 5.3k). Lateral to the notochord, other mesoderm grows ventrally and eventually surrounds the gut tract. This mesoderm then hollows out to form the body cavity, or coelom. On each side of the notochord, mesoderm proliferates to form segmental blocks of tissue called **somites.** Somites are visible externally as swellings in the body wall. Mesoderm eventually forms muscle, connective tissue, and coelomic lining.

▼ Examine a neurula cross section and locate the notochord, mesodermal somites, and coelom. ▲

Neural Tube Formation

During late gastrulation, external changes along the upper surface of the amphibian embryo begin the formation of the neural tube—a process called **neurulation** (fig. 5.3j–l). Neurulation begins in response to the formation of the notochord by chordamesoderm. This stimulation of neurulation is an example of induction. **Induction,** the stimulation of development of a group of cells as a result of cellular changes in adjacent cells, occurs many times during animal development.

After gastrulation is completed, an oval-shaped area on the dorsal side of the embryo marks the presumptive neural tube. This region is called the **neural plate.** Microfilaments in neural-plate cells cause a flattening and thickening of the neural plate. Edges of the neural plate roll up and over the midline of the neural plate. These longitudinal ridges, called **neural folds,** meet dorsally to form the neural tube.

▼ Examine a cross section of a neurula stage under low power. Identify the ectoderm, endoderm, archenteron, and neural tube. ▲

Tadpole Stage

After further development, the embryo becomes tadpolelike with suckers, gills, a tail, and eyes. Movements of the tadpole result in hatching from the jellylike medium within which the embryo has been developing. Eventually, legs are formed and the tail is absorbed as the tadpole metamorphoses into the adult amphibian.

▼ Examine a tadpole. ▲

6. What is the influence of yolk of the amphibian egg on early embryonic cleavages? _____ _____

7. What is involution? _____ _____

8. What is epiboly? _____ _____

9. What is chordamesoderm? _____ _____

10. On each side of the notochord of the amphibian embryo, blocks of mesoderm called _____ form and are visible externally as swellings in the body wall.

KEY TERMS

blastula   75
chordamesoderm   77
cortical reactions   74
differentiation   72

ectoderm   72
embryology   72
endoderm   72

gastrula   75
mesoderm   72
morula   75

WORKSHEET 5.1   Embryology

Egg Types, Cleavage Patterns, and Embryonic Germ Layers

1. Compare and contrast the early cleavages of the sea urchin (sea star) and the frog. What is largely responsible for the differences in early cleavages that you observed?

2. Name each of the three embryonic germ layers and give an example of a tissue that develops from each.
   a.
   b.
   c.

Echinoderm Embryology

3. What is the blastopore, and what processes are responsible for its formation in sea urchin embryos?

4. At what point in sea urchin embryology does the embryo begin to enlarge?

5. Mesoderm formation begins at what stage of sea urchin development? Describe how it occurs.

Amphibian Embryology

6. Contrast the method of mesoderm formation in amphibian embryos to that in echinoderm embryos.

7. What is the yolk plug?

8. What is the fate of the blastocoel in the amphibian gastrula?

9. Describe neurulation in the amphibian embryo.

Adaptations of Stream Invertebrates— A Scavenger Hunt

Learning Objectives

After completing this exercise, you should be able to do the following:

1. Explain the concept of evolutionary adaptation.
2. Describe the physical and chemical characteristics of freshwater streams.
3. Describe some of the community interactions that occur in freshwater streams.
4. Describe adaptations of stream invertebrates.
5. Use aquatic nets and other equipment to sample invertebrates in a stream.

Prelaboratory Quiz

Study this week's laboratory exercise and then complete the following quiz to assess your preparation for the laboratory.

1. The term that refers to all organisms living in a certain area and the organisms' physical environment is
   a. population.
   b. ecosystem.
   c. community.
   d. niche.

2. Which of the following is an accurate definition of "evolutionary adaptation?"
   a. any genetic change than improves the organism's ability to survive
   b. any genetic change that improves the organism's ability to exploit food resources
   c. any genetic change that improves the organism's reproductive success in a specified environment
   d. any genetic change that improves the organism's reproductive success in any environment

3. All of the following statements are teleological EXCEPT one. Which one is the exception?
   a. Genetic change and evolution in a species occur because the species needs a change in order to survive.
   b. Evolution occurs as a result of random genetic changes in a species that may or may not be adaptive.
   c. Evolution directs change in a species.
   d. Evolution always results in progress. Species improve through evolutionary change.

4. A thin, 2–3 mm layer of nonmoving water at the surface of objects in a stream is called the
   a. neuston.
   b. benthos.
   c. boundary layer.
   d. hyporheos.

81

5. Current velocity in a stream is usually greatest
   a. near the bottom substrate.
   b. near the shore.
   c. throughout the stream cross section.
   d. near the center of the stream.

6. When a chemical parameter of a stream falls outside of the range that supports an animal and the animal dies, this parameter is said to be outside of the animal's
   a. range of the optimum.
   b. tolerance limit.
   c. trophic level.

7. The source of most primary production for streams is
   a. vegetation growing along the stream bank—especially autumn leaf fall.
   b. algae growing in the stream.
   c. attached aquatic vegetation.
   d. All of the above contribute equally to primary production for stream food webs.

8. Herbivory in streams begins with animals classified as
   a. shredders.
   b. filterers.
   c. gatherers.
   d. predators.

9. Animals, like the larvae of the water penny beetle, feed on fine organic material attached to or settled on substrate materials. These animals are classified as
   a. shredders.
   b. filterers.
   c. scrapers and gatherers.
   d. predators.

10. Animals living in or on the streambed are living in the
    a. neuston.
    b. boundary layer.
    c. neckton.
    d. benthos.

11. True/False   A pH of 5.0 is harmful to most aquatic animals.

12. True/False   Very soft water is required for the formation of exoskeletons of arthropods and the shells of molluscs.

13. True/False   Dissolved oxygen levels in warm streams tend to be higher than those in cold streams.

14. True/False   Flattening of some stream insects allows them to live in the neuston of a stream.

15. True/False   Some stream insects use silk to attach themselves to the substrate of a stream.

EVOLUTIONARY ADAPTATION

Evolution has resulted in the great variety of living organisms that we see today. A goal of the next unit of this laboratory manual is to study that diversity in a survey of the animal phyla. The survey approach helps one appreciate common themes of animal organization that result from common evolutionary ancestry. Another approach to studying animal diversity is to study particular ecosystems. An **ecosystem** consists of all of the organisms living in a certain area and the organisms' physical environment. The ecosystem approach helps one appreciate the interactions that occur between animals and their environment. The goal of this exercise is to study a freshwater stream in order to appreciate the evolutionary adaptations of invertebrates to the stream ecosystem.

An **evolutionary adaptation** is any change in an animal's genetic makeup that increases the animal's chance of successful reproduction in a specified environment. Changes that could be considered adaptations can directly affect an animal's ability to produce large numbers of eggs, care for large numbers of offspring, or reproduce more often. Other adaptations can indirectly increase an animal's reproductive potential. These changes can increase an animal's ability to feed more efficiently and store more energy for egg production. Other changes allow animals to live in places where other animals are unable to live, thus reducing competition for resources and making it more likely that young will survive. Still other changes allow an animal to withstand harsh climatic conditions or potentially harmful interactions with other individuals. These changes allow an animal to survive and reproduce in the future.

Notice that the definition of adaptation specifies a particular environment. This qualification is very important. A trait that is adaptive in one environment may result in decreased reproductive success in another environment. If an animal migrates to another location or if the environment changes, what were formerly adaptive traits could result in extinction.

When studying animal adaptations, it is tempting to think of evolution as a directed process that produces change in response to the needs of an animal. This kind of goal-oriented thinking is said to be **teleological.** Evolution is simply a result of random genetic changes (mutations) that result in some individuals being more effective at reproducing than others in the population. There is no force that directs when a mutation will occur or what the nature of the change will be.

A second pitfall in thinking is to attribute adaptive significance to every process or structure that an animal possesses. Mutations are random occurrences. They are usually neutral rather than harmful or beneficial for an animal. This means that evolution may result in structures that are useless for an animal. These structures may be retained because they are coded by genes that may be linked to chromosomes containing genes that code for adaptive traits.

STREAM ECOLOGY

Freshwater stream ecosystems are a part of nearly all landscapes. Their physical and chemical characteristics are relatively restrictive, and animals that live in streams often

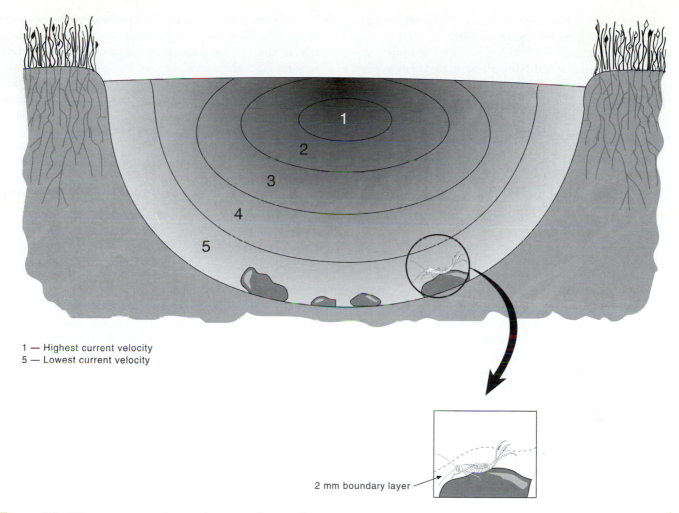

1 — Highest current velocity
5 — Lowest current velocity

2 mm boundary layer

Figure 6.1 This cross-sectional view of a stream illustrates how current velocity varies depending on where the measurements are taken. As one moves from the center of the stream toward the banks or bottom substrate, current velocity is reduced because of friction between the water and the streambed. The inset shows the boundary layer, which is a small layer of still water on the surface of objects in the streambed. Many benthic stream invertebrates are adapted for life in the boundary layer.

show body structures that are adaptations to stream habitats. Many of these structural (morphological) adaptations are easily recognized by careful observation. Before one can understand the adaptive significance of the morphological traits, one needs to know something about the stream ecosystem. While streams can be very different from one another in terms of depth, type of substrate, current velocity, and so forth, there are certain characteristics that streams share. Some of these are discussed next.

Physical Characteristics

The most obvious characteristic of streams is the current. Animals that live in streams must be able to maintain themselves in, or seek shelter from, moving water. Some members of a species must withstand periods of flooding, in which current velocity increases to the point that it threatens to scour animals from the streambed. On the other hand, many stream animals have an inherent requirement for moving water. Moving water delivers nutrients and dissolved oxy-

gen to stream animals and aids in animal dispersal. The current that we experience when we stand in a stream, however, is not necessarily what is experienced by an insect in the stream (fig. 6.1). As one moves away from the center of a stream, current velocity decreases because of friction between the water and the stream bank or streambed. Within 2–3 mm of the surface of a rock, or other object on the substrate, the current disappears and water is stationary. This layer of standing water is called the **boundary layer.** Many invertebrates that live on the streambed **(benthos)** are adapted to living in this layer of stationary water.

Other physical characteristics that influence animal life in streams include water temperature, substrate type, and turbidity. Temperature influences where an animal lives, the stage of its life cycle, and the availability of oxygen for the animal. Substrate type also influences stream animals. The particle size of substrate materials may vary from meters in diameter (boulders) to fractions of a millimeter (silt). The color of substrates varies from very light to very dark. Although the reasons are not always entirely clear, a certain species may be consistently found associated with a particular substrate. Streams are often fairly turbid, especially larger rivers with silty substrates. Fine particles are often suspended in moving water. Animals adapted to larger rivers are usually not affected by turbid conditions. On the other hand, the runoff of topsoil from construction sites and farming operations can result in sedimentation that may foul gills, cover habitats in the spaces between rocks and gravel, or foul food supplies.

Chemical Characteristics

Streams also have chemical characteristics that influence animal life. Most important among these are the amount of dissolved oxygen, pH, and hardness. Animals are most successful when these chemical characteristics are within certain ranges of values. These ranges are species-specific and are called **ranges of the optimum.** Animals can survive when a chemical parameter falls out of the range of the optimum. Extreme variations in the parameter, however, make conditions unsuitable for life, and the animal dies. This is referred to as the **tolerance limit** of the animal. Different species of animals have different ranges of optimum for various chemical characteristics of streams.

Oxygen levels of streams vary from low values of two or three parts per million (p.p.m.) to saturation levels of 14 p.p.m. The amount of oxygen present in a stream depends partly on water temperature. The solubility of oxygen in water is greater at lower temperatures, so cold water usually has more dissolved oxygen than warm water. Oxygen is rarely a limiting factor in clean streams. Streams have a larger surface area relative to their volume than lakes. This large surface area provides for adequate exchange of gases between the water and the air. Gas exchange is enhanced by the fact that water is moving in streams. Water rushing over boulders or tree limbs causes turbulence that mixes air and water, providing faster rates of gas exchange between air and water. Streams can be naturally low in oxygen for two reasons: (1) heavy leaf fall in autumn can enhance rates of decomposition, and oxygen in the stream can become depleted and (2) continuous ice cover can limit the exchange of oxygen between the air and the water and can result in low oxygen levels. Pollution, on the other hand, can deplete streams of oxygen. Heated discharge from factories can raise water temperature and reduce the solubility of oxygen in water, and organic pollutants enhance rates of decomposition and deplete the oxygen supply.

The pH (acidity) also influences animals of a stream. The pH of a stream is determined by characteristics of the surrounding landscape. Streams that run through regions where the bedrock is limestone (calcium carbonate) will normally have a pH close to neutral. In other regions where there is little calcium in the surrounding soils, the pH can be naturally lower. Acid deposition has lowered the pH of many streams. Although there is much variation in the tolerance levels of stream animals, most evidence suggests that pH levels below 5.5 are harmful to many common stream animals.

The hardness of a stream is a measure of the dissolved minerals in the water. Dissolved calcium, magnesium, chloride, sulfate, and sodium ions are important for at least three reasons. First, these ions are used by an animal for various physiological processes. Second, the presence of these ions in the water reduces the osmotic stress on a freshwater animal. This means that the animal will take on less water by osmosis and expend less energy expelling excess water. Lastly, in molluscs and many arthropods, calcium is used to harden the shell (or exoskeleton in arthropods). Reduced availability of these ions (soft water) may cause problems for an animal in any of these three areas. Hardness often varies between 50 and 300 p.p.m.

Stream Food Webs

The sequence of organisms through which energy moves in an ecosystem is a **food chain.** In most ecosystems, it is more accurate to envision complexly interconnected food chains called **food webs.** There are relatively few attached aquatic plants in most streams and very little floating algae (phytoplankton). The source of most plant material (primary production) that provides energy for stream animals comes from outside the stream. Plants on the stream bank supply primary production in the form of leaves that fall into the stream, especially in autumn (fig. 6.2).

The feeding (trophic) levels of a stream can be classified as follows. Herbivory begins with animals that serve as **shredders.** Many mayfly nymphs, some caddis fly nymphs, and crustaceans (amphipods), for example, feed on entire leaves. As they feed, smaller pieces of leaves are freed and carried downstream by the current. This shredding continues until the particle size becomes very small and the food is prepared for the next set of users. **Filterers** are animals with specializations for trapping suspended food particles. Some mayfly nymphs have fine hairlike setae on their forelegs that are used as filters. Some caddis fly nymphs spin silk nets that are used to trap food particles. Black fly larvae possess fan-

Figure 6.2 Primary production for most streams comes from vegetation surrounding the stream. Leaf fall in autumn may supply enough nutrients for stream food webs for the entire year.

like appendages that trap food. Bivalves filter food using gill filaments. Some filtering animals are specialized for filtering coarse food particles from the water. The smallest particles that they miss are filtered by animals with finer filtering surfaces. Other animals, called **scrapers and gatherers,** scrape fine organic material that has settled on, or is attached to, substrate materials. Scrapers and gatherers include snails, some mayflies, and beetle larvae like the water penny. There are also many predators in freshwater streams. Some are invertebrates that prey on other invertebrates. Stone fly larvae and helgramites are invertebrate predators. Fishes and salamanders are examples of vertebrate predators. Animals like crayfish are **detritivores** and feed on a variety of dead and decaying animal and plant matter.

STOP AND ASK YOURSELF

1. What is an evolutionary adaptation? _____
_____
_____

2. What is an ecosystem? _____
_____
_____

3. What is the benthos of a stream? _____
_____
_____

4. What is the boundary layer of a stream? _____
_____
_____

5. What is a tolerance limit? _____
_____
_____

6. Why is dissolved oxygen rarely a limiting factor in clean streams? _____
_____
_____

7. What is the primary source of energy for a stream ecosystem? _____
_____
_____

8. What is the role of shredders in the stream ecosystem?
_____
_____
_____

A STUDY OF A STREAM ECOSYSTEM

The purpose of the following activities is to study the ecology of a freshwater stream. You will first study the physical and chemical characteristics of the stream you are visiting. Following that exercise, you will be using collecting equipment to study adaptations of animals living in the stream. Before going into the stream, a few safety precautions should be followed.

1. Use a buddy system. Work in groups of two or more and stay with your group. If one member of the group encounters trouble, the other members will be there to summon help.
2. Never get out of visual and calling range of other groups.
3. Do not exceed knee depth when wading.
4. Watch for underwater obstacles. Use the handle of your net to steady yourself and explore the substrate.
5. Rocky substrates are often very slippery. Be very careful when walking across them.
6. Avoid extremely fast currents.

Physical and Chemical Characteristics

▼ When you arrive at the stream you will be studying, take some time to look around. How wide is the stream? How deep is the stream in your location? Look up and down the length of the stream. What do you think is the primary source of energy for animals living in the stream? What makes up the substrate of this stream? What other factors are there that may influence the lives of animals in the stream (e.g., farm pastures, industrial establishments, highways, subdivisions)? Record the date and location of your field trip and your observations on worksheet 6.1.

Determine the current velocity in the stream. Your instructor will give you directions for using a current meter if one is available. Alternatively, measure a convenient distance (perhaps 15 m) and time the movement of a float across that distance. Repeat your measurement three times and calculate an average. Express your results in meters per second and record the results on worksheet 6.1. ▲

There are a variety of chemical tests that can be performed with stream water. The three parameters described in this exercise—dissolved oxygen, pH, and hardness—are important for the reasons described. Your instructor will provide you with reagents from a water analysis kit and give you specific instructions on their use. Perform the chemical tests and record your results in worksheet 6.1. If additional reagents are available, your instructor may request that more tests be performed. Record your conclusions regarding the chemical characteristics of your stream on worksheet 6.1.

Stream Invertebrates—A Scavenger Hunt

Invertebrates are used in this exercise because they are easily collected, usually quite numerous, and show remarkable diversity. They are not often observed in casual visits to

TABLE 6.1	Adaptations of Stream Insects

Adaptation	Adaptive Significance
Morphological adaptations	
Flattening	Flattening of the body allows an animal to utilize the boundary layer, thus avoiding current. It also allows animals to crawl under objects and within the spaces between substrate particles (figs. 6.3 and 6.7).
Streamlining	Streamlining lessens resistance to fluids moving around the animal. The animal tends to orient upstream (figs. 6.3 and 6.8).
Suckers	Suckers help an animal adhere to the substrate. They are not common in stream animals except for leeches and a few fly larvae.
Friction pads and marginal contact	A flattened ventral surface and domed upper surface increase frictional resistance with the substrate and force the margin of the animal against the substrate. This seals the ventral side and prevents moving water from getting under the animal (fig. 6.7).
Hooks and grapples	Claws on the legs or other appendages of animals are used to hold onto the roughness of stone surfaces (figs. 6.5 and 6.6).
Small size	Small animals can utilize the boundary layer and spaces between substrate particles.
Silk and sticky secretions	Silk is used to attach animals to their substrate (figs. 6.6 and 6.9).
Ballast	Ballast may be in the form of stone cases constructed with the help of silk (fig. 6.4).
Neuston	Some invertebrates live on top of the water, supported by water's surface tension. Hairlike setae that repel water may be present on legs or other body parts and help prevent the animal from breaking through the water surface (fig. 6.10).
Behavioral adaptations	
Avoidance	Many animals seek refuge from the current by living under stones or under the banks of streams.
Living in the hyporheos	Some stream animals burrow below the surface of the substrate. Some spend their entire life histories in this way—completely hidden from the current.

streams, however, because they are quite small. Stream invertebrates show a number of morphological adaptations to the physical characteristics of the stream ecosystem. Table 6.1 lists adaptations that are easily observed in stream invertebrates. Many of these are adaptations to living in moving water. Others are behavioral traits that allow invertebrates to avoid the current (figs. 6.3–6.10).

▼ Each group of students will be given a D-frame net, a collecting pan, vials, forceps, a squeeze bottle of 70% ethanol, a pencil, and vial labels. Your goal is to collect as many different kinds of invertebrates as you can find in the time allotted. Approach your collecting site from downstream so you do not disturb the substrate where you will be working. Collecting simply involves sweeping your net through vegetation along the stream bank or through debris on the bottom of the stream. After collecting in sandy or muddy substrates, wash the contents of the net by gently dipping the net into the water. In swift water with rocky substrate, place the flat edge of the net on the stream bottom so that the net opens upstream. Disturb the substrate in front of the net with your foot and allow the current to carry debris into the net. Fill your collecting pan about one-half full of water and empty the contents of the net into the pan. Examine the net and the pan for invertebrates. Preserve one or two specimens of each kind of animal in a separate vial of ethanol. Use a pencil and a vial label to number the animals in your collection. Place the label inside the vial with the specimen. Determine whether or not any of the adaptations listed in table 6.1 applies to the animal. A specimen may have more than one adaptation. If so, indicate all that apply. Other specimens may have no adaptations visible. Indicate that as well. Use worksheet 6.2 to record the specimen by number and location. The location record should include information on substrate type, water depth, current velocity, and location within the substrate. Your instructor may have set up dissecting microscopes so that you can view specimens more closely. It is not necessary to know the names of the animals you are collecting to complete this exercise. However, your instructor may want to take time to help you learn to recognize a few of your specimens by name. Record the name in worksheet 6.2 if the specimen has been identified.

At the end of today's laboratory period, you will hand in your collection and worksheets 6.1 and 6.2. Your instructor will grade your worksheets and assign points based on the number of different adaptations that you found. Based on your observations in your stream and the information in this manual, fill in the food web in figure 6.11 in worksheet 6.3. ▲

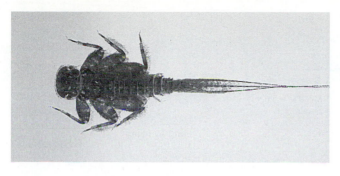

(a)   (b)

Figure 6.3   (a) Mayfly nymphs (order Ephemeroptera) are recognized by the presence of two or three tail filaments and dorsal or lateral abdominal gills. Mayflies usually serve as shredders or filterers in stream ecosystems and are often streamlined or flattened. A member of the family Heptageniidae is shown here. (b) An adult mayfly, *Blasturus intermedius*. The adult mayfly's life span ranges from only a few hours to one to two days. They do not feed. They mate, the females lay eggs on the water's surface, and they die.

Figure 6.5   Helgramites or dobson fly larvae (*Corydalus cornuta*) are predators that feed on other invertebrates. They are often found under, or clinging to, rocks in swift currents. Hooks on prolegs at the posterior tip of the abdomen help the animal to maintain its position in the current.

Figure 6.4   Caddis fly larvae (order Trichoptera) often build cases of stone or twigs. These materials are cemented together with silk secretions.

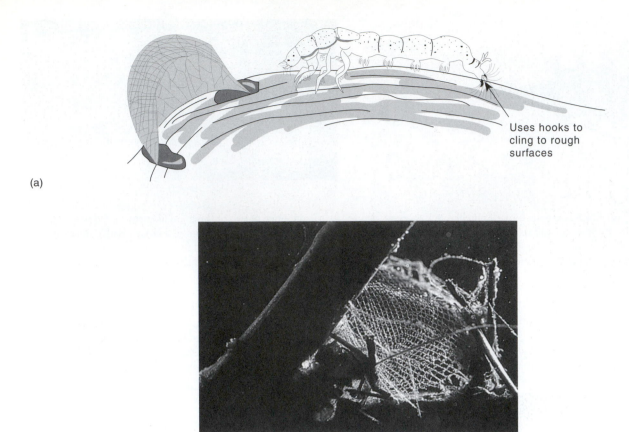

(a)

Uses hooks to
cling to rough
surfaces

(b)

Figure 6.6 Some caddis fly larvae (order Trichoptera) do not build cases but may construct silken retreats on the undersurface of rocks or nets used to capture debris drifting downstream. *(a) Hydropsychae* is shown here. *(b)* Food-filtering net of *Hydropsychae* (×8).

Figure 6.7 This flattened, scale-like beetle larva (order Coleoptera) is a water penny (*Psepheneus lecontel*). Flattening allows it to live within the boundary layer. The doming of its dorsal surface and the tight marginal contact of its body with rocks that it crawls over prevent it from being swept away. Water pennies are scrapers and gatherers (×10).

Figure 6.8 This insect is a stone fly nymph (order Plecoptera). Stone flies are recognized by two tail filaments and gills that attach at the ventral surface of the thorax. They are often streamlined insects and prey on other stream invertebrates.

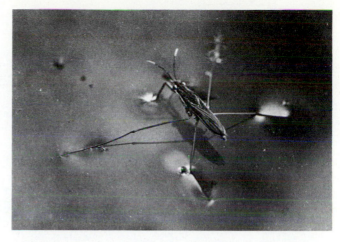

Figure 6.10 This water strider (*Gerris*) is a member of the insect order Hemiptera. The water-repelling setae on its legs and the surface tension of the water allow water striders to live on top of the water (neuston). Note the "dents" in the water's surface where the insect's legs contact the water. Water striders are usually found in quiet water in pool areas or along the bank. Water striders prey on invertebrates swimming just below the water surface.

Figure 6.9 This black fly larva (*Simulium*) secretes silk from its salivary glands. The silk is attached to rocks or other firm substrates. Hooks at the tip of the abdomen allow the black fly larva to cling to the silk and the substrate in spite of very fast currents. The mouthparts of a black fly larva are fan shaped and filter its food—fragments of plant material—from the water. Adult black flies are annoying biters.

KEY TERMS

benthos   83

ecosystem   82

evolutionary adaptation   82

food chain   84

food web   84

ranges of the optimum   84

tolerance limit   84

WORKSHEET 6.1   Physical and Chemical Characteristics

Name and location of the stream _____

Date of the scavenger hunt _____

1. Summary of the physical characteristics of the stream.

   a. The approximate width of this stream is

   b. The average depth of this stream is about

   c. The current velocity is

   d. The major sources of energy for animals living in this stream are

   e. The substrate of this stream consists of

   f. Other factors that may influence animal life in this stream (list)

2. Summarize the results of your physical and chemical measurements. Are there any parameters that you measured that seem to be out of the ordinary?

WORKSHEET 6.2

Adaptations of Stream Animals

1. Fill in the following table by placing the number of the specimen beside the adaptation. Indicate where each specimen was collected (midstream under rocks, under bank, clinging to log, etc.)

Adaptation	Specimen Number	Location	Specimen Name (optional)
Flattening			
Streamlining			
Suckers			
Friction pads and marginal contact			
Hooks and grapples			
Small size			
Silk and sticky secretions			
Ballast			
Neuston			
Avoidance			
Hyporheos			
No apparent adaptations present			

WORKSHEET 6.3

Evolutionary Adaptations and Stream Ecology

1. Name two common problems regarding how people sometimes think about evolutionary adaptations.

2. Name three chemical characteristics of streams that can influence animal life in a stream.

3. Describe one factor that can cause a natural reduction in dissolved oxygen levels in a stream.

4. Describe one factor that results from human influences that can cause a reduction in dissolved oxygen in a stream.

5. What is the food of the following classes of stream animals? Give one example of each.
   a. shredders

   b. filterers

   c. scrapers and gatherers

6. How are the following features adaptations for living in a stream?
   a. flattening

   b. streamlining

   c. silk and sticky secretions

   d. living in the hyporheos

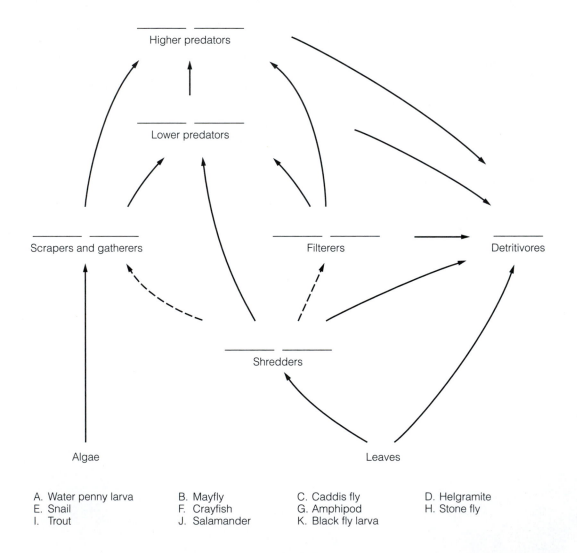

A. Water penny larva   B. Mayfly   C. Caddis fly   D. Helgramite
E. Snail   F. Crayfish   G. Amphipod   H. Stone fly
I. Trout   J. Salamander   K. Black fly larva

Figure 6.11   A simplified food web of a hypothetical stream. Dashed lines going through the shredders to scrapers, gatherers, and filterers indicate that the shredders tear up plant material and the smaller pieces that are left are used by other classes of herbivores. Solid lines indicate energy flow between different trophic levels. Fill in the food web by matching the organism with its location in the food web. Did you collect any organisms not listed here? Consult with your laboratory instructor and additional references to determine where they fit in this food web. How many trophic levels are present in this food web?

UNIT II

An Evolutionary Approach to the Animal Phyla

Unit II is similar to the traditional survey approach of generations of zoology courses. The goal of this unit is to give you a broad exposure to the protozoans, the animal phyla, and the evolutionary relationships that unite the diversity found in the animal kingdom. Unit II begins with exercise 7—an introduction to animal classification and cladistics. The "Evolutionary Perspective" at the beginning of exercises 8 through 17 is meant to help you relate the phylum being studied to those already discussed and those that will be studied in subsequent laboratory exercises. The introduction in exercises 8 through 17 is designed to introduce you to the unique characteristics of the phylum (phyla) studied in each exercise. That is followed by studies of class representatives and processes unique to particular animal groups. Most exercises in unit II end with an exercise on class-level taxonomy of the phylum studied.

The Classification of Organisms

Learning Objectives

After completing this exercise, you should be able to do the following:

1. Describe a classification system as a reflection of the order present in nature.

2. Describe systematics, or taxonomy, as the study of the kinds of organisms and their evolutionary relationships.

3. Describe a species as a single kind of organism that is given a distinctive name by taxonomists.

4. Describe other levels of classification as reflecting the relatedness among different species.

5. Interpret and describe the usefulness of cladograms in modern taxonomic studies.

6. Construct a cladogram using simple block animals.

7. Assign names to simple block animals using a few basic rules of zoological nomenclature.

Prelaboratory Quiz

Study this week's laboratory exercise and then complete the following quiz to assess your preparation for the laboratory.

1. An inherited trait used in determining taxonomic relationships among a group of animals is a(n)
   a. characteristic.
   b. character.
   c. anamorphy.
   d. symplesiomorphy.

2. A character that has been derived since a group diverged from a common ancestor is called a(n)
   a. symplesiomorphy.
   b. anamorphy.
   c. synapomorphy.
   d. anapomorphy.

3. The next level of taxonomy more inclusive than order is
   a. class.
   b. family.
   c. genus.
   d. phylum.

4. A subset of taxa that shares a certain derived character is called a
   a. polyphyletic group.
   b. paraphyletic group.
   c. species group.
   d. clade.

5. Which of the following is the correct way of designating a species name?
   a. Musca domestica
   b. *Musca domestica*
   c. *domestica*
   d. *Musca Domestica*

6. True/False   If all information about the evolution of a group is known, the group will always be represented as a monophyletic group.

7. True/False   A paraphyletic group can be traced to more than one ancestral species.

8. True/False   The simplest, most economical explanation for an evolutionary lineage is a parsimonious explanation.

9. True/False   A group of organisms that is used to help decide what character is ancestral for a group being studied is called an ingroup.

10. True/False   The goal of animal systematics is to study the diversity of animals and organize them into groupings that reflect evolutionary relationships among the groups.

Biologists estimate that there are in excess of 30 million different kinds of organisms on the earth. Most of these different organisms are undescribed and reside in tropical rain forests. As scientists work to find these undescribed organisms, the organisms must be named. Our need to name organisms stems from the fact that we have a language, and that language allows us to encode and classify concepts. This need to encode and classify is a natural consequence of being human. For example, the first question that any young child asks when seeing an unfamiliar animal in a zoo is, "What is that?"

Scientists go about naming and classifying organisms in ways that have helped to make sense out of the diversity in nature. A potpourri of millions of names is of little use to anyone. To be useful, a naming system must reflect the order that is present in nature. This means that before names are assigned, a scientist must study the organism to determine the relationships of the organism to other previously described organisms. These interrelationships are reflections of the evolutionary pathways that lead to modern organisms. The study of the kinds and diversity of organisms and the evolutionary relationships among them is called **systematics,** or **taxonomy.** (Some biologists distinguish between taxonomy and systematics. Taxonomy is often thought of as the work involved with the original description of species, and systematics as the assignment of species into evolutionary groups.) The assignment of a distinctive name to each species is called **nomenclature.**

A TAXONOMIC HIERARCHY

Our modern classification system is founded on the work of Karl von Linne (Carolus Linnaeus, 1707–1778). In this classification system, each different kind of organism is called a **species.** The species concept is actually quite difficult to define precisely. In fact, there is no single definition that can be used in all cases. It will be sufficient for our purposes to think of a species as a group of organisms that produces more of its own kind through sexual reproduction. Realize, however, that there are many instances in which this definition does not work well.

Karl von Linne used a two-part name for each species; therefore, our naming system is referred to as a **binomial system.** The first part of a species name is the **genus** name. It always begins with a capital letter. The second part of a species name is the **species epithet.** It always begins with a lowercase letter and is never used without the genus name. The entire name is written in italics, or underlined, because it is Latinized. For example, the species name (often called the scientific name) of humans is *Homo sapiens,* and the species name of the housefly is *Musca domestica.* When the genus name is understood, it can be abbreviated (*H. sapiens*). An abbreviation is used only after the entire name has been written earlier in a paper.

Taxonomic levels above the species are used to describe evolutionary relationships. These taxonomic levels, or taxa (sing. taxon), are arranged hierarchically (from more specific to broader): **species, genus, family, order, class, phylum, kingdom,** and **domain** (table 7.1). These groupings are based upon increasingly divergent characteristics and, ideally, reflect evolutionary relationships. The housefly (*Musca domestica*), for example, has its own set of unique characteristics that makes it a distinctive species. It also shares certain characteristics with other true flies—the horsefly, for example (fig. 7.1). The most important of these shared characteristics is a single pair of wings. This shared characteristic is a reflection of the relatively close evolutionary ties between these species. They are both members of the order Diptera. These flies also share characteristics with a broader group of animals—butterflies, beetles, and wasps. These animals all have three pairs of legs and are called insects (class Hexapoda). Insects are all related, but not all insects are as closely related as houseflies are to horseflies. Insects, in turn, share important characteristics with other animals—ticks, spiders, crayfish, and crabs. They all possess a jointed exoskeleton. Thus, the housefly and all of its relatives belong to the phylum Arthropoda.

Above the species level, there are no precise definitions of what makes up a particular taxon. Disagreements among scientists as to whether two species should be included in the same or different taxonomic groups are common. Disagreements like these are the fuel that keeps taxonomy a vital, lively scientific field of study.

TABLE 7.1   Taxonomic Categories of a Human and a Housefly

Taxon	Human	Housefly
Domain	Eukarya	Eukarya
Kingdom	Animalia	Animalia
Phylum	Chordata	Arthropoda
Class	Mammalia	Hexapoda
Order	Primates	Diptera
Family	Hominidae	Muscidae
Genus	*Homo*	*Musca*
Species	*Homo sapiens*	*Musca domestica*

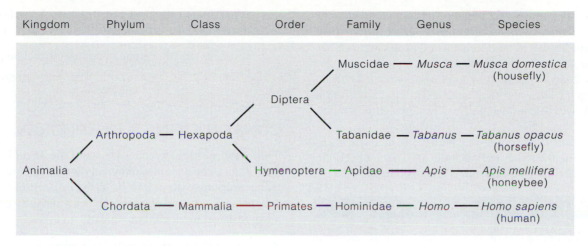

Kingdom	Phylum	Class	Order	Family	Genus	Species

Figure 7.1 A taxonomic hierarchy. The classification of a housefly, horsefly, honeybee, and human.

Students often ask the question, "Why are scientific names more useful than common names?" The answer to this question is easily understood. As pointed out previously, scientific names reflect evolutionary relationships, common names do not. Just as importantly, scientific names make sense out of a chaotic world of common names. Many species have common names that vary from region to region and country to country. For example, a common sparrow is called an English sparrow, barn sparrow, and house sparrow in different regions of the country. It has a single scientific name that is recognized by scientists throughout the world—*Passer domesticus*. In addition, common names often refer to more than one species. It is impossible to distinguish between species of mosquitoes or crayfish with superficial examination. Common names, therefore, often do not designate a particular species. On the other hand, species names refer to only one kind of organism. No two species have the same scientific name.

EVOLUTIONARY TREE DIAGRAMS

The goal of systematic studies is to arrange animals into **monophyletic groups,** which include a single ancestor and all of its descendants. Insufficient knowledge of a particular group of animals sometimes results in lineages that appear to have more than one ancestor. This kind of lineage is called a **polyphyletic group.** Polyphyletic interpretation is always considered erroneous, and a search for additional data usually ensues. Sometimes a lineage includes some, but not all members of the group. This kind of lineage is called a **paraphyletic group** and also reflects insufficient knowledge of the group.

Evolutionary tree diagrams are used to depict ideas regarding family relationships. One of the most commonly used tree diagrams is called a **cladogram.** Cladograms are a result of one approach to systematics called **phylogenetic systematics (cladistics).** Cladograms depict a sequence in the origin of taxonomic characteristics. Figure 7.2 is a cladogram that shows five hypothetical taxa (1–5) and the char-

acteristics (A–H) used in deriving the taxonomic relationships. Characteristics that have a genetic basis, that can be measured, and that indicate relatedness are called **characters.** To decide what character is ancestral for a group of organisms, cladists look for a related group of organisms that is not included in the study group. This related group is called an **outgroup.** Notice that taxon 5 is the outgroup for the other four taxa. Character A is common to all members of the lineage. This character is called a **symplesiomorphy.** It indicates that all members of the study group are related because it is present in ancestors of all members of the study group. It cannot be used to distinguish between members of the study group.

A character that has arisen since common ancestry with the outgroup is called a **derived character,** or **synapormorphy.** Taxa 1–4 in figure 7.2 share character B. Character B

Outgroup

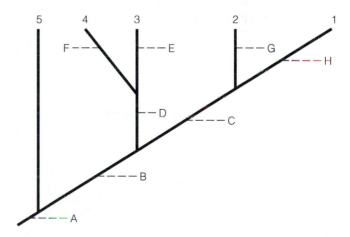

Figure 7.2 Interpreting cladograms. This hypothetical cladogram shows five taxa (1–5) and the characters (A–H) used in deriving the taxonomic relationships. Character A is symplesiomorphic for the entire group. Taxon 5 is the outgroup because it shares only that ancestral character with taxa 1–4. All other characters are more recently derived.

separates taxa 1–4 from the outgroup and is a synapomorphy. Similarly, characters C and D arose more recently than B. Character C is shared by closely related taxa 1 and 2. Character D is shared by closely related taxa 3 and 4. Taxa that share a certain synapomorphy form a subset called a **clade.** Taxa 1 and 2 form a clade characterized by C. The entire cladogram can be considered a hypothesis regarding a monophyletic lineage. New data (newly investigated characters) are used to test the hypothesis. Figure 7.3 is an actual cladogram showing the taxonomic relationships of the vertebrates (fishes, amphibians, reptiles, birds, and mammals).

S T O P   A N D   A S K   Y O U R S E L F

1. Why is our system of naming kinds of organisms called a binomial system? _____

_____

_____

2. What is the next, more inclusive, taxonomic level following "order"? _____

3. We call a group of animals that are all descended from a common ancestor a _____ grouping.

4. What do we call characters that have been derived since their common ancestry? _____

5. What do we call a group of animals that share a certain character? _____

CONSTRUCTING A CLADOGRAM

This exercise will help you understand the hierarchical nature of the classification system and the cladograms that you will be working with in unit II. In the first part of this exercise, you will use a set of wooden block animals to construct a hypothetical monophyletic lineage.

▼ Examine the block animals on display (fig. 7.4). Complete the following observations:

1. One of the animals has a character that clearly sets it apart from all other animals present. This animal is a member of the outgroup that we will use to establish the ancestral character for the study group. What is the synapomorphy that unites all other animals with members of the outgroup? Record that character in blank A of the cladogram (fig. 7.5) in worksheet 7.1.

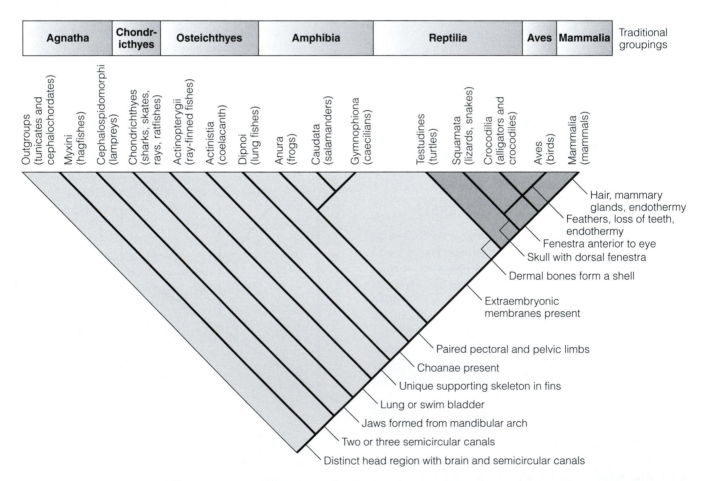

Figure 7.3 A cladogram showing vertebrate phylogeny.

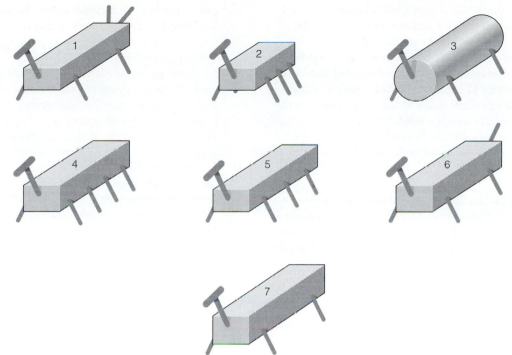

Figure 7.4 Block animals.

2. Group animals other than the outgroup together on the table. What character distinguishes these taxa from the outgroup? Record that character in blank B1 and the contrasting outgroup character in blank B2 in figure 7.5. Write the number 3 in the outgroup blank at the tip of the outgroup lineage of the cladogram.

3. Group animals 1, 6, and 7 together on the table in front of you. They constitute a clade. What character distinguishes this clade from animals 2, 4, and 5? Record this character in blank C1, and the contrasting character in blank C2, of figure 7.5.

4. Animal 7 can be easily distinguished from animals 1 and 6. Record the pair of contrasting characters in blanks D1 and D2 of figure 7.5.

5. In a similar way, distinguish between animals 1 and 6 and record the contrasting characters. Write the number of these animals in their positions at the tips of the branches of this clade.

6. Now return to animals 2, 4, and 5. What are the characters that can be used to group these animals? Record the characters and the numbers designating animals 2, 4, and 5 at the ends of their branches. ▲

You have now constructed one hypothesis that depicts a monophyletic grouping. In completing the following exercise, you will assign names to the six animals in the study group. There are rules of zoological nomenclature that are used in assigning scientific names. The rules are very lengthy, and we will mention only a few.

1. Species names are Latinized by adding Latin endings. A few of these endings are shown in table 7.2. Notice that some of the endings are feminine and some are masculine. In naming a species, both the genus name and the species epithet should reflect the same gender.

2. The name of a genus is one word, which is a singular Latinized noun. It begins with a capital letter, and the word is italicized. The species epithet is also a singular Latinized noun. It begins with a lowercase letter and is also italicized.

3. Above the genus level, names are Latinized, capitalized, plural nouns. They are not italicized.

4. There are conventional endings for superfamily (-oidea), family (-idae), and subfamily (-inae) taxa. You will be using the family-level taxonomic category in the exercise that follows. Higher taxonomic categories do not officially have standard endings for their name.

5. Scientific names are often descriptive of a character used in classifying the animal, although this is not

TABLE 7.2   Case Endings of a Few Latin Nouns

	Feminine	Masculine
Singular	-a	-us
	-am	-er
	-ae (possessive)	-um
Plural	-ae	-i
	-as	-a
		-os
	-arum (possessive)	-orum (possessive)

required. It is impossible for a single class or order name to be descriptive of all animals in the class or order. In the past, species names have been given in honor of the person describing the species or another person who has contributed to knowledge of the animal or group. This practice is less common today. The purpose of a name is not to describe a taxon or to honor a person. It is only a label.

6. A taxon can bear only one correct name.
7. The correct name for an animal is the first correctly published name.

▼ Assume that all animals, 1–7, are separate species and members of a single class united by the common ancestral character—a wooden body. We will assign the class name "Wooda" to the class. We will also assume that animals other than the outgroup do not constitute a paraphyletic grouping—there are no species missing from the group. Complete the following observations:

1. Species that are most closely related are more recently diverged from a common ancestor, and their branch points will be nearer the top of the cladogram. Consider animals 1, 6, and 7. There are two animals that are more closely related to each other than to the third. What are the numbers designating these two closely related animals? _____ This pair of animals should be named as two species within a common genus. The other member of this clade should be given a separate genus name and species epithet. Notice, however, that there is no other species in the genus containing animal 7. In this case, any

species epithet can be used. Using the naming rules described previously, assign species names to animals 1, 6, and 7. Record these names in the taxonomic hierarchy in worksheet 7.1 (fig. 7.6).

2. All three species of the 1, 6, 7 clade are united by a single character. Assign a family name to the clade and record it in figure 7.6. Remember that family names use the "-idae" suffix.

3. Repeat the naming procedure for animals 2, 4, and 5. In this case, you need to decide whether or not animals 2 and 5 belong to the same family as animal 4. Unfortunately, above the species level, assignment of taxonomic names can be somewhat subjective. Taxonomists often struggle with these kinds of decisions. Once you have made your decision, write your reasoning for it at the bottom of figure 7.6. Add to the hierarchy in figure 7.6 by drawing lines connecting the genus level with the order level, including either one or two additional family designations. Write in the family name(s).

4. Finally, complete the hierarchy by adding an order name for the entire group of six species.

One of the last steps in systematic studies is one that we will not complete. Systematists ask the following question: Is this cladogram the simplest cladogram that can be constructed using the data available? This question employs the principle of **parsimony**. As with many natural occurrences, systematists have found that the simplest, most economical explanation for a set of observations is usually the most accurate explanation. This determination usually involves computerized statistical analysis. ▲

KEY TERMS

binomial system   100
cladogram   101
class   100
domain   100
family   100

genus   100
kingdom   100
nomenclature   100
order   100
phylum   100

species   100
symplesiomorphy   101
synapormorphy (derived character)   101
systematics   100
taxonomy   100

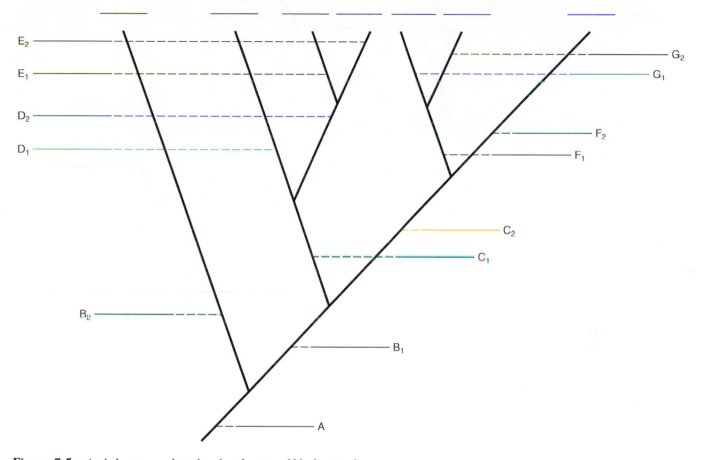

Figure 7.5   A cladogram used in the classification of block animals.

| Class | Order | Family | Genus | Species epithet | Block animal |

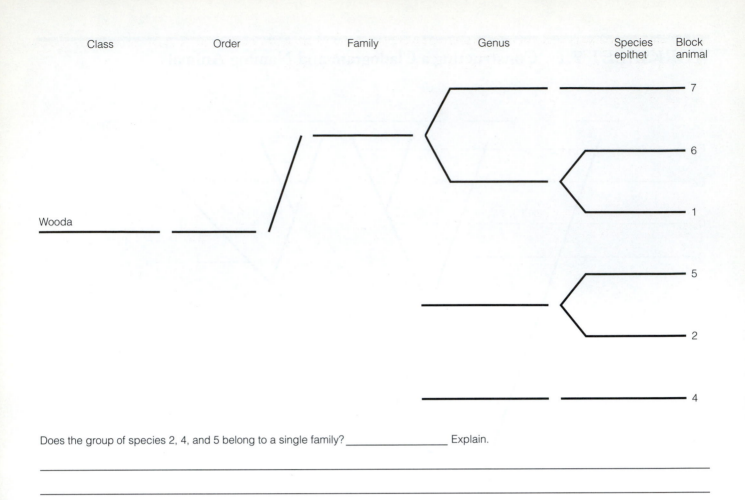

Wooda

Does the group of species 2, 4, and 5 belong to a single family? _____ Explain.

_____

_____

Figure 7.6 A taxonomic hierarchy of block animals.

WORKSHEET 7.2   The Classification of Organisms

1. What is taxonomy?

2. True/False   The goal of classifying organisms is giving them names that show their family relationships.

3. What is a species?

4. Are animals belonging to the same family but different genera more closely related than animals belonging to different classes?

5. Name the levels of classification between kingdom and species.

6. Why do scientists use scientific names when referring to a species?

7. What is a cladogram?

8. Why is the description of synapomorphies important in the work of systematists?

Exercise 8

Animal-like Protists

Learning Objectives

After completing this exercise, you should be able to do the following:

1. Describe current ideas regarding the relationship of members of the kingdom Protista to other kingdoms.

2. Summarize the classification of the animal-like protists.

3. Recognize members of each phylum and subphylum studied.

4. Describe the function of organelles studied in each phylum or subphylum.

Prelaboratory Quiz

Study this week's laboratory exercise and then complete the following quiz to assess your preparation for the laboratory.

1. The Protozoa perform all functions within the confines of a single plasma membrane. Thus, they are said to display
   a. tissue-level organization.
   b. unicellular organization.
   c. organelle-level organization.
   d. organ-level organization.

2. Members of the phylum Sarcomastigophora may have all of the following characteristics EXCEPT
   a. flagella.
   b. pseudopodia.
   c. a single type of nucleus.
   d. spores.

3. Members of the subphylum Sarcodina include
   a. euglenoids.
   b. *Paramecium*.
   c. amoebas.
   d. all intracellular parasitic protists.

4. All of the following are characteristic of members of the phylum Apicomplexa EXCEPT
   a. intracellular parasites.
   b. apical complex used in penetrating host cells.
   c. cilia and flagella present in all stages.
   d. most form spores in their life cycles.

5. Vertebrates acquire malaria as a result of the parasite being introduced into the vertebrate through
   a. the bite of a mosquito.
   b. the bite of a fly.
   c. drinking contaminated water.
   d. eating contaminated meat.

6. Members of which of the following phyla possess both a macronucleus and a micronucleus?
   a. Sarcomastigophora
   b. Apicomplexa
   c. Ciliophora

7. True/False   The Protozoa constitute a monophyletic grouping.

8. True/False   Members of the subphylum Mastigophora possess cilia.

9. True/False   The ectoplasm is a gellike layer of cytoplasm next to the plasma membrane of an amoeba and other protists.

10. True/False   Contractile vacuoles are used in osmoregulation (water regulation).

EVOLUTIONARY PERSPECTIVE

Ancient members of the Archaea were the first living organisms on this planet. The Archaea gave rise to the Protista during the Precambrian period about 1.5 billion years ago. Most biologists agree that the protists probably arose from the archaeans by way of more than one ancestral group. Various phyla of protists, therefore, represent separate evolutionary lineages. The protists are **polyphyletic.** Some protists are plantlike because they are primarily autotrophic. (They produce their own food.) Others are animallike because they are primarily heterotrophic. (They feed upon other organisms.) In this exercise, you will study three phyla of animallike protists.

UNICELLULAR AND COLONIAL EUKARYOTES

To the animal biologist, the animallike Protista (those without chlorophyll) have been known as the Protozoa (formerly a phylum designation). The designation Protozoa is now considered a subkingdom consisting of seven phyla of protists. The protozoans are undoubtedly descended from those forms of life that gave rise to multicellular organisms. Protozoans present today, however, have had a long evolutionary history and may be considerably different from the ancestral stock. It is misleading, therefore, to describe the Protozoa as "primitive," as is commonly done. As a group, however, they display many of the features that probably characterized the earliest protists.

In the Protozoa, all functions are performed within the limits of a single plasma membrane. Protozoans, therefore, display **unicellular (cytoplasmic) organization.** Even though there are no tissues or organs in protozoans, there is a division of labor within the organism through specialized **organelles.** Individual protozoans may also associate together to form colonies, and there may be division of labor between members of a colony.

CLASSIFICATION

The classification of the Protista is unsettled. There are numerous phyla described, and you will study representatives of the following three phyla:

Phylum Sarcomastigophora
　　Subphylum Mastigophora
　　Subphylum Sarcodina
Phylum Apicomplexa
Phylum Ciliophora

PHYLUM SARCOMASTIGOPHORA

Characteristics: Flagella, pseudopodia, or both; single type of nucleus; no spores formed.

Subphylum Mastigophora

Characteristics: One or more flagella. Autotrophic (cl. Phytomastigophora), heterotrophic (cl. Zoomastigophora), or both; reproduction usually by fission.

Euglena

▼ *Euglena* is a common freshwater phytomastigophoran found in ponds and slow-moving streams. Live protozoans can be slowed for study using methylcellulose or a similar product. Place a drop of methylcellulose on a clean microscope slide. Spread it over the middle third of the slide. Next, obtain live specimens of *Euglena* from a culture dish, add them to the slide, and cover with a coverslip. Observe under high power, and compare what you see to figure 8.1. Note the green color, the red eyespot, and movements of the flagellum. After viewing your specimens alive, remove the slide from the microscope and add a drop of iodine-potassium iodide (IKI) to the edge of the coverslip. This will kill the euglenoids but will make the flagellum easier to see. Repeat these observations using a prepared *Euglena* slide.

Knowledge about the details of the *Euglena* structure has increased due to the electron microscope. Note the following structures in figure 8.1a, but do not expect to see these details in the light microscope. ▲

flagella	locomotion
eyespot and **photoreceptor**	photoreception
contractile vacuole	osmoregulation
flagellar pocket	contains the bases of the flagella; aids in discharge of waste and water via the contractile vacuole
kinetosome	base of the flagellum; development and control of the flagellum
chloroplasts	photosynthesis
pyrenoid	starch storage

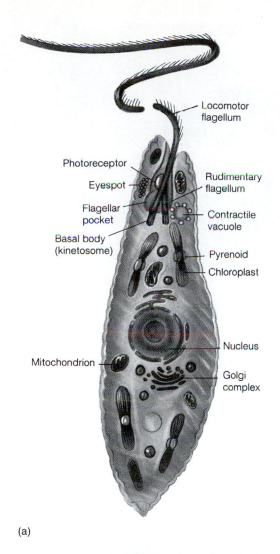

Locomotor
flagellum

Photoreceptor

Eyespot

Rudimentary
flagellum

Flagellar
pocket

Contractile
vacuole

Basal body
(kinetosome)

Pyrenoid

Chloroplast

Mitochondrion

Nucleus

Golgi
complex

(a)

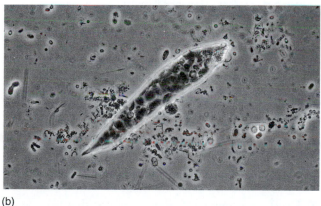

(b)

Figure 8.1 *Euglena.* (*a*) The ultrastructure of *Euglena* as would be observed in electron micrographs. (*b*) A light photomicrograph of *Euglena* (×250).

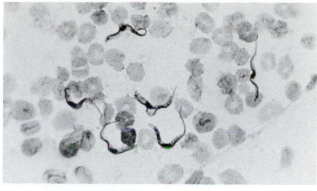

(a)

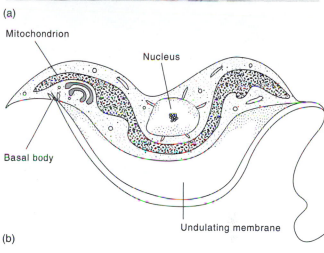

Mitochondrion

Nucleus

Basal body

Undulating membrane

(b)

Figure 8.2 *Trypanosoma.* (*a*) A photomicrograph of *Trypanosoma* in vertebrate blood. (*b*) The structure of *Trypanosoma.* The flagellum of *Trypanosomes* often trails behind the protist and is connected to the protist by an undulating membrane.

Other Mastigophora

▼ Examine other Mastigophora as available. Note that most Zoomastigophora are symbionts, including *Trypanosoma* (fig. 8.2), *Trichonympha* (fig. 8.3), and *Trichomonas*.

If termites are available, make a preparation of their hindgut, mastigophoran symbionts. Place a drop of insect saline on a slide. Working in saline and using fine-tipped forceps, grasp the terminal segments of the abdomen of the termite and pull the hindgut free of the remainder of the body. Discard the head, thorax, and remaining abdomen. Cover the hindgut with a coverslip, and gently squash with the eraser end of a pencil. View under high power.

Volvox is a phytomastigophoran that shows colonial organization. If live organisms are available, place a sample from the culture dish in the concavity of a depression slide. Fill the concavity of the slide with culture medium and cover with a coverslip. Be sure there is no air trapped below the coverslip. Note the graceful rolling of the colony through the water, which results from the activity of the pair of flagella present on each individual (zooid) in the colony (fig. 8.4).

While most zooids in the colony are unspecialized, reproduction is dependent upon certain individuals being

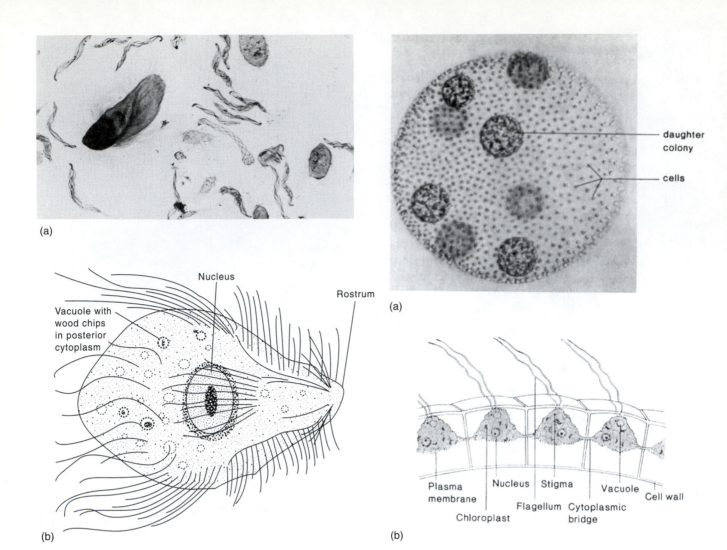

(a)

(b)

Figure 8.3 *Trichonympha* is a very complex mastigophoran found in the gut of termites. It possesses thousands of flagella. (*a*) Photomicrograph (×400). (*b*) Drawing based on an electromicrograph.

(a)

(b)

Figure 8.4 *Volvox,* a colonial flagellate. (*a*) A *Volvox* colony showing asexually produced daughter colonies. (*b*) An enlargement of a portion of a body wall.

specialized for that function. Asexual reproduction occurs when certain cells withdraw to the watery interior of the colony and form daughter colonies. When the parental colony dies and ruptures, daughter colonies are released. Sexual reproduction occurs in autumn when specialized cells differentiate into macrogametes or microgametes. Macrogametes are large, filled with nutrient reserves, and nonmotile. Microgametes form as a packet of flagellated cells that leave the parental colony and swim to a colony containing macrogametes. Microgametes then break apart and syngamy (fertilization) occurs. The zygote develops into a zygospore by secreting a resistant wall around itself. It is an overwintering stage that is released when the parental colony dies. A small colony is released from the zygospore in the spring. Observe any preparations available that demonstrate these features. ▲

Subphylum Sarcodina

Characteristics: Pseudopodia present; flagella occasionally present (in developmental stages). The amoebas.

Amoeba

▼ *Amoeba* is a common freshwater protist that lives on the bottom of ponds and slow-moving streams. Examine living specimens. Obtain a sample of water from the bottom of a culture dish, place one or two drops of the sample on a microscope slide, and cover with a coverslip. The amoeba will appear transparent, granular, and irregular in outline. If you have trouble finding a specimen, ask your instructor for help. Compare what you see to figure 8.5 and locate as many of the following as possible on your specimen:

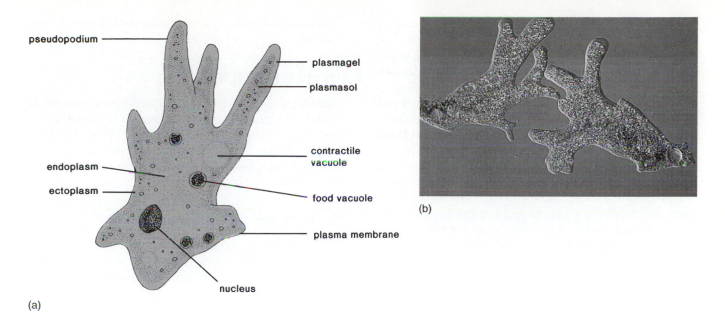

(a)

(b)

Figure 8.5 *Amoeba proteus.* (a) A diagrammatic representation of this protist's structure. (b) A light photomicrograph (×100).

ectoplasm	nongranular, gellike layer of cytoplasm next to the plasma membrane
endoplasm	granular, fluid layer of cytoplasm in the interior of the amoeba
pseudopodia	lobelike extensions of the amoeba used in locomotion and feeding
food vacuole	food storage and digestion
contractile vacuole	osmoregulation

Save this wet mount for the next observation in which you will watch the streaming of the cytoplasm and the formation of pseudopodia. ▲

Amoeboid Movement

▼ Sarcomastigophorans as well as other cells (e.g., white blood cells) display a type of movement known as **amoeboid movement.** As you watch your living amoeba, note that the fluid interior (endoplasm) flows forward, forming a pseudopodium. As the endoplasm reaches the tip of the pseudopod, it flows to the outside and changes state to form ectoplasm. At the posterior of the amoeba, ectoplasm is converted into endoplasm and begins flowing toward the advancing pseudopod. Microfilaments, similar to the actin of muscle cells, are believed to be responsible for this movement. (See cell structure, exercise 2.)

Examine a prepared slide of *Amoeba* and compare it to the living specimen. ▲

Other Sarcodina

▼ Examine other sarcodines as available. Not all amoebas are naked. Some, including the freshwater *Arcella, Difflugia,* and *Actinosphaerium,* and marine radiolarians and fora-
minifera, form tests. Tests can be formed from sand grains (*Difflugia*) or can be secreted in the form of calcium carbonate (foraminifera) or silica (radiolarians). Radiolarians are known to have existed for millions of years because of the fossilized tests found in limestone and chalk beds. ▲

STOP AND ASK YOURSELF

1. What does it mean to say that the protists are polyphyletic? _____

   _____

2. Protists have a(n) _____ level of organization.

3. In what sense is there division of labor in protists?

   _____

   _____

   _____

4. What are two subphyla of sarcomastigophorans? _____ and _____

5. What is the difference between ectoplasm and endoplasm? _____

   _____

   _____

PHYLUM APICOMPLEXA

Characteristics: All parasites. Apical complex used for penetrating host cells. Lack cilia and flagella, except in certain reproductive stages. Coccidians or apicomplexans are named based upon the presence of the **apical complex.**

The most important coccidians are members of the class Sporozoea. They are intracellular parasites of animals (including humans) and form spores or oocysts following sexual reproduction. All sporozoans have complex life cycles that often involve both vertebrate and invertebrate hosts.

Plasmodium, the sporozoan that causes malaria, is a parasite of red blood cells and liver cells of vertebrates. Multiple fission (schizogony) occurs in liver and red blood cells. Some of the products of this fission enter blood or liver cells for more asexual reproduction. Others enter red blood cells and form gametocytes (gametogeny), which are taken into a mosquito during a blood meal and fuse. The zygote penetrates the gut of the mosquito and is transformed into an oocyst. Meiosis and multiple fission (sporogeny) produce haploid sporozoites that enter a vertebrate host when the mosquito bites (fig. 8.6).

▼ Examine a demonstration of *Plasmodium* from vertebrate blood. ▲

PHYLUM CILIOPHORA

Characteristics: Cilia, macronuclei, and micronuclei usually present. The ciliates are the largest, most complex, and most diverse group of protozoans. They occupy nearly all aquatic habitats. Some are symbiotic. Reproduction can be asexual through fission or sexual through a process called conjugation.

Paramecium

▼ *Paramecium* is a common freshwater ciliate. Obtain a drop of culture containing *Paramecium* and place it on a clean microscope slide along with a drop of methylcellulose.

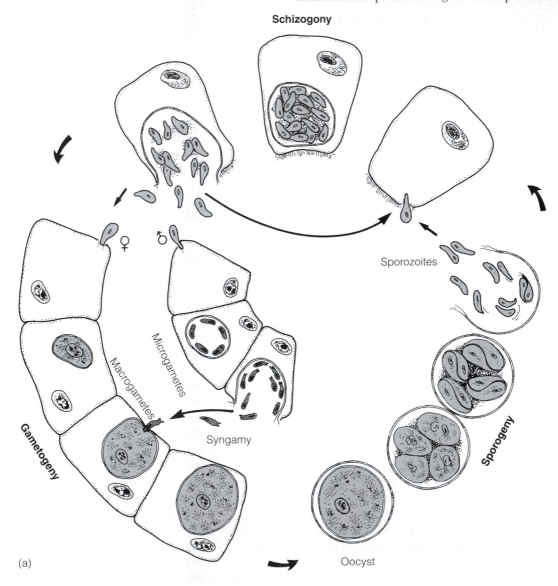

Schizogony

Sporozoites

Sporogeny

Oocyst

Syngamy

Microgametes

Macrogametes

Gametogeny

♀ ♂

(a)

Figure 8.6 *(a)* A generalized life cycle of a coccidian involves three phases. Schizogony is multiple fission of asexual organisms within host cells. Gametogeny occurs when some individuals produce flagellated microgametes and others produce macrogametes. Syngamy produces an oocyst, which is ingested by, or otherwise introduced into, a new host. Sporogeny results in the formation of many haploid sporozoites. *(b)* The life cycle of *Plasmodium*.

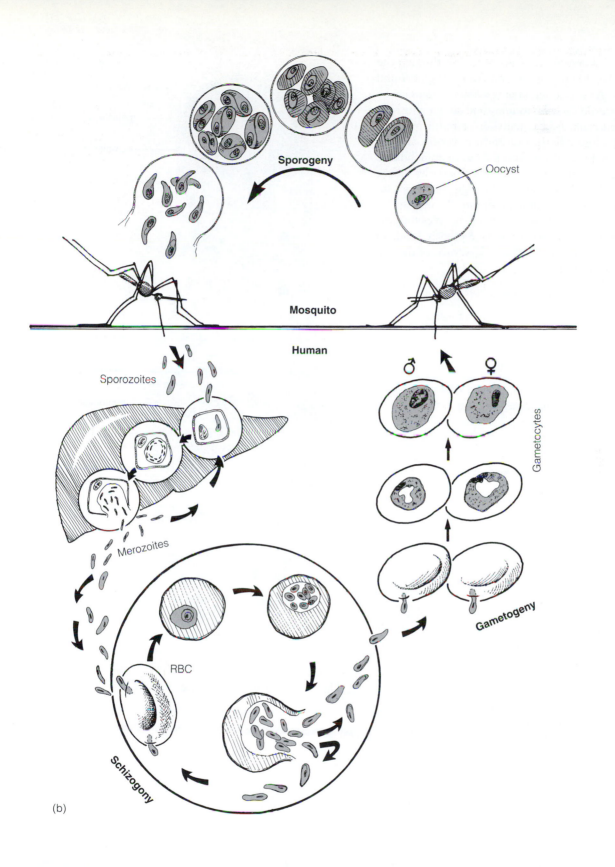

Sporogeny

Oocyst

Mosquito

Human

Sporozoites

♂ ♀

Gametocytes

Merozoites

Gametogeny

RBC

Schizogony

(b)

Note its "slipper" shape. As it swims, it rotates through the water. You will be able to see the oral groove (figs. 8.7 and 8.8) as it swims. As you watch your specimens, you may be able to see contractile vacuoles forming and discharging near the ends of *Paramecium*. Add a small drop of freshly prepared yeast/congo red solution to the side of the coverslip. (If your slide is starting to dry, or if your specimens are immobilized, make a new preparation.) As stained yeast particles diffuse under the coverslip, *Paramecia* should begin feeding. Observe cilia of the oral groove draw yeast into the cytostome. After a few minutes, you should see food vacuoles forming in the region of the cytostome. Observe a prepared slide of *Paramecium* and compare it to figure 8.7. Which of the following structures can you see on your slide? ▲

pellicle	protective outer covering
cilia	locomotion and feeding
trichocysts	attachment
macronucleus	controls metabolic processes
micronucleus	hereditary units that participate in sexual reproduction
oral groove, cytostome, and **cytopharynx**	feeding complex
food vacuoles	food storage and digestion
contractile vacuoles	osmoregulation

Other Ciliophora

The presence of at least one macronucleus and micronucleus is characteristic of ciliates. Stained preparations of a small ciliate, called *Colpidium*, show these structures clearly.

▼ Examine a slide of *Colpidium* under high power. Examine any other specimens available for an appreciation of the diversity present in this phylum. They may include *Vorticella* (fig. 8.9), *Stentor* (fig. 8.10), and ciliates from the rumen of cattle. ▲

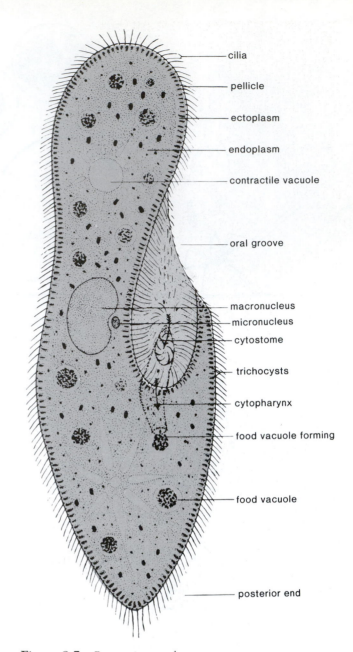

Figure 8.7 *Paramecium caudatum.*

STOP AND ASK YOURSELF

6. Where does asexual reproduction of *Plasmodium* occur? _____

7. What phylum of protists are parasitic, have spores or oocysts in their life histories, and lack cilia or flagella (except in gametes)? _____

8. What is the function of the oral groove of *Paramecium*? _____

9. What are contractile vacuoles? _____

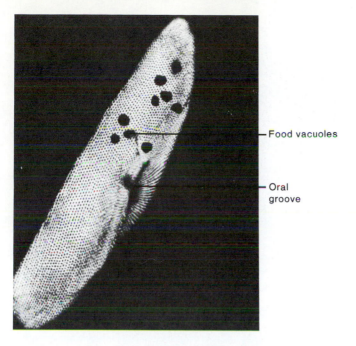

Figure 8.8 *Paramecium* (nigrosin stain shows sculpturing of pellicle) (×450).

Food vacuoles

Oral groove

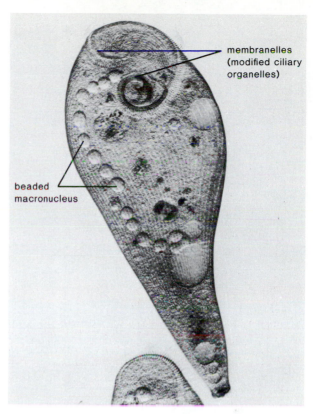

membranelles (modified ciliary organelles)

beaded macronucleus

Figure 8.10 *Stentor.*

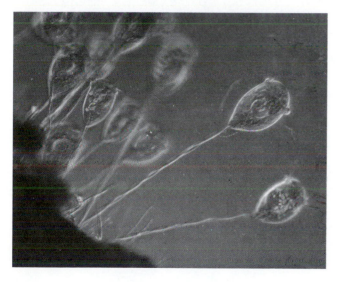

Figure 8.9 *Vorticella*, differential interference optical photomicrograph (×200).

KEY TERMS

amoeboid movement   113

apical complex   113

cilia   116

contractile vacuole   110 & 113

ectoplasm   113

endoplasm   113

food vacuole   113

macronucleus   116

micronucleus   116

pellicle   116

polyphyletic   110

pseudopodia   113

unicellular (cytoplasmic)
   organization   110

WORKSHEET 8.1   Animal-like Protists

Evolutionary Perspective

1. Describe what is known of the evolutionary origin(s) of the kingdom Protista.

Phylum Sarcomastigophora

2. Characterize members of the phylum Sarcomastigophora.

3. Describe amoeboid movement.

Phylum Apicomplexa

4. What is the apical complex?

5. Briefly describe the life cycle of *Plasmodium*. What part of this life cycle did you observe?

Phylum Ciliophora

6. What features of the Ciliophora did you observe in lab?

7. Describe the method of feeding of *Paramecium*.

Exercise 9

Porifera

Learning Objectives

After completing this exercise, you should be able to do the following:

1. Describe the evolutionary relationships of the Porifera with the protists and other members of the Animalia.
2. Describe the cellular level of organization.
3. Name and describe the function of the cell types found in sponges.
4. Describe the three sponge colony forms and the evolutionary significance of each.
5. Describe the function of structures studied in the laboratory and recognize each in the specimens studied.

Prelaboratory Quiz

Study this week's laboratory exercise and then complete the following quiz to assess your preparation for the laboratory.

Name the three cell types that make up the body wall of a sponge and list their function(s).

Name	Function
1. _____	_____
2. _____	_____
3. _____	_____

4. True/False   Asconoid sponges have a spongocoel. Water enters the spongocoel through radial canals.
5. True/False   Asconoid sponges have the simplest body organization.
6. True/False   Incurrent canals of a leucon sponge are lined by choanocytes.
7. True/False   The pathway for water leaving a sponge is through the osculum.
8. True/False   The largest sponges have the leuconoid body form.
9. True/False   In the leuconoid sponges, incurrent canals lead to flagellated chambers.
10. True/False   Sponges have a tissue-level organization.

EVOLUTIONARY PERSPECTIVE

The sponges are members of the phylum Porifera and, like all animals, they are multicellular. The origins of multicellularity are shrouded in mystery. Many zoologists believe that multicellularity could have arisen as cells divided but remained together as a colony of protists. (Recall the structure of *Volvox*.) Although there are a number of variations of this hypothesis, they are all treated here as the **colonial hypothesis.** Ancient flagellated protists are

121

often described as possible ancestral stocks because flagellated protists form colonies. A second proposed hypothesis is the **syncytial hypothesis.** A syncytial cell is a large, multinucleate cell. The formation of plasma membranes within the cytoplasm of a syncytial protist could have formed a small, multicellular organism. This hypothesis is supported by the fact that syncytial organization occurs in some protist phyla.

A fundamental question concerning animal origins is whether animals are monophyletic (derived from a single ancestor) or polyphyletic (derived from more than one ancestor). Many zoologists believe that sponges derived from the protists, but separately from other animal phyla, and that the animal kingdom is at least diphyletic. The presence of flagellated cells in the sponges makes the colonial hypothesis an attractive explanation of their origin. Other zoologists disagree. They believe that the Porifera and other animal phyla can be traced to a single protistan ancestor.

THE CELLULAR LEVEL OF ORGANIZATION

The Porifera consist of loosely organized cells with no well-defined tissues. They are asymmetrical or radially symmetrical, sessile, and mostly marine. (See the discussion of symmetry on page xii.) Even though sponges do not have tissues, their bodies are more than colonies of independent cells. Sponge cells are specialized for specific functions. This is often referred to as division of labor. The major cell types and their functions are as follows (fig. 9.1):

pinacocytes   Thin, flat cells covering the outer surface and some of the inner surface. May be mildly contractile, thus having the ability to change the shape of the sponge. May be specialized into porocytes (myocytes) that surround pores and can contract to regulate circulation of water to the interior of the sponge.

choanocytes   Flagellated cells lining sponge chambers. Create water currents and filter suspended food particles from the water. Sometimes called collar cells.

mesenchyme cells   Amoeboid cells embedded in the noncellular matrix. May store, digest, and distribute food, or be specialized to form reproductive cells (archaeocytes), secrete spicules (scleroblasts), secrete spongin (spongioblasts), or secrete fibrils (collenocytes and lophocytes). Spicules and spongin give structural support to the sponge colony. Cells are loosely arranged in jellylike material called the **mesohyl.** Amoeboid cells, spicules or spongin, and various fibrils are embedded in the mesohyl.

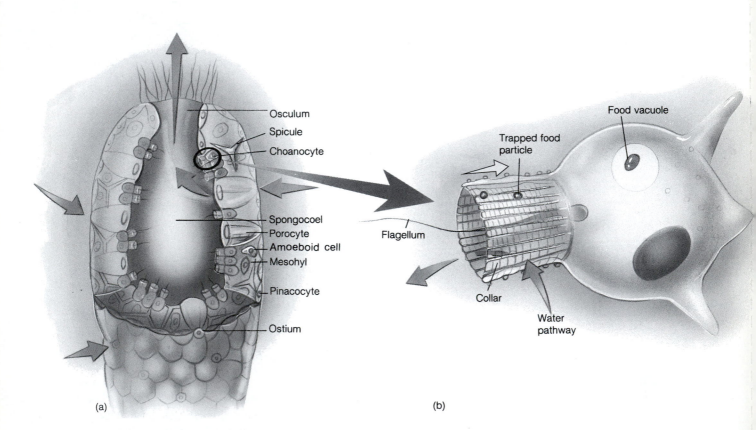

Figure 9.1   Morphology of a simple sponge, including body-wall organization. (a) In this example, pinacocytes form the outer body wall, and amoeboid cells and the spicules are found in the mesohyl. Ostia are formed by porocytes. (b) Choanocytes are cells in which the flagellum is surrounded by a collar of microvilli that traps food particles. Food is moved toward the base of the cell, where it is incorporated into a food vacuole and passed to amoeboid cells where digestion takes place. Solid arrows show water flow pattern. The open arrow shows the direction of movement of trapped food particles.

CLASSIFICATION

Class Calcarea
> Calcium carbonate spicules in the form of needles or three- or four-rayed stars. Asconoid, syconoid, or leuconoid. Marine.

Class Hexactinellida
> Siliceous spicules in the form of six-rayed stars. Syconoid or leuconoid. Marine, mostly deep water.

Class Demospongiae
> Spicules siliceous (but not six-rayed) and/or spongin. Leuconoid. Freshwater (one family) and marine.

Rather than trying to characterize each poriferan class, we will examine the three body forms characteristic of sponges. Body forms are described based upon the arrangement of canals. Recall from exercise 1 that when observing preserved specimens under a dissecting microscope, it is best to immerse them completely in water to prevent the refraction of light off moist surfaces. Failure to do so will result in a distorted image.

THE ASCONOID SPONGE— *LEUCOSELENIA* (CALCISPONGIAE)

▼ **Asconoid sponges** have the simplest organization. Choanocytes line the spongocoel, drawing water through small openings called **ostia** and expelling it through the osculum (fig. 9.2a). Examine *Leucoselenia* slides using the low power of a compound microscope. Locate the following: ▲

osculum	water outlet
spicules	triradite, protruding from the body wall
spongocoel	central cavity lined by choanocytes

THE SYCONOID SPONGE—*SCYPHA* (CALCISPONGIAE)

▼ **Syconoid sponges** have a tubular design similar to the ascon sponge, but the body wall is folded. The "folds" form

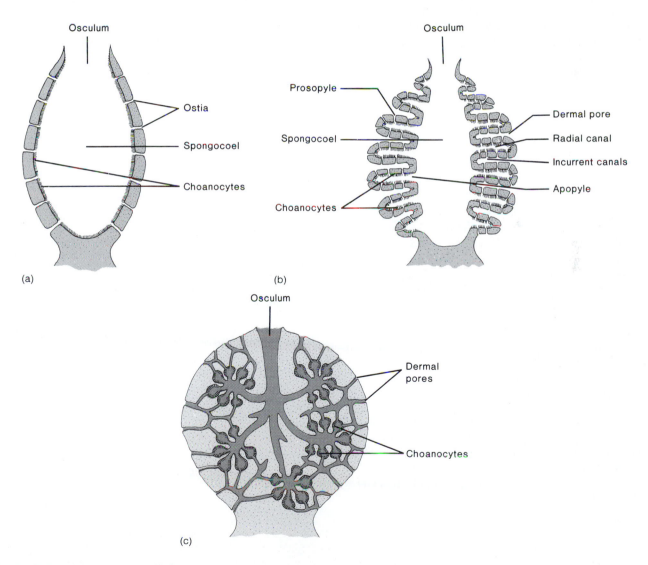

Figure 9.2 Sponge body forms: (*a*) ascon sponge; (*b*) sycon sponge; (*c*) leucon sponge.

radial canals and incurrent canals (fig. 9.2b). Choanocytes line the radial canals rather than the spongocoel. Examine preserved specimens of *Scypha* under a dissecting microscope. Externally locate the osculum, which opens at the top of the sponge, and the many small dermal pores along the sides. Dermal pores open into the canal system of the sponge. Choanocytes create water currents that draw water through dermal pores to the interior of the sponge. Eventually water is discharged through the osculum.

With a sharp scalpel or razor blade make a longitudinal section through the specimen. Locate internal openings to the radial canals, the apopyles. The large cavity into which the incision was made is the spongocoel.

Examine a cross section of *Scypha* using a compound microscope. Examine first under lower power (fig. 9.3). Note the central spongocoel. Radial canals radiate from the spongocoel. They connect through small openings (prosopyles) to incurrent canals that open to the outside of the sponge. Recall that the dermal pore is the external opening to the incurrent canal and the apopyle is the internal opening of the radial canal. Return to the radial canal and change to high power. Examine the inner cell layer of the radial canal. Choanocytes make up this layer (figs. 9.4 and 9.5). Some slides are prepared with the calcium carbonate spicules intact; in other slides, spicules are dissolved before mounting. If your slide shows spicules, note their triradiate structure. If your slide does not show them, obtain and examine a separate preparation of spicules (fig. 9.6). ▲

THE LEUCONOID SPONGE

Leuconoid sponges represent the most complex body form. The canal system is extensively branched (fig. 9.2c). Small incurrent canals lead to flagellated chambers lined by choanocytes. Flagellated chambers discharge water into excurrent canals that eventually lead to an osculum. Usually there are many oscula in each sponge. The "bath sponge" is an example of a leuconoid sponge. The skeleton of this sponge is made of a soft protein, called spongin, rather than calcium carbonate or silica.

▼ Examine demonstration materials showing the leuconoid body form. ▲

SPONGE EVOLUTION

The ascon-to-sycon-to-leucon progression probably does not reflect an exact sequence in the evolution of sponge body forms. (The vast majority of sponges, however, are leuconoid.) What seems clear is that natural selection resulted in increased body size and greater surface area for choanocytes.

▼ Look again at demonstrations of the three sponge body forms. Which of the three body forms has the largest overall size? _____

How is canal organization correlated with body size? _____

_____

Which of the three body forms has the most surface area for choanocytes? _____

Why would this be advantageous? _____

_____

_____

_____ ▲

STOP AND ASK YOURSELF

1. What is the function of each of the following:

   a. osculum _____

   _____

   b. ostium _____

   _____

   c. spongocoel _____

   _____

   d. apopyles _____

   _____

   e. radial canals _____

   _____

   f. choanocytes _____

   _____

2. How could you distinguish between a dermal pore and an apopyle in a slide preparation of a sycon sponge? _____

3. Describe the organization of the filtering surfaces of a syconoid sponge. _____

   _____

   _____

4. Why can one wash a car with a natural bath sponge and not harm the car's finish? _____

   _____

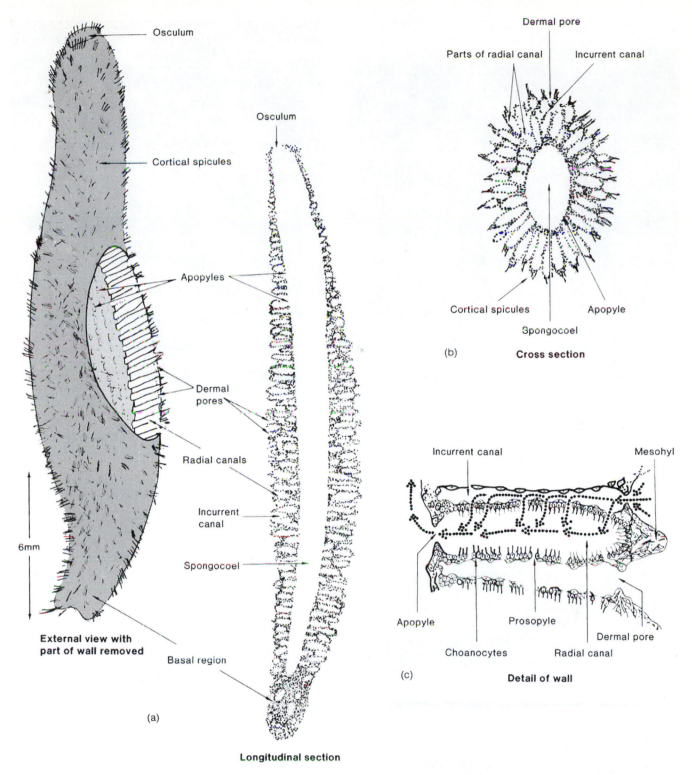

Figure 9.3 *Scypha:* (*a*) external structure and longitudinal section; (*b*) cross section; (*c*) structure of the body wall. The dotted arrows in (*c*) show the path of water movement.

flagellum   collar

cell body

Figure 9.4   Choanocyte (scanning electron micrograph) (×9,400).

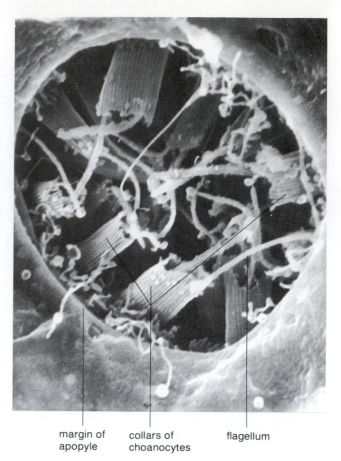

margin of   collars of   flagellum
apopyle   choanocytes

Figure 9.5   Choanocytes as viewed through an apopyle (scanning electron micrograph) (×6,200).

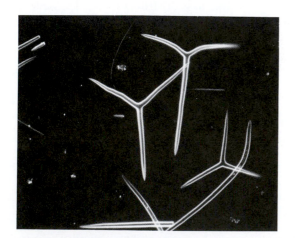

Figure 9.6   Sponge spicules (×120).

KEY TERMS

WORKSHEET 9.1 Porifera

Evolutionary Perspective

1. Contrast colonial and syncytial hypotheses of the origin of multicellularity in animals.

2. Why do many zoologists believe that sponges arose from ancient flagellated protists?

Cellular Level of Organization

3. Name three general cell types found in the Porifera and describe their functions.

4. Describe three sponge body forms and the selective advantage of the more complex sponge body form.

Exercise 10

Cnidaria
(Coelenterata)

Learning Objectives

After completing this exercise, you should be able to do the following:

1. Describe the evolutionary significance of the phylum Cnidaria and appreciate evolutionary relationships within the phylum.
2. Describe diploblastic, tissue-level organization.
3. Describe cnidarian polymorphism.
4. Define radial symmetry and describe its advantages for sedentary organisms.
5. Describe the cnidarian body plan.
6. Recognize class representatives.
7. Recognize and describe the function of those structures studied in the laboratory.

Prelaboratory Quiz

Study this week's laboratory exercise and then complete the following quiz to assess your preparation for the laboratory.

1. Members of the phyla Cnidaria and Ctenophora possess
   a. radial or biradial symmetry.
   b. diploblastic, tissue-level organization.
   c. oral and aboral ends.
   d. All of the above are correct.

2. The life cycle of many cnidarians involves an alternation between _____ and _____ stages.
   a. aquatic and terrestrial
   b. polyp and medusa
   c. bilateral and radial
   d. diploblastic and triploblastic

3. Members of the class Hydrozoa include
   a. sea anemones.
   b. true jellyfish.
   c. corals.
   d. colonial hydroids and *Hydra*.

4. Members of the class Anthozoa include
   a. *Hydra*.
   b. corals and sea anemones.
   c. true jellyfish.
   d. freshwater cnidarians.

5. The larval stage of cnidarians is called the
   a. planula.
   b. hydranth.
   c. polyp.
   d. medusa.

6. True/False   The epidermis of cnidarians contains epithelio-muscular cells and cnidocytes.

7. True/False   The phylum name Cnidaria is derived from the presence of specialized cells called cnidocytes, which are used in reproduction.

8. True/False   With a few exceptions, members of the phylum Cnidaria are monoecious.

9. True/False   Members of the class Anthozoa do not display alternation of generations.

10. True/False   The predominate life-history stage in members of the class Scyphozoa is the asexually reproducing polyp.

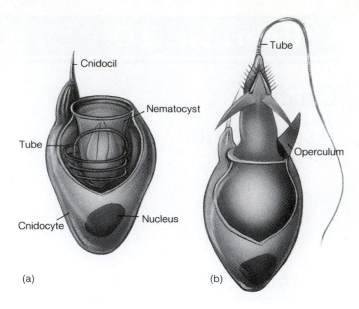

Figure 10.1 Cnidarian nematocysts: (*a*) undischarged; (*b*) discharged.

EVOLUTIONARY PERSPECTIVE

Members of the phylum Cnidaria are radially symmetrical. (See the discussion of symmetry on page xii.) The Cnidaria, along with another lesser known phylum (Ctenophora), are called the radiate phyla. The evolutionary significance of the radiate phyla is debated. Zoologists who think the animal kingdom is polyphyletic contend that the radiate phyla had a separate origin from most other groups of animals. Other zoologists think that the radiate phyla are closely related to the ancestors of all other animal groups. They contend that radially symmetrical animals could have evolved from a colonial flagellate, which may have looked similar to *Volvox*. Still other zoologists think that bilateral symmetry was the ancestral form of symmetry, and they contend that the radiate phyla are further removed from the base of the evolutionary tree. Currently it is impossible to determine which, if any, of these interpretations is correct. As in all discussions of the evolutionary relationships among higher taxonomic groups, one must be careful to avoid picturing modern cnidarians, or any modern animals, as being ancestral. Modern species are the result of hundreds of millions of years of evolution and may be very different from ancestral forms.

THE RADIATE ANIMALS— TISSUE-LEVEL ORGANIZATION

Radial animals have no anterior and posterior (see terms of direction on page xii). As a result, terms of direction are defined based on the position of the mouth opening. The end of the organism that contains the mouth is the **oral end,** and the opposite end is the **aboral end.** The radial symmetry of some cnidarians is modified into **biradial symmetry.** In biradial symmetry, a single plane, passing through a central (oral/aboral) axis, divides the organism into mirror images. Biradial symmetry occurs when a single, or a paired, structure (e.g., feeding or sensory tentacles) occurs in a basically radial animal. It differs from bilateral symmetry in that there is no distinction between dorsal and ventral surfaces. Radial symmetry is advantageous for sedentary animals because sensory receptors are evenly distributed around the body. These organisms can respond to stimuli that come from all directions.

Cnidarians also possess **diploblastic, tissue-level organization.** This means that similar cells are organized into tissues, and all cells are derived from two embryological layers

(see exercise 5). Ectoderm of the embryo gives rise to the outer layer of the body wall, called the **epidermis,** and the endoderm of the embryo gives rise to the inner layer of the body wall, called the **gastrodermis.** Cells of the epidermis and gastrodermis are specialized to perform functions such as protection, food gathering, coordination, movement, digestion, and absorption.

PHYLUM CNIDARIA—NEMATOCYSTS

The phylum name comes from the presence of specialized cells used in defense, feeding, and attachment. The **cnidocytes** contain stinging organelles called **nematocysts** (fig. 10.1).

Alternation of Generations

The life cycle of a typical cnidarian displays **alternation of generations**—it alternates between an often sessile **polyp** (hydroid) stage and a swimming **medusa** (jellyfish) stage (fig. 10.2). The polyp stage is often an asexual stage, and the medusa is often the sexual stage. In some cnidarian classes, either the polyp or medusa is reduced or missing.

With a few exceptions, the Cnidaria are **dioecious**—the sexually reproducing stage has individuals that are either male or female. In the Cnidaria, it is usually impossible to distinguish sexes by superficial examination.

Gastrovascular Cavity

The gastrovascular cavity is a large central cavity that serves to receive and digest food. The single opening on the oral surface serves as both mouth and anus. Tentacles surrounding the oral opening aid in feeding.

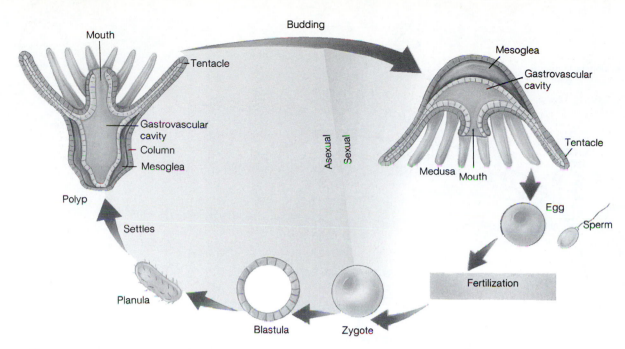

Figure 10.2 A generalized cnidarian life cycle showing alternation between medusoid and polypoid forms.

The Body Wall

The body wall of cnidarians consists of two cellular layers. The epidermis consists of **epithelio-muscular cells** and cnidocytes. Epithelio-muscular cells are contractile and aid in movement. The inner gastrodermis lines the gastrovascular cavity and contains **gland cells** for the production and release of enzymes. The gastrodermis also has flagellated, **nutritive-muscular cells** that contain food vacuoles produced from phagocytosis of partly digested food particles. Between the epidermis and gastrodermis is the **mesoglea.** It is a noncellular gel and is abundant in the medusa.

CLASSIFICATION

Class Hydrozoa
Class Scyphozoa
Class Cubozoa
Class Anthozoa

Class Hydrozoa

Characteristics: Polyp and medusa stages present, though one may be reduced in importance; cnidocytes present in the epidermis; gametes produced epidermally; no wandering mesenchyme cells in the mesoglea; medusa with velum; freshwater and marine. Examples: *Obelia, Hydra, Gonionemus,* and *Physalia.*

Obelia

We will first examine the colonial hydroid *Obelia* because it clearly illustrates alternation of generations (best represented in the Hydrozoa). *Obelia* is a marine hydrozoan whose colonial hydroid grows attached to some substrate.

Because it is colonial, the hydroid will have multiple polyps. *Obelia* has two types of polyps. Other species have up to five.

▼ Examine a slide of the *Obelia* hydroid under a dissecting microscope. Using figure 10.3, locate the following structures:

coenosarc	cellular body wall surrounding the gastrovascular cavity
perisarc	nonliving, protective, chitinous sheath surrounding coenosarc
hydranth	feeding polyp with mouth, tentacles, and hydrotheca
gonangium	reproductive polyp, observe: blastostyle with medusa buds, gonotheca.

Next, obtain a slide of the *Obelia* medusa. The *Obelia* medusa is very small and should be observed with a compound microscope. You should be able to locate the exumbrellar and subumbrellar surfaces, manubrium, tentacles, and, at the margin of the medusa, clear, round statocysts (equilibrium and balance receptors).

Reexamine figure 10.3 to study the life cycle of *Obelia.* Sexual reproduction in the medusa results in the formation of a **planula** larval stage. The planula settles and grows into the asexual polyp. This life cycle is believed by many to represent the primitive cnidarian life cycle. If so, it is probable that the ancestral hydrozoans represent the cnidarian stock from which the other two classes and modern Hydrozoa evolved. Some biologists, however, maintain that the anthozoan life cycle is the more primitive life cycle and, therefore, believe the Anthozoa represent the ancestral stock. The controversy remains and may never be resolved. ▲

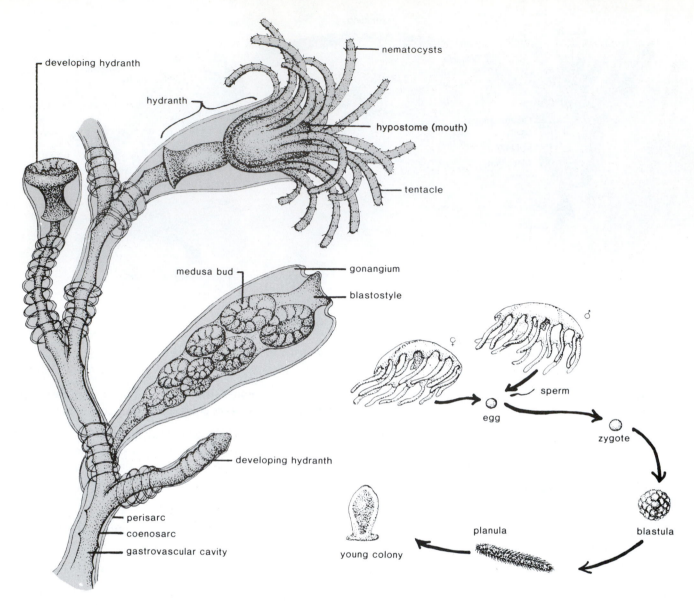

Figure 10.3 *Obelia* life cycle.

Gonionemus

▼ Obtain a preserved *Gonionemus* medusa. *Gonionemus* is a hydrozoan with a well-developed medusa and a minute polyp.

Examine *Gonionemus* using a dissecting microscope. Compare your observations to figures 10.4 and 10.5. The convex surface of the medusa is referred to as the **exumbrella** and the concave surface the **subumbrella.** Place your *Gonionemus,* subumbrellar surface up, in a dish of water. Note the consistency of the medusa. This is due to the large quantity of mesoglea present. The mouth opening is contained on the **manubrium** that hangs from the middle of the oral surface. Four sets of **gonads** radiate from the manubrium to the margin of the medusa. Note that a membranous shelf extends inward from the margin of the umbrella. This is the **velum** and is diagnostic of hydrozoan medusae. It is believed to aid in swimming. At the margin of the medusa, **tentacles**

with nematocysts and adhesive pads can be seen. At the base of each tentacle are photoreceptors and equilibrium and balance receptors. We will not attempt to locate these.

Internally, the gastrovascular cavity divides into four **radial canals** that connect at the margin of the medusa with a **ring canal.** Radial canals are located just aboral to the gonads and can be seen through the exumbrellar surface. These canals distribute food throughout the medusa. ▲

Hydra

Hydra is a freshwater hydrozoan that lacks the medusa stage (fig. 10.6a). (Its sexual and asexual reproductive cycles will not be described here. Consult your text if instructed to do so.)

▼ Obtain a prepared slide mounted with a cross section of *Hydra.* Note the diploblastic organization of the body wall. Locate epidermis, gastrodermis, mesoglea, and gastrovascular cavity (fig. 10.6b). Examine living *Hydra* by placing a

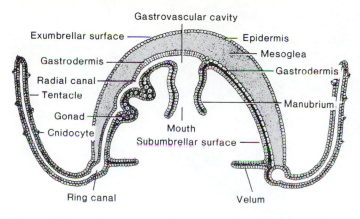

Figure 10.4 The structure of *Gonionemus* (oral/aboral section).

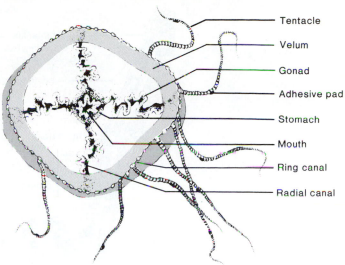

Figure 10.5 *Gonionemus*, oral view.

specimen in a dish and observing the *Hydra* under a dissecting microscope. Allow the *Hydra* to recover from the shock of handling, then gently touch it with a needle. Observe the response. Transfer it to a depression slide, cover with a coverslip, and examine the tentacles closely. The cnidocytes can be seen as small "bumps." Add a drop of Bouin's fluid to cause nematocysts to discharge. Examine under high power (reduce diaphragm aperture). How many kinds of nematocysts do you see? ▲

Other Hydrozoans

Another freshwater cnidarian belongs to the genus *Craspedacusta*. This hydrozoan has a small polyp stage that lives in temperate freshwater lakes during the spring and summer. In the fall, the medusa is produced. Zygotes overwinter and develop into a polyp the following spring.

▼ Examine any specimens available. ▲

Physalia is called the Portuguese man-of-war. Although not evident from looking at a specimen, *Physalia* represents a colony containing both polyps and medusae. Different types of polyps are specialized for feeding, protection, and asexual reproduction. Medusae are specialized for swimming, forming gas-filled floats, forming oil-filled floats, defense, or sexual reproduction. *Physalia* and closely related hydrozoans normally float or swim in deep-water areas of oceans.

▼ Examine a demonstration specimen. ▲

STOP AND ASK YOURSELF

1. What is biradial symmetry? _____

_____

2. Why is radial symmetry advantageous for sedentary animals? _____

_____

_____

3. What is diploblastic organization? _____

_____

4. How does the term "alternation of generations" apply to members of the phylum Cnidaria?

_____

_____

5. Describe the location and function of the following:

a. nutritive-muscular cells _____

_____

b. manubrium _____

_____

c. radial canals _____

_____

d. planula _____

_____

Class Scyphozoa

Characteristics: Polyp reduced or absent. Medusa prominent without velum; gametes produced gastrodermally; cnidocytes present in gastrodermis and epidermis; mesoglea with wandering mesenchyme cells of epidermal origin. Marine. *Aurelia*, *Rhizostoma*.

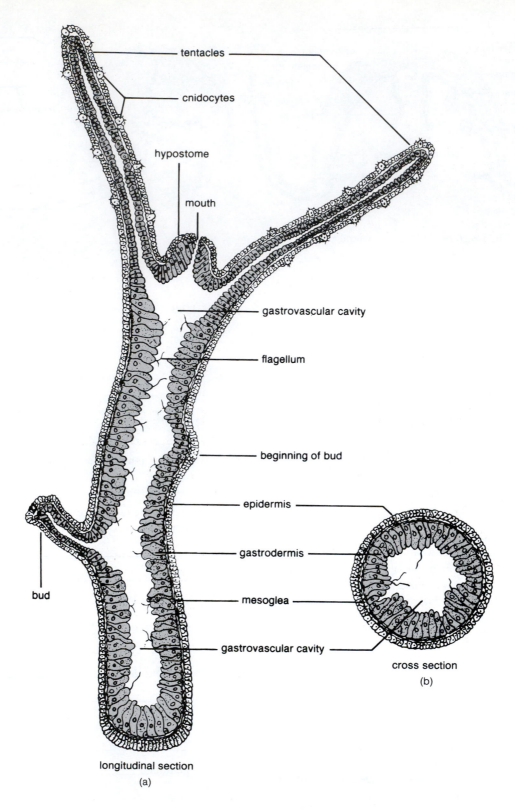

tentacles

cnidocytes

hypostome

mouth

gastrovascular cavity

flagellum

beginning of bud

epidermis

gastrodermis

mesoglea

gastrovascular cavity

cross section

(b)

bud

longitudinal section

(a)

Figure 10.6 The structure of *Hydra:* (*a*) longitudinal section; (*b*) cross section.

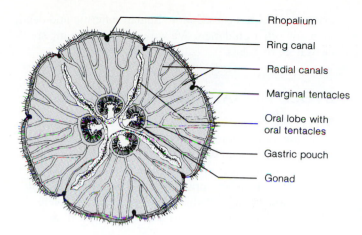

- Rhopalium
- Ring canal
- Radial canals
- Marginal tentacles
- Oral lobe with oral tentacles
- Gastric pouch
- Gonad

Figure 10.7 *Aurelia* medusa, oral view.

Aurelia

▼ The Scyphozoa are referred to as the "true jellyfish." Obtain a specimen of *Aurelia* and place it in a water-filled bowl. Spread it aboral-side up. Note the eight notches at the margin of the jellyfish. These are the location of **rhopalia,** sense organs for photoreception (ocellus) and equilibrium and balance (statocysts). Note also the **radial canals, ring canal,** and **tentacles.** Next, turn the specimen oral-side up (fig. 10.7). Find the mouth that is surrounded by four **oral lobes** (modifications of the manubrium). These are armed with nematocysts and are used in food capture. Sexes are separate, and **gonads** lie within modifications of the gastrovascular cavity called **gastric pouches.** Gonads are the horseshoe-shaped structures lining the margin of each gastric pouch. ▲

The zygote develops into a planula that settles and forms a small polyp. The polyp is first called a **scyphistoma,** and then a **strobila.** The strobila produces miniature medusae, called **ephyrae,** by budding. The ephyrae are stacked one on top of the other on the strobila, the top one being closest to maturity. As ephyrae mature, they break loose from the strobila and take up a free-swimming lifestyle. They then grow to sexually mature adults.

▼ Examine any materials available that demonstrate the life cycle of *Aurelia*. ▲

Class Anthozoa

Characteristics: Polyp only, no medusa; solitary or colonial; gastrovascular cavity divided by mesenteries; cnidocytes present in the gastrodermis. Sea anemones and corals.

Metridium

▼ Examine the sea anemone *Metridium*. The body is cylindrical with the mouth and tentacles located on the **oral disk.** The **pedal disk** is the point of attachment to the substrate. Between the pedal and oral disks is the **column.** Inside the mouth note numerous ridges and one, or occasionally more, smooth grooves called **siphonoglyphs.** This ciliated groove directs water into the **pharynx.**

If your instructor directs you to do so, find the following internal structures by making a longitudinal incision through the body wall (fig. 10.8). ▲

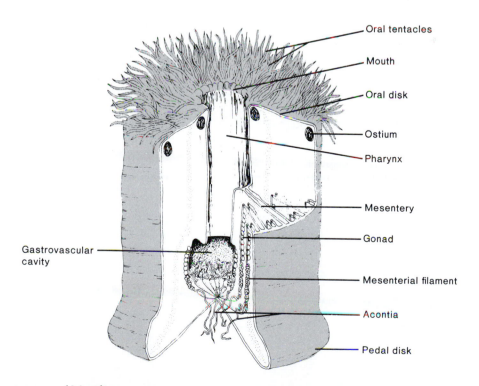

- Oral tentacles
- Mouth
- Oral disk
- Ostium
- Pharynx
- Mesentery
- Gonad
- Mesenterial filament
- Acontia
- Pedal disk
- Gastrovascular cavity

Figure 10.8 The structure of *Metridium*.

pharynx	tubular passageway from the oral opening to the gastrovascular cavity
gastrovascular cavity	cavity below and around the gullet
mesenteries	extensions of the body wall into the gastrovascular cavity; increase surface area for absorption and secretion
gonads	on the margins of incomplete septa (those not connecting to the pharynx)
acontia	filaments bearing nematocysts at the aboral end of the gastrovascular cavity

The Anthozoa lack a medusa stage. Gonads in the dioecious polyp produce gametes that are released externally, where fertilization takes place. After settling to the substrate, the planula grows into the adult polyp. Reproduction also occurs asexually by **fragmentation** (regeneration of individuals from pieces of the polyp).

Other Anthozoa

▼ Examine demonstrations of other Anthozoa, particularly the stony and soft corals. Stony corals are responsible for the formation of coral reefs. They secrete a calcium carbonate exoskeleton around their pedal disk and column. Note the small openings in a piece of hard coral where an anthozoan once lived. Soft corals are common in warm waters and are often very colorful. They secrete an internal skeleton of calcium carbonate or protein. They include sea fans, sea whips, red corals, and pipe organ corals. ▲

Class Cubozoa

Characteristics: Cuboidal medusa; tentacles hang from corners of the medusa; polyps very small or unknown. Sea wasps.

Cubozoans are found in the warm tropical waters off the coast of Australia. *Chironex fleckeri* has extremely potent ne-matocysts and has caused more human suffering and death than any other cnidarian. We will not examine any specimens from this class.

EVOLUTIONARY RELATIONSHIPS

▼ A cladogram showing a traditional interpretation of the evolutionary relationships among the cnidarian classes is shown in worksheet 10.1 (fig. 10.9). Examine the synapomorphies depicted on the cladogram. Based upon your studies in this week's laboratory, you should be able to determine the position of each of the classes studied. Complete the cladogram and answer the questions that follow it. ▲

STOP AND ASK YOURSELF

6. Describe the location and function of the following:

   a. rhopalium _____

   _____

   b. oral arms _____

   _____

   c. strobila _____

   _____

   d. siphonoglyph _____

   _____

7. Members of what cnidarian class lack a medusa stage? _____

8. Members of what cnidarian class have a reduced polyp stage? _____

9. What is fragmentation? _____

   _____

KEY TERMS

alternation of generations 130	epidermis 130	nematocysts 130
cnidocytes 130	gastrodermis 130	planula 131
diploblastic 130	gastrovascular cavity 136	tissue-level organization 130

WORKSHEET 10.1   Cnidaria and Ctenophora

Evolutionary Perspective

1. What are three ideas regarding the evolutionary position of the radiate phyla?

The Radiate Animals—Tissue-Level Organization

2. What is diploblastic, tissue-level organization?

Phylum Cnidaria

3. How would you characterize the four cnidarian classes studied in this exercise?

4. What cell type is unique to the Cnidaria?

5. How is the life cycle of an anthozoan different from that of most members of the other two classes?

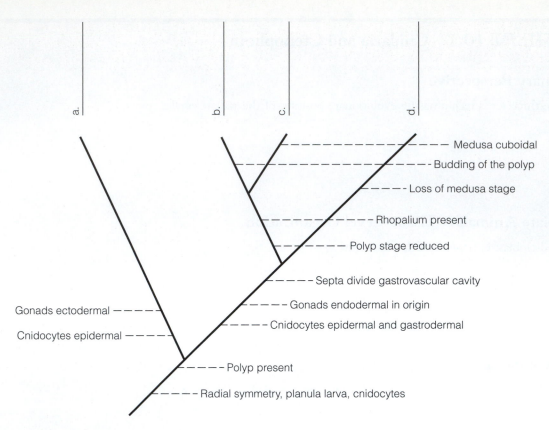

Figure 10.9 Evolutionary relationships among the Cnidaria. Fill in the class names.

6. What symplesiomorphies are common to all members of the phylum Cnidaria?

7. What derived character(s) distinguishes the scyphozoan/anthozoan clade from the hydrozoans?

8. What derived character distinguishes the anthozoans from all other cnidarians?

9. Members of what class are probably most closely related to ancestral cnidarians?

Exercise 11

Platyhelminthes

Learning Objectives

After completing this exercise, you should be able to do the following:

1. Describe the evolutionary significance of the Platyhelminthes.

2. Describe the triploblastic acoelomate body plan.

3. Describe bilateral symmetry and the advantages it gives active animals.

4. Define and give examples of animals with organ-level organization.

5. Describe the classification of the Platyhelminthes and recognize class representatives.

6. Describe life histories for some members of each of the classes studied.

7. Describe the functions of those structures studied in the laboratory and recognize them in the specimens studied.

Prelaboratory Quiz

Study this week's laboratory exercise and then complete the following quiz to assess your preparation for the laboratory.

1. Members of the phylum Platyhelminthes are characterized by all of the following EXCEPT
   a. organ-level organization.
   b. triploblastic body plan.
   c. coelomate body plan.
   d. bilateral symmetry.

2. In members of the phylum Platyhelminthes, muscles and excretory structures arise from the embryonic germ layer called the
   a. ectoderm.
   b. mesoderm.
   c. endoderm.

3. Animals that are bilaterally symmetrical usually have a head where nervous tissue and sensory structures are found. This development of a head in a bilaterally symmetrical animal is called
   a. organ organization.
   b. cephalization.
   c. diploblastic organization.
   d. centralization.

4. Members of the class Turbellaria include the
   a. free-living flatworms.
   b. monogenetic flukes.
   c. tapeworms.
   d. roundworms.

5. Members of the class Cestoidea include the
   a. free-living flatworms.
   b. monogenetic flukes.
   c. tapeworms.
   d. roundworms.

6. True/False   A host that harbors the immature stage of a parasite is called the definitive host.

7. True/False   Snails are the usual intermediate host of members of the class Trematoda.

8. True/False   Members of the class Cestoidea have a blind-ending digestive tract.

9. True/False   The definitive host of a tapeworm may be infected by eating flesh of an intermediate host containing a cysticercus larva.

10. True/False   Mature or gravid proglottids are found near the scolex of a tapeworm.

EVOLUTIONARY PERSPECTIVE

The members of the phylum Platyhelminthes, the flatworms, exhibit bilateral symmetry and triploblastic organization. Two of the three classes within this phylum are parasitic, and the third class contains the free-living flatworms.

The evolutionary relationships between the Platyhelminthes and other phyla are controversial. The classical interpretation is that the triploblastic acoelomate body plan is intermediate between diploblastic and triploblastic coelomate plans. The flatworms would thus be an evolutionary side-branch from a hypothetical triploblastic, acoelomate ancestor. Evolution from radial ancestors could have resulted from a planulalike larval stage becoming sexually mature.

Other zoologists think that the radiate phyla are not in the main evolutionary lineage. They prefer to envision the evolution of flatworms from a bilateral ancestor.

A third hypothesis, supported by recent evidence, is that the acoelomate body plan is a secondary characteristic. If so, the ancestors of the Platyhelminthes may have had a body cavity that was eliminated in the evolution of the flatworms.

According to the first two hypotheses, ancestral triploblastic acoelomates represent a major point of divergence within the animal kingdom. Their descendants include an annelid-arthropod lineage, a mollusc lineage, and possibly pseudocoelomate and echinoderm-chordate lineages.

Within the phylum, it is generally agreed that ancient free-living flatworms (class Turbellaria) were ancestral to modern turbellarians and members of the two classes of parasites (Trematoda and Cestoidea).

THE TRIPLOBLASTIC ACOELOMATE ANIMALS—THE ORGAN LEVEL OF ORGANIZATION

The Platyhelminthes include free-living flatworms, like the planarians, and the parasitic tapeworms and flukes. The term "flatworm" refers to the fact that the body is dorsoventrally flattened. Flatworms are the first organisms to have tissues organized into organs and the first to demonstrate bilateral symmetry. **Bilateral symmetry** means that one plane passing through the longitudinal axis of an organism divides it into right and left halves that are mirror images. It is characteristic of active, crawling, or swimming organisms and usually results in the formation of a distinct head (**cephalization**) where nervous tissue and sensory structures accumulate. This reflects the importance to the organism of monitoring the environment it is meeting—rather than that through which it has just passed—and results in the presence of definite anterior and posterior ends. The Platyhelminthes and all phyla above them on the evolutionary tree are bilaterally symmetrical or have evolved from bilaterally symmetrical ancestors.

Although there was some interdependence of tissues in the Cnidaria, interdependence is more highly developed in the Platyhelminthes. Thus, the flatworms display the **organ level of organization.** Three major sets of organs characterize the phylum. The excretory system consists of flame cells and their associated ducts. The nervous system consists of a pair of anterior ganglia, usually with two ventral nerve cords running the length of the organism. Nerve cords are interconnected by transverse nerves to form a ladderlike structure. The digestive tract is incomplete. A single opening serves for ingestion of food and elimination of wastes.

The Platyhelminthes are **triploblastic** and **acoelomate.** Figure 12.2 shows a diagrammatic cross section of a triploblastic, acoelomate animal. There are three embryonic germ layers: ectoderm, endoderm, and mesoderm. As with the Cnidaria, ectoderm gives rise to the outer epithelium, and the endoderm gives rise to the lining of the gut tract. The third germ layer, **mesoderm,** gives rise to the tissue between ectoderm and endoderm, including muscle, excretory structures, and undifferentiated cells referred to as parenchyma. The term "acoelomate" refers to the fact that there is no body cavity (fluid-filled space) between or within any of the embryonic germ layers.

CLASSIFICATION

Class Turbellaria
Class Monogenea
Class Trematoda
Class Cestoidea

Class Turbellaria

Characteristics: Free-living, ciliated epidermis containing rodlike rhabdites, ventral mouth opening, monoecious, both sexual reproduction and asexual fission common. Example: *Dugesia* (planaria).

▼ Examine a prepared whole mount of *Dugesia* using a dissecting microscope (fig. 11.1). Note the presence of a definite anterior end. Two sets of specialized sensory receptors should be obvious. Photoreceptors can be seen dorsally. **Auricles** can be seen as lateral extensions of the head. Chemoreceptors are located on the auricles.

Most turbellarians are carnivores and feed on live invertebrates or scavenge on larger dead animals. If your specimen has the digestive cavity (**gastrovascular cavity**) stained, note that it is in the form of a single tube anteriorly and then divides posteriorly. The mouth opening is located

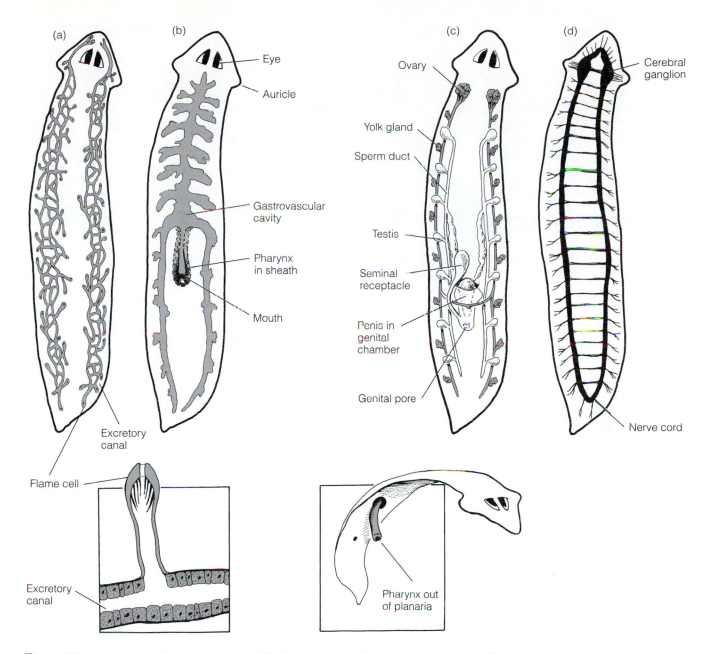

Figure 11.1 *Dugesia:* (*a*) excretory system; (*b*) digestive system; (*c*) reproductive system; (*d*) nervous system.

midventrally. In feeding, a protrusible **pharynx** is extended through the mouth opening. Digestion is both extracellular and intracellular.

Examine a living specimen in a water-filled dish. Observe the gliding movement. Epidermal mucous glands lay down a sheet of mucus as cilia and body-wall musculature propel the animal. Examine a demonstration of *Dugesia* feeding on fresh beef liver.

Planarians have remarkable powers of regeneration. This is easily demonstrated. Place a planarian on a cube of ice. Allow it to become immobilized and then cut it in half (or some other section) with a razor blade. Rinse the pieces into a dish containing pond or spring water. Cover to prevent evaporation. Change the water in the dish every second day until regeneration is completed.

The excretory system of a freshwater planarian functions primarily in osmoregulation. Planarians live in a hypotonic environment and, therefore, are continually taking on water by osmosis. Protonephridia consist of networks of fine tubules that originate as tiny enlargements called **flame cells.** Flame cells remove this excess water from the tissues. Flickering cilia within the flame cells create currents that drive excess fluids through the tubule system to the outside. Observe a demonstration showing planarian flame cells. It was prepared by squashing the posterior third of a planarian between a slide and a coverslip. Look for the flickering tuft of cilia from which flame cells get their name.

Obtain a prepared slide showing a cross section of *Dugesia* (figs. 11.2 and 11.3). Examine it under low power of a compound microscope. A section through the middle of the

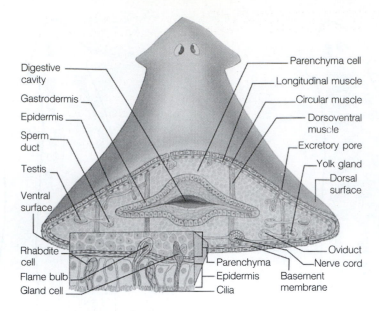

Figure 11.2 *Dugesia.* Cross section through anterior third.

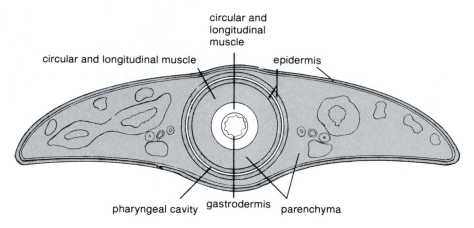

Figure 11.3 *Dugesia.* Cross section through the pharyngeal region.

planarian will show the pharynx; sections through the anterior and posterior thirds will not show the pharynx. In an anterior or posterior section, note the acoelomate organization and the derivatives of the three embryonic germ layers. The **epidermis** is ectodermal in origin, the undifferentiated **parenchyma (mesenchyme)** that fills the interior is mesodermal in origin, and **gastrodermis** that lines the gastrovascular cavity is endodermal in origin. In locating the latter, look for the space of the gastrovascular cavity and then look for the cells lining that cavity. In the upper epidermis, locate the small, rodlike **rhabdites.** Rhabdites are characteristic of the class and can be discharged to form a protective mucous coat around the planarian. Muscle cells can also be seen in this slide. Just inside the epidermis is a layer of **circular muscle,** and just inside that is a layer of longitudinal muscle. The **longitudinal muscle** will be seen as small dots because cells are cut in cross section. Strands of **dorsoventral muscle** run between the dorsal and ventral surfaces. Examine a section through the pharynx. Compare what you see to figure 11.3. ▲

Reproductive structures will not be studied because they are difficult to see in most preparations. The Turbellaria are, however, **monoecious.** Recall that monoecious species have both ovaries and testes in one individual. In the Turbellaria, as in most monoecious taxa, cross-fertilization is the rule. After copulation, fertilized eggs are deposited in the substrate of freshwater ponds and streams. Development may be direct (young resemble the adults), or in some marine turbellarians, larval stages may hatch from the eggs. As mentioned previously, the regenerative powers of turbellarians are well developed, and asexual reproduction is common.

▼ Exercise 1 of this laboratory manual had an activity investigating phototaxis of planarians. If you did not complete that exercise earlier, your instructor may want you to turn there at this time. In addition to investigating phototaxis, experiments can be designed to study the response of planaria to temperature and the direction of the pull of gravity. ▲

1. What is cephalization? _____

_____

2. What is bilateral symmetry, and why is it characteristic of most actively crawling or swimming animals?

_____

_____

_____

3. What embryonic germ layer is present in the Platyhelminthes but was absent in the Cnidaria?

_____

What kinds of tissues develop from that third germ layer? _____

4. Turbellarians are monoecious. What does that mean? _____

_____

Class Monogenea

Characteristics: Monogenetic flukes; mostly ectoparasites on vertebrates (usually on fishes, occasionally on turtles, frogs, copepods, squids); bear opisthaptor; body wall with tegument. Examples: *Gyrodactylus* and *Polystoma*.

▼ Members of the Monogenea, Trematoda, and Cestoidea possess an outer epidermis modified as a tegument. It is non-ciliated and syncytial (a continuous layer of fused cells). It is adaptive for parasites in these classes because it aids in the transport of nutrients, gases, and wastes across the body wall and protects against enzymes and the host's immune system. Monogenetic flukes have a single generation in their life cycle, that is, one adult develops from one egg. Embryonic development leads to a larval stage called an oncomiracidium. Most monogenetic flukes are external parasites of fishes, where they attach to gill filaments and feed on epithelial cells, mucus, or blood. Observe any demonstration slides that are available. Note the large attachment structure called an **opisthaptor**. ▲

Class Trematoda

Characteristics: Flukes; body wall with tegument; leaflike or cylindrical; usually with oral and ventral suckers; adults, parasites of vertebrates. Examples: *Fasciola*, *Clonorchis* (*Opisthorchis*), *Schistosoma*.

▼ Obtain a prepared slide of the human liver fluke, *Clonorchis* (*Opisthorchis*) *sinensis*. It illustrates trematode structure clearly (fig. 11.4). Using a dissecting microscope, note the tapered anterior and more rounded posterior ends. At the anterior tip is the **oral sucker.** It serves as a holdfast and surrounds the mouth opening. Posterior to the oral sucker is the **ventral sucker,** or **acetabulum.** It also serves as

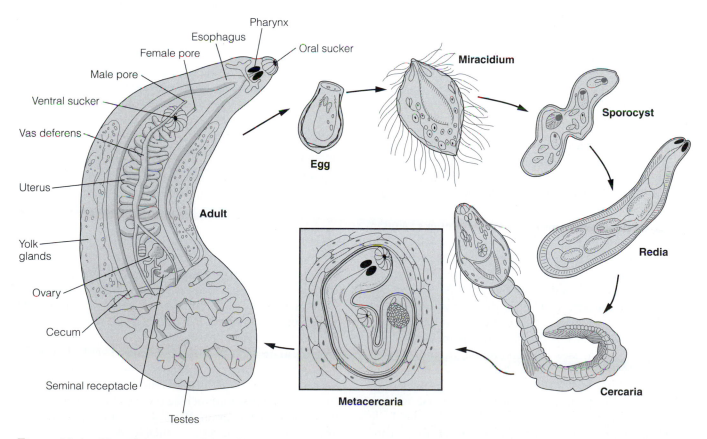

Figure 11.4 *Clonorchis sinensis*, adult and larval stages.

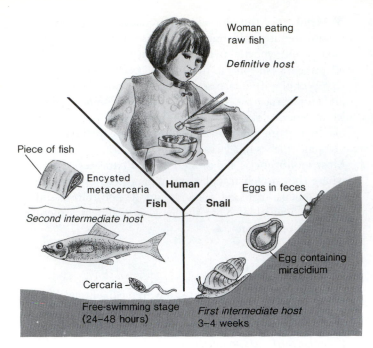

Woman eating
raw fish

Definitive host

Piece of fish

Encysted
metacarcaria

Human

Fish   **Snail**

Eggs in feces

Second intermediate host

Egg containing
miracidium

Cercaria

Free-swimming stage
(24–48 hours)

First intermediate host
3–4 weeks

Figure 11.5 The life cycle of *Clonorchis sinensis*. Humans become infected by eating uncooked or poorly cooked fish containing metacercariae. Eggs passed with feces contain miracidia, which are eaten by snails. Miracidia develop into sporocysts within the snail. Sporocysts develop rediae, which develop cercariae. Hundreds of cercariae leave the snail and encyst in fish muscle as metacercariae.

a holdfast. Extending posteriorly from the mouth is the blind-ending digestive tract. The muscular **pharynx** is just posterior to the mouth. The pharynx is followed by a short **esophagus.** The digestive tract divides to become Y-shaped. Each arm of the "Y" is called a **cecum.** The remaining internal structures are components of the reproductive complex. Most trematodes are monoecious, with cross-fertilization the rule. ▲

Trematodes typically have a two- or three-host life cycle. The host that harbors the adult stage of the parasite is the **definitive host.** The host that harbors the immature stage is the **intermediate host.** The life cycle of trematodes can be extremely complex. The life cycle of *Clonorchis sinensis* (fig. 11.5) represents one of the many variations found in the class.

Clonorchis sinensis adults live in the bile passageways of the human liver, where they feed on human cells and cell fragments. Mature eggs pass through the host via the bile duct into the intestine and out of the host with the feces. **Miracidia** hatch after eggs are eaten by the first intermediate host, the snail. Miracidia develop into **sporocysts.** Sporocysts develop rediae asexually, and **rediae,** in turn, develop **cercariae.** Hundreds of cercariae leave for each miracidium that infected the snail. Cercariae penetrate the second intermediate host, the fish, and develop into **metacercariae,** which encyst in muscle. Metacercariae have the adult body form but are sexually immature. When metacercariae are in-

gested by humans eating incompletely cooked fish, metacercariae develop into adults.

▼ Examine any slide of preserved or living materials that illustrate the life cycle of *Clonorchis sinensis.* Examine other trematodes available for study. ▲

Class Cestoidea

Characteristics: Tapeworms; body wall with tegument bearing microtriches; body consisting of scolex with hooks and/or suckers and a long tapelike strobila divided into proglottids; no digestive tract; monoecious, parasites of vertebrates. Examples: *Taenia, Hymenolepis.*

Cestodes, with some exceptions, have two hosts in their life cycles. Figure 11.6 shows stages in the life cycle of a typical cestode, the dog tapeworm, *Taenia pisiformis.* Adults generally live in the digestive tract of their vertebrate host, where they absorb nutrients from the host's intestine. The **scolex** is a holdfast structure. Anterior proglottids contain **immature** reproductive structures, middle proglottids contain **mature** reproductive structures, and posterior proglottids contain eggs and are referred to as **gravid** proglottids. As proglottids are produced behind the scolex, they move posteriorly and begin accumulating fertilized eggs. Cestodes are monoecious. Cross-fertilization with another individual or proglottid is the rule. When a proglottid is gravid, reproductive organs degenerate, and the proglottid breaks off the end of the **strobila.** These "bags" of eggs then pass out of the host with the feces.

Fertilized eggs develop into an encapsulated, six-hooked larva called a **hexacanth,** or **oncosphere.** The hexacanth is ingested by the intermediate host and uses its hooks to penetrate the wall of the digestive tract. In the intermediate host, the hexacanth encysts as a second larval stage. The second larval stage is either a **cysticercus** or **cysticercoid** larva. When the intermediate host is eaten, the second larval stage enters the definitive host.

▼ Obtain a slide of the dog tapeworm *Taenia pisiformis* (fig. 11.6). (These slides usually have representative regions of the tapeworm mounted under one coverslip.) Examine the scolex with its suckers (fig. 11.7). Just behind the scolex note a number of immature proglottids. Mature proglottids are indicated by the presence of reproductive structures and some fertilized eggs. Gravid proglottids are full of fertilized eggs. These segments are ready to break free and pass out of the host. Note the absence of a digestive tract. Nutrients are absorbed from the host's intestine across the tapeworm body wall. Note the **microtriches** that increase the surface area for absorption (fig. 11.8). Examine any materials available that illustrate aspects of the life cycle described here. In *T. pisiformis,* the intermediate host is the rabbit, and the larval stage in that host is a cysticercus.

Human tapeworms include the beef tapeworm, *T. saginata;* the pork tapeworm, *T. solium;* and others. Examine any demonstration materials available. ▲

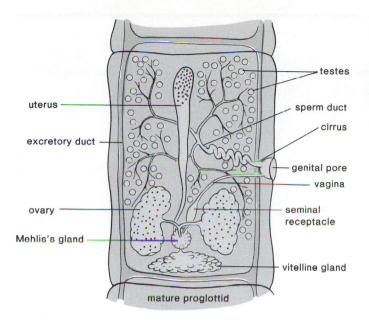

testes

sperm duct

cirrus

genital pore

vagina

uterus

excretory duct

ovary

Mehlis's gland

seminal receptacle

vitelline gland

mature proglottid

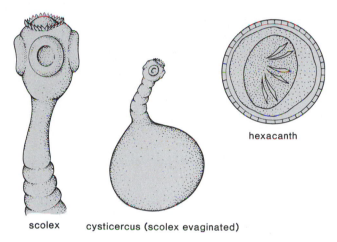

scolex

cysticercus (scolex evaginated)

hexacanth

Figure 11.6 *Taenia pisiformis*, adult and larval stages.

EVOLUTIONARY RELATIONSHIPS

▼ A cladogram showing a traditional interpretation of the evolutionary relationships among the platyhelminth classes is shown in worksheet 11.1 (fig. 11.9). Examine the synapomorphies depicted on the cladogram. Based upon your studies in this week's laboratory, you should be able to determine the position of each of the classes studied. Complete the cladogram and answer the questions that follow it. ▲

STOP AND ASK YOURSELF

5. What is a definitive host? _____

_____

6. What is (are) the intermediate host(s) in the life cycle of *Clonorchis sinensis*? _____

_____

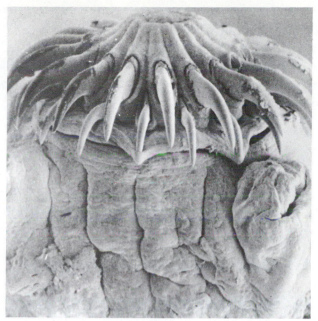

Figure 11.7 *Taenia taeniaeformis*, scolex with hooks.

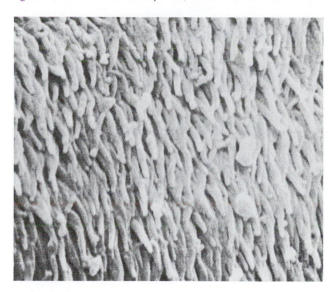

Figure 11.8 The tegument of *Hymenolepsis diminuta* showing microtriches.

7. Where would you expect to find sporocysts of *Clonorchis sinensis*? _____

_____

8. What body function(s) is (are) carried out within a tapeworm proglottid? _____

_____

9. Where would you find the cysticercus of *Taenia pisiformis*? _____

_____

10. The chain of proglottids of a tapeworm is called the

_____

KEY TERMS

acoelomate 140
cephalization 140
definitive host 144

intermediate host 144
monoecious 142

organ level of organization 140
triploblastic 140

WORKSHEET 11.1   Platyhelminthes

Evolutionary Perspective

1. What are three ideas regarding the evolutionary significance of ancestral, triploblastic acoelomate animals?

The Triploblastic Acoelomate Animals

2. What are the three major sets of organs that characterize members of the phylum Platyhelminthes?

Classification

3. Characterize members of the class Trematoda.

4. Characterize members of the class Cestoidea.

5. How would you describe the excretory structures found in planarians?

6. What are rhabdites?

7. What procedures do you think would be most effective for control of *Clonorchis sinensis* in endemic regions? Explain your answer based on specific knowledge of the parasite's life cycle.

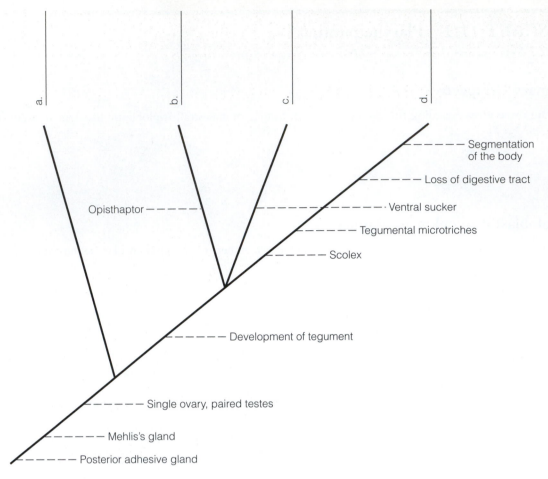

a.

b.

c.

d.

— — — — Segmentation of the body

— — — — Loss of digestive tract

Opisthaptor — — — — — — — — — — — — — — — Ventral sucker

— — — — Tegumental microtriches

— — — — Scolex

— — — — Development of tegument

— — — — Single ovary, paired testes

— — — — Mehlis's gland

— — — — Posterior adhesive gland

Figure 11.9 Evolutionary relationships among the Platyhelminthes. Fill in the class names.

8. What single derived character distinguishes members of the classes Cestoidea, Monogenea, and Trematoda from all other members of the phylum Platyhelminthes? _____

9. Name at least one derived character for each of the following flatworm classes:

   a. Monogenea _____

   b. Trematoda _____

   c. Cestoidea _____

10. Name one class of the phylum Platyhelminthes for which there are no known synapomorphies. _____

11. According to figure 11.9, which of the four classes of Platyhelminthes is most closely related to ancestral members of this phylum? _____

12. Explain why an absence of synapomorphies in a lineage leads systematists to the conclusion that members of that lineage are closely related to the ancestral members of the taxon.

Exercise 12

The Pseudocoelomate Body Plan: Aschelminths

Learning Objectives

After completing this exercise, you should be able to do the following:

1. Describe the evolutionary significance of the pseudocoelomate phyla.

2. Describe the organ-system level of organization.

3. Describe the nature of body cavities and the advantages they give organisms that possess them.

4. Name six pseudocoelomate phyla and describe the diversity in structure, function, and habitats within these phyla.

5. Carry out a dissection of a nematode worm.

6. Identify and describe the structure and function of the pseudocoelomate organ systems studied in the laboratory.

Prelaboratory Quiz

Study this week's laboratory exercise and then complete the following quiz to assess your preparation for the laboratory.

1. A body cavity that is entirely lined by mesoderm is a
   a. coelom.
   b. pseudocoelom.
   c. gut.
   d. cloaca.

2. All of the following are true of body cavities EXCEPT
   a. they are usually filled with parenchyma.
   b. they provide room for organ development.
   c. they provide area for storage.
   d. they act as hydrostatic skeletons.
   e. they provide more surface area for the diffusion of gases, nutrients, and wastes.

3. All of the following are aschelminths EXCEPT
   a. nematodes.
   b. rotifers.
   c. flukes.
   d. nematomorphs.

4. In this week's laboratory, we will perform a dissection of
   a. an earthworm.
   b. a nematode.
   c. a rotifer.
   d. a gastrotrich.

5. The development of an individual from an unfertilized egg is called _____ . It is encountered in this week's laboratory during observations of the phylum _____ .
   a. vitellogenesis/Nematoda
   b. parthenogenesis/Nematoda
   c. vitellogenesis/Rotifera
   d. parthenogenesis/Rotifera
   e. parthenogenesis/Nematomorpha

6. True/False   The aschelminths include a diverse group of closely related phyla.

7. True/False   All nematodes are parasites.

8. True/False   A female *Ascaris* is smaller than the male and has a hooked posterior tip.

9. True/False   Mesenteries are mesodermal sheets that suspend the gut tract of pseudocoelomate animals within their body cavities.

10. True/False   In temperate regions, most rotifers spend the winter as dormant zygotes.

EVOLUTIONARY PERSPECTIVE

The aschelminths include a group of phyla that share one common feature, the pseudocoelom. They include the phyla Nematoda, Rotifera, Kinorhyncha, Acanthocephala, Gastrotricha, Nematomorpha, and others (fig. 12.1). Except for their one common feature, they are a diverse group of animals, only distantly related. Some zoologists think that ancestral members of these phyla diverged separately from the main evolutionary line. Lumping these phyla together, therefore, is artificial, yet convenient for purposes of study in the biology laboratory. Because the evolutionary relationships among the aschelminths is uncertain, no cladogram is included in worksheet 12.1.

THE TRIPLOBLASTIC PSEUDOCOELOMATE ANIMALS— ORGAN-SYSTEM LEVEL OF ORGANIZATION

The pseudocoelomates are the first animals we will study that have organs associated together to form **organ systems.** Different organs functioning together make up systems. All remaining phyla show organ-system organization.

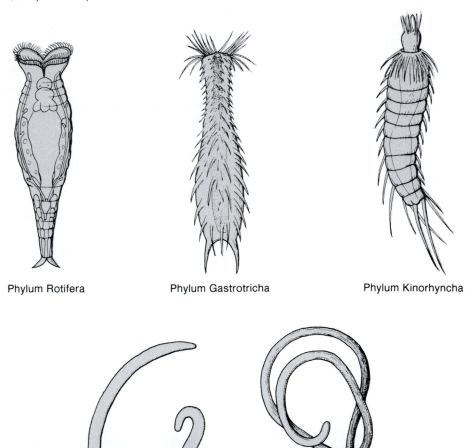

Phylum Rotifera   Phylum Gastrotricha   Phylum Kinorhyncha

Phylum Nematoda   Phylum Nematomorpha

Figure 12.1   Representatives of the pseudocoelomate phyla.

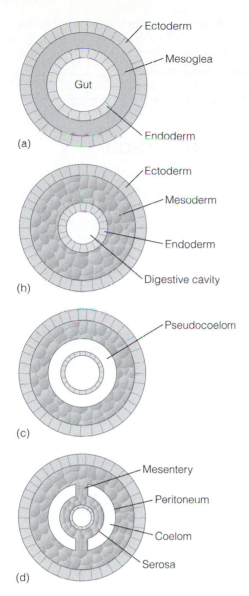

Figure 12.2   Animal body plans: (*a*) diploblastic;
(*b*) triploblastic acoelomate; (*c*) triploblastic pseudocoelomate;
(*d*) triploblastic coelomate.

The Nature of Body Cavities

A body cavity is characteristic of all bilateral animals
above the acoelomates. (Fig. 12.2 shows diagrammatic
sections of the body walls of diploblastic and triploblastic
animals. Diploblastic and triploblastic acoelomate plans
were covered in exercises 10 and 11 and are included for
comparative purposes.) Body cavities are advantageous
because they

1. provide more room for organ development;
2. provide more surface area for diffusion of gases,
   nutrients, and wastes into and out of organs;
3. provide area for storage;
4. often act as hydrostatic skeletons;
5. provide a vehicle for eliminating wastes and
   reproductive products from the body; and
6. facilitate increased body size.

Of these, the **hydrostatic skeleton** deserves further comment.
Body-cavity fluids confined by the body wall give support but,
at the same time, the body remains flexible. Body-wall muscles acting on this incompressible fluid facilitate movement.

Internal organs are separated from the body wall by, and
suspended in, a fluid-filled cavity. A body cavity may be one
of two types. A **pseudocoelom** is a body cavity not entirely
lined by mesoderm (see fig. 12.2c). Therefore, no muscular
or connective tissues are associated with the gut tract, no
mesodermal sheet covers the inner surface of the body wall,
and no membranes suspend organs within the body cavity.
Embryologically, the pseudocoelom is derived from the blastocoel of the embryo (see exercise 5).

A **coelom** is a body cavity that is completely surrounded
by mesoderm. The inner body wall is lined by a thin mesodermal sheet, the **peritoneum,** and visceral organs are lined on
the outside by a mesodermal sheet, the **serosa.** The peritoneum
and the serosa are continuous and suspend visceral structures
in the body cavity. These suspending sheets are called **mesenteries.** Having mesodermally derived tissues, such as muscle
and connective tissue, associated with internal organs en-
hances the function of virtually all internal body systems.

PHYLUM ROTIFERA

Rotifers are mostly freshwater pseudocoelomates that derive
their name from a ciliated **corona.** When cilia are beating,
the impression is of rotating wheels. The body is elongate,
consisting of a **head,** a **trunk,** and a **foot.**

We will examine a living specimen of *Philodina* (fig. 12.3).
Philodina is frequently used in introductory courses because it
clearly shows details of internal and external anatomy.

▼ Obtain a drop of water from the bottom of a culture dish
and make a wet mount. Observe the cylindrical body with
head, trunk, and foot. The foot is used in attaching to the
substrate. The ciliated corona can be seen anteriorly. Note
the beating cilia. Cilia create water currents used in feeding.
Note the periodic retraction of the corona. This is also associated with feeding. Just below the corona, look for a muscular grinding structure called the **mastax.** It is associated
with the gut tract and is easily seen in living rotifers because
of its rhythmical contractions. ▲

Rotifers are dioecious. Frequently, as in *Philodina*, reproduction occurs by the development of unfertilized diploid
eggs. This is called **parthenogenesis.** In fact, males are unknown in *Philodina*. In other species, parthenogenesis occurs
in the spring and summer. In autumn, parthenogenic males
are produced. Fertilization results in a zygote that overwin-
ters and hatches into a female in the spring. This is an adaptation to withstand the harsh winter conditions of temperate
lakes. Other rotifers exhibit normal sexual reproduction.

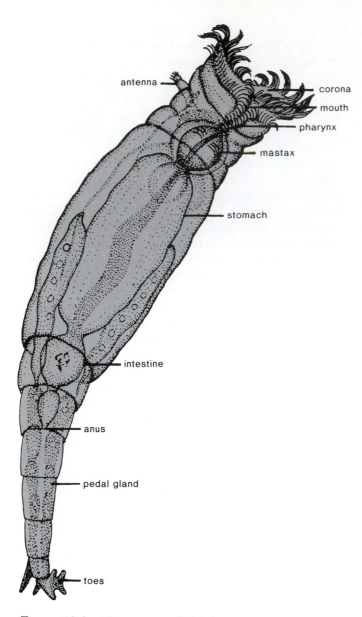

Figure 12.3 The structure of *Philodina*.

Labels on figure:
antenna
corona
mouth
pharynx
mastax
stomach
intestine
anus
pedal gland
toes

STOP AND ASK YOURSELF

1. What is an organ system? _____
_____

2. What is a coelom? _____
_____
_____

3. What is a hydrostatic skeleton? _____
_____
_____

4. Describe the feeding mechanism of *Philodina*. _____
_____

5. What is parthenogenesis? _____
_____

DISSECTION METHODS

In the exercise that follows, you will be performing the first of a number of dissections called for in this manual. It is, therefore, appropriate to give some instructions for dissection. Contrary to what might be one's first impression, dissection involves only limited cutting. The purpose of dissection is to expose body parts for viewing in as natural a state as possible. This usually involves the separation of body parts along natural parting lines. For this you need at least a scalpel, scissors, forceps, dissecting needle, and dissecting pins. The scalpel is used for entry into the body of an organism. The initial cuts are shallow so the thickness and consistency of the tissue can be determined. Once the body cavity is opened, scissors can be used to continue cutting. This is usually done on larger specimens where one blade of the scissors can easily fit into the body cavity. It is important to keep the lower blade of the scissors parallel to and next to the inner surface of the body wall. This avoids cutting into the internal structures. Once the body wall is pinned open, use forceps for cleaning fat or other adhering tissue from the surface of the structure. The dissecting needle will be one of your most useful instruments. It is used for gently lifting, moving, and separating tissues and organs. A heavy metal probe, or the handle of your scalpel, can be used for separating and lifting larger, heavier structures.

Before starting your dissection, read through the entire description, looking at any figures available. Instructions for dissection are always given based on the anatomical terminology noted on pages xii–xiii. Review that terminology before beginning. All terms are in reference to the animal, not the dissector. For example, if the dissection is carried out on the ventral aspect of the organism, a structure described as being on the right will be on the specimen's right, but the student's left. If one is dissecting a small specimen with delicate internal structures, it is advisable to carry out the dissection with the specimen completely submerged in water. Water makes delicate structures easier to move around because they have a tendency to float. Remember also that if you want to observe a dissected specimen under a dissecting microscope, the specimen must be pinned near the edge of the dissecting pan to fit into the field of view. Finally, figures and photographs are for purposes of orienting the student to the location of specific structures. Visual aids cannot be used in guiding the dissection, and they are not a substitute for the dissection experience. Follow the text closely in dissecting. All instructions and precautions are given there.

PHYLUM NEMATODA

Characteristics: Round in cross section, covered by a flexible, nonliving cuticle; lack motile cilia and flagella; longitudinal muscles only. Examples: *Ascaris*, *Trichinella*, *Necator*.

Ascaris—External Structure

▼ Obtain a specimen of *Ascaris*. It will most likely be *Ascaris suum*, the pork ascarid. A species indistinguishable from the pork ascarid occurs in humans, *Ascaris lumbricoides*. Nematodes are dioecious. Males are smaller, and the posterior tip is hooked. Females are larger and lack the hook. Determine the sex of your worm. Examine the external structure of the worm using a dissecting microscope or hand lens. Note the cylindrical appearance of the worm. Distinguish anterior from posterior. The anterior tip of both sexes is less pointed than the posterior. In looking straight down on the anterior tip, note the three lips that surround a triangular **mouth.** Near the posterior tip one can see a slitlike **anus.** The anus is ventral and can be used to distinguish dorsal from ventral. Find lateral lines running down both sides of the worm. Within each is an excretory canal.

Internal Structure

Place your worm dorsal surface up along one side of a dissecting pan (so you can fit the specimen under a dissecting microscope) and place a pin in each end. Internal structures in preserved *Ascaris* are delicate, so it is best to submerge the specimen entirely during dissection and observation. Water will tend to float internal organs and result in less damage. Using a sharp scalpel, make a longitudinal incision along the length of the worm. Cut just through the body wall, being careful not to damage internal structures. Pin the body wall to the bottom of the dissecting pan as you proceed.

Internal structure is shown in figure 12.4. Note the **pseudocoelom.** Running the length of the body cavity is a flat, ribbonlike **intestine.** Anteriorly, the digestive tract is modified into the **pharynx.** This structure is muscular and has a pumping function. The remainder of the body cavity is filled with reproductive structures. In looking at reproductive structures, examine your worm first and then trade with someone who has the opposite sex. You are responsible for both.

Female

The female reproductive tract is Y-shaped. Note the mass of coiled and folded tubules. The duct system is single near its attachment at the genital opening (located within the anterior one-third of the worm) and double at the opposite extreme. **Ovaries** are the smallest diameter tubules at the free ends of the "Y." As you trace the tubules proximally, note that their diameters increase. Find a region where the tubule diameter is intermediate. This region is the **oviduct.** There

is no grossly observable transition between either ovary and oviduct or oviduct and **uterus,** the largest diameter tubules. The oviduct carries eggs between the ovary and the uterus. Fertilization occurs in the uterus, and eggs are held here prior to their release. The two uteri unite to form the **vagina.** Copulatory spicules of the male are inserted into the vagina for sperm transfer. The position of the **genital pore** can be located internally where the vagina attaches to the body wall.

Male

The reproductive tract of the male is a single continuous tube. It ends at the posterior tip where it unites with the digestive tract to form a **cloaca** (an opening common to digestive and reproductive tracts). The **testis** is the region at the smaller, free end of the tubule. Sperm formed in the testis travel down the intermediate-diameter **vas deferens** to the **seminal vesicle,** where sperm are stored. The region of the tubule near the cloaca is the **ejaculatory duct.** The ejaculatory duct propels sperm into the female vagina during copulation. Locate these regions. As with the female, there is no grossly observable boundary between them.

Ascaris Cross Section

Obtain a prepared *Ascaris* slide that shows male and female cross sections (figs. 12.5 and 12.6). The larger section is the female. Examine it under low power. The outer body wall is covered by a nonliving, protective cuticle. Below the cuticle, note the body wall, the pseudocoelom, and other internal structures. Recall that the pseudocoelomates are triploblastic. Derivatives of the three germ layers can be seen in your cross section. Just below the cuticle is a layer of cells called the epidermis. It is ectodermal. Locate lateral lines, lateral thickenings of the epidermal layer. Within each lateral line is an excretory canal.

The flattened intestine is endodermal. There is no mesoderm associated with the intestine to help it maintain its shape. Refer back to figure 12.2 and recall how this organization differs from that of coelomates.

Muscles and reproductive structures are mesodermal. Observe **longitudinal muscles** just inside the epidermis. When muscles contract, they act against the hydrostatic skeleton. Because there are no circular muscles, a thrashing movement results. You will observe this later. Cross sections of females are usually taken through the middle third of the body and thus show two sections of uteri. Each uterus is a large-diameter tubule containing eggs. Oviducts are smaller in diameter with a central cavity (lumen) for the passage of eggs from the ovary to the uterus. Sections through the ovary are nearly solid.

Examine the cross section of a male. In addition to the body wall and intestine, note cross sections through the reproductive tract. The testis will be represented in numerous small-diameter sections. The cells inside are amoeboid sperm cells (recall there are no cilia or flagella in the nematodes). You should also see several sections through the

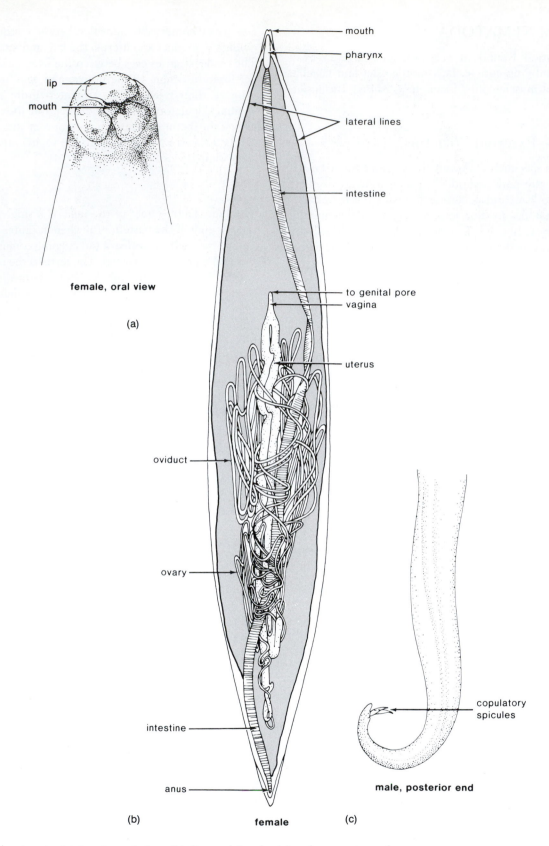

female, oral view

(a)

lip

mouth

mouth

pharynx

lateral lines

intestine

to genital pore

vagina

uterus

oviduct

ovary

intestine

anus

(b)

female

copulatory spicules

male, posterior end

(c)

Figure 12.4 *Ascaris:* (*a*) female, oral view; (*b*) dissected female; (*c*) male, posterior end.

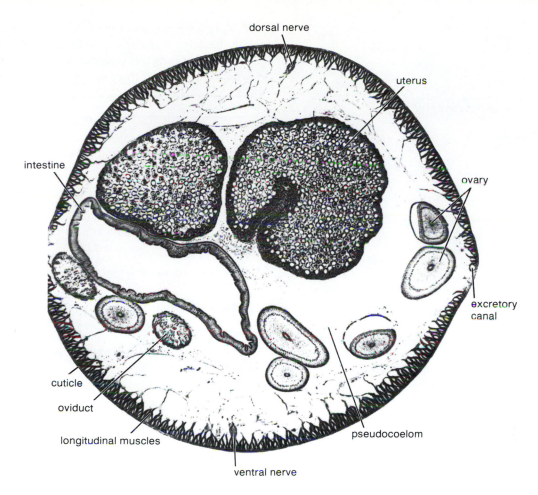

Figure 12.5 Cross section of an *Ascaris* female.

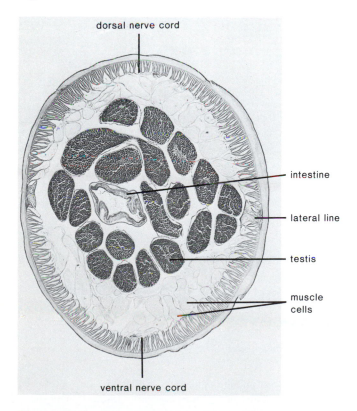

Figure 12.6 Cross section of an *Ascaris* male.

larger vas deferens and one section through the very large seminal vesicle. Sperm should also be present in these sections. ▲

Life Cycles

After fertilization, eggs of nematodes are usually released to the environment, where development takes place. (There are a number of exceptions.) Larvae look similar to adults but are smaller and sexually immature. Larvae molt as they grow to adulthood. In *Ascaris*, eggs are deposited with feces. Development through the initial larval stage occurs within the very resistant egg capsule. When ingested, larvae hatch and migrate through the intestinal wall and then, via the bloodstream, to the lungs. They finally migrate up the respiratory tract and are swallowed, thus making their way back to the small intestine, where they mature (fig. 12.7).

Trichinella spiralis is a nematode that causes trichinosis in nearly all meat-eating animals. Humans usually become involved by eating incompletely cooked pork. Eggs hatch within the female, and larvae are released from the female worm into the intestine of the host. These larvae migrate through the intestinal wall to the lymphatic system and then the circulatory system. They encyst in active striated muscle, like the diaphragm and intercostal muscles, where they remain viable for many years (fig. 12.8).

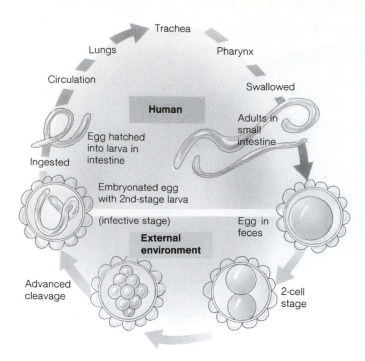

Figure 12.7 The life cycle of *Ascaris lumbricoides*.
Source: Redrawn from original from Centers for Disease Control and Prevention, Atlanta, GA.

Many other life-cycle variations occur in parasitic nematodes. These nematodes include the hookworms like *Ancylostoma* and *Necator*, whose adults live attached to the intestinal wall, where they feed on blood. Eggs hatch after being released into the soil, and the larvae penetrate the skin of the host. Filarial nematodes depend upon arthropod vectors for transmission to new hosts. The adult dog heartworm, for example, lives in the ventricles of the heart. Larvae move into the peripheral circulation, where they can be picked up by a mosquito. When the mosquito bites another dog, they are introduced into a second host.

▼ Examine demonstrations of the following parasites as they are available: ▲

Trichinella spiralis
    Adult stages from the intestine
    Larvae encysted in muscle
Necator americanus and *Ancylostoma duodenale*
    Hookworms
Enterobius vermicularis
    Pinworm

Free-Living Nematodes

In the preceding activities we have emphasized parasitic nematodes. By far most nematodes are free-living. They occur in virtually every aquatic and terrestrial habitat. One commonly used in the laboratory is the vinegar eel, *Tubatrix aceti* (fig. 12.9). It is relatively transparent, thus showing internal structure in living specimens.

▼ Make a wet mount of the fluid from the top of a culture dish. Examine it under low and high power. Note particularly the thrashing movements characteristic of nematodes. What is the basis for these movements?

_____

_____ ▲

A handful of rich garden soil contains thousands of nematodes living off decaying organic matter. These nematodes are important in the decomposition of organic matter and serve as a source of food for soil arthropods, earthworms, and some fungi. Soil nematodes are essential to the energy flow and nutrient cycling in soil ecosystems. Soil nematodes can be collected using a Baermann apparatus.

▼ Make a wet mount of fluid drawn from a Baermann apparatus and examine under low and high power. ▲

OTHER PSEUDOCOELOMATE PHYLA

▼ Observe demonstrations of the following phyla. Be able to recognize them and describe where they are found (see fig. 12.1). ▲

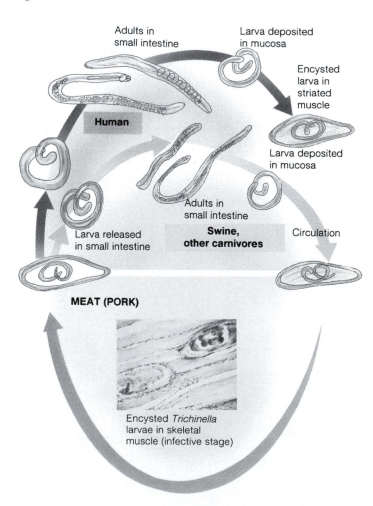

Figure 12.8 The life cycle of *Trichinella spiralis*.
Source: Redrawn from original from Centers for Disease Control and Prevention, Atlanta, GA.

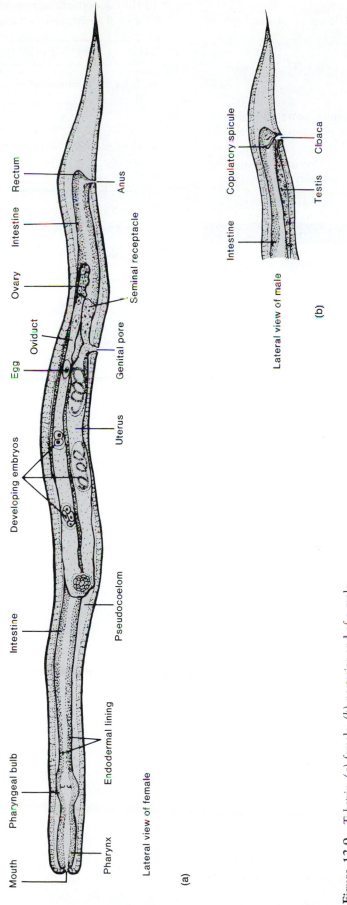

Figure 12.9 *Tubatrix*: (*a*) female; (*b*) posterior end of a male.

Mouth

Pharyngeal bulb

Pharynx

Endodermal lining

Intestine

Pseudocoelom

Developing embryos

Uterus

Egg

Oviduct

Genital pore

Ovary

Seminal receptacle

Intestine

Rectum

Anus

Lateral view of female

(a)

Intestine

Copulatory spicule

Testis

Cloaca

Lateral view of male

(b)

Phylum Gastrotricha

Freshwater and marine. Cuticle with scales and spines. Move via ventral cilia.

Phylum Kinorhyncha

Marine. Retractile head with spines. The apparent segmentation is superficial.

Phylum Nematomorpha

Horsehair worms. Elongate adults are free-living in streams and ponds. Larvae are parasites of arthropods.

Phylum Acanthocephala

Spiny-headed worms. Parasites of vertebrates as adults. Anterior with spine-covered holdfast. Common in fish, turtles, and pigs.

S T O P   A N D   A S K   Y O U R S E L F

6. In terms of overall structure, how is the reproductive tract of a male *Ascaris* different from that of a female *Ascaris?* _____

_____

7. How does the arrangement of muscles in the body wall of a nematode influence its form of locomotion?

_____

_____

8. Why is the intestine of a nematode in the shape of a flat ribbon? _____

_____

9. How does one become infected with *Trichinella spiralis?* _____

_____

10. What can one do to avoid being infected with hookworms in regions where they are endemic?

_____

_____

KEY TERMS

coelom   151

hydrostatic skeleton   151

organ systems   150

parthenogenesis   151

pseudocoelom   151 & 153

WORKSHEET 12.1   The Pseudocoelomate Body Plan: Aschelminths

Evolutionary Perspective

1. Explain why lumping together the aschelminths is an artificial grouping.

Triploblastic Pseudocoelomate Animals

2. List four advantages of having a body cavity.

Phylum Rotifera

3. In what way is parthenogenesis advantageous for some animals? Are there disadvantages as well? If so, what are they?

Phylum Nematoda

4. If given two unlabeled cross sections of *Ascaris*, could you determine which was the male and which was the female? Explain the differences.

5. Assess the importance of nematodes in terms of their impact on humans and numbers of individuals.

Other Pseudocoelomate Phyla

6. Where could one expect to find members of the following pseudocoelomate phyla?
   a. Gastrotricha

   b. Kinorhyncha

   c. Acanthocephala

Exercise 13

Mollusca

Learning Objectives

After completing this exercise, you should be able to do the following:

1. Describe the evolutionary significance of the molluscs.
2. Describe the typical molluscan body plan.
3. Describe the diversity in structure, function, and habitat within the phylum.
4. Characterize each of the classes studied including habitat, feeding and reproductive habits, and distinctive features of body organization.
5. Recognize class representatives.
6. Describe the structure and function of organ systems studied in each class.

Prelaboratory Quiz

Study this week's laboratory exercise and then complete the following quiz to assess your preparation for the laboratory.

1. Members of all of the following phyla are protostomes EXCEPT
   a. Annelida.
   b. Arthropoda.
   c. Platyhelminthes.
   d. Mollusca.

2. One function of the mantle of most molluscs is to
   a. filter feed.
   b. secrete the shell.
   c. house visceral organs.
   d. prevent desiccation.

3. A rasping, tonguelike structure found in many molluscs is called the
   a. radula.
   b. pen.
   c. beak.
   d. siphon.

4. Snails and slugs are members of the class
   a. Monoplacophora.
   b. Bivalvia.
   c. Gastropoda.
   d. Polyplacophora.
   e. Scaphopoda.

5. Chitons are members of the class
   a. Monoplacophora.
   b. Bivalvia.
   c. Gastropoda.
   d. Polyplacophora.
   e. Scaphopoda.

6. True/False   A glochidium is a larval stage of most marine bivalves.

7. True/False   In most members of the class Cephalopoda, the mantle is modified to promote jetlike propulsion.

8. True/False   In this week's laboratory, we will be carrying out a dissection of a freshwater mussel.

9. True/False   All molluscs possess an open circulatory system.

10. True/False   Blood, in molluscs, serves as a hydraulic skeleton in addition to carrying nutrients.

EVOLUTIONARY PERSPECTIVE

The phyla described in the following exercises are divided into two large groups, sometimes called "superphyla." Members of the phylum Mollusca, along with the annelids and arthropods, are referred to as **protostomes.** Zoologists are not certain of the relationship of the molluscs to other animal phyla, in spite of their protostome affinities. At one time, the molluscs were thought to be closely related to the annelids and arthropods because of the suggestion of segmentation (metamerism) in members of one molluscan class. Now, however, most zoologists agree that this segmentation is very different from that seen in annelids and arthropods. This has led some to suggest that the molluscs diverged from ancient triploblastic animals independently of any other phylum. Others maintain that, in spite of the absence of annelidlike segmentation in molluscs, protostomate affinities still tie the molluscs to the annelid-arthropod lineage. Whichever is the case, the relationships of the molluscs to any other animal phylum are distant.

The molluscs are a very diverse group. The fossil record indicates that representatives of all modern classes were present in prehistoric seas 500 million years ago. Since then they have undergone adaptive radiation into freshwater and terrestrial habitats, including tropical rain forests and deserts. Members of the class Cephalopoda are often cited as the most highly evolved invertebrate group. Their complexity of structure and function rivals that found in many chordates.

THE TRIPLOBLASTIC COELOMATE ANIMALS—MOLLUSCA

The molluscs make up a very large group (second only to the arthropods in number of species) containing the snails, bivalves, chitons, squid, octopuses, and others. Along with the remaining phyla, the molluscs are triploblastic and coelomate. The coelom, however, is secondarily reduced in size and importance in molluscs, forming only the pericardial cavity surrounding the heart and cavities of the gonads and nephridia.

The molluscan body consists of three regions: the **head-foot,** the **visceral mass,** and the **mantle** (fig. 13.1). The head-foot functions largely in locomotion and retracting the body into a shell. It is muscular and relatively fast in responding to stimuli. The visceral mass contains organ systems devoted to digestion, reproduction, excretion, and other visceral functions. In a generalized mollusc, the visceral mass is carried dorsal to the head-foot. The mantle is a fleshy tissue that is attached to the visceral mass and secretes a **shell.** The shell is made up of calcium carbonate and protein. With a few exceptions, the shell can be considered characteristic of the phylum. Between the mantle and the visceral mass is the **mantle cavity.** The mantle cavity contains gills and openings to excretory, digestive, and reproductive systems. The mantle and mantle cavity may be variously modified for sensory reception and locomotion.

A rasping tonguelike structure, the **radula,** is present in all mollusc classes except the bivalves. It consists of a chitinous belt of rasping teeth. This belt overlies a fleshy, tonguelike structure supported by a cartilaginous odontophore. As the radula is moved back and forth over the odontophore, food is rasped into small particles and passed back to the digestive tract.

Most molluscs have an **open circulatory system.** The heart pumps blood through vessels to large tissue spaces called blood sinuses, where the blood bathes the tissues. Blood is then returned to the gills (or lungs) and the heart.

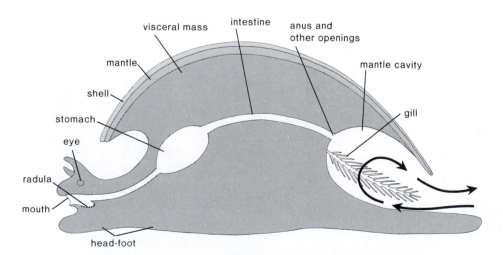

Figure 13.1   The generalized mollusc.

The cephalopods are the sole exception to this generalization. Their circulatory system is closed; blood is confined to vessels. Blood in molluscs performs two functions: carrying nutrients and oxygen and serving as a **hydraulic skeleton.** A hydraulic skeleton serves much the same purpose as the hydrostatic skeleton of pseudocoelomates. In the hydraulic skeleton, fluids are moved through tissue spaces rather than the body cavity. The muscles associated with these movements are often some distance from the body part being moved. Oxygen is carried by the respiratory pigment, hemocyanin. Rather than being confined to blood cells, hemocyanin is carried in solution in the blood. Additionally, molluscs have more complex digestive, excretory, nervous, and sensory systems than previous phyla. Considerable specialization of systems for individual life-styles is seen in all the above systems. Molluscs are usually dioecious, although some are monoecious. Many aquatic molluscs have larval stages that will be described later.

CLASSIFICATION*

Class Caudofoveata
Class Aplacophora
Class Polyplacophora
Class Monoplacophora
Class Gastropoda
Class Cephalopoda
Class Bivalvia
Class Scaphopoda

*This taxonomic listing reflects a phylogenetic arrangement. The observations that follow, however, begin with molluscs that are familiar to most students.

Class Gastropoda—Snails, Slugs, Limpets

Characteristics: Shell, when present, usually coiled; body symmetry distorted; some monoecious. Examples: *Physa, Helix, Busycon.*

▼ Obtain a living snail. If the specimen is aquatic, be sure to keep the snail submerged for most of the following set of observations. Allow your snail to recover from being handled, then compare it to figure 13.2. Identify the head-foot, the mantle, and the shell. The mantle can usually be observed around the lower edge of the shell. The visceral mass is covered by the shell. It will be observed later on a preserved snail that has had the shell removed.

Hold the shell apex up with the opening **(aperture)** toward you. Is the aperture on the right or left? _____ If on the right, the shell is **dextral;** if on the left, the shell is **sinistral.** This is an important taxonomic characteristic. Note the whorls (complete turns) of the shell and fine lines of growth. The apex is the oldest part of the shell.

If the specimen is aquatic, the following must be done with the snail submerged in a dish of water. In an undisturbed living snail, note the head with its two pairs of **tentacles. Eyes** may be located at the tip or base of the more dorsal tentacles. Touch one of the tentacles of your living snail with a needle. Notice the quick response that draws the tentacle back. This response is the result of fast-acting retractor muscles. Watch as the tentacle is reextended. It is a much slower process because it is accomplished by a hydraulic mechanism. Blood is shifted from other parts of the body by the contraction of distant muscles. Blood gradually fills the tissue spaces of the tentacle, thus extending the tentacle.

Place your snail on a microscope slide. Allow the snail to attach itself firmly. Invert and support the ends of the slide so the ventral surface of the foot can be observed with a hand lens under the dissecting microscope. Note the mouth opening and the method of locomotion. The gliding movement is accomplished by waves of muscular contraction passing over the foot. Observe a prepared slide showing a gastropod radula. Details of radular structure are taxonomically important.

(a)

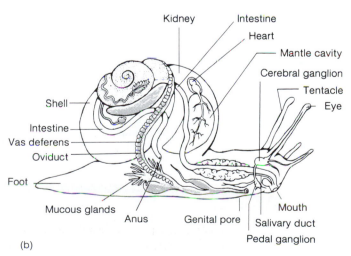

(b)

Figure 13.2   Class Gastropoda: (*a*) a land (pulmonate) gastropod; (*b*) internal structure of a generalized gastropod.

Study a specimen of the large marine snail, *Busycon*, that has been removed from its shell. Locate the head-foot, mantle, and visceral mass. Note the calcareous **operculum** on the dorsal aspect of the posterior half of the foot. This protective plate closes over the aperture of the shell when the snail withdraws into the shell. On the head find the tentacles and eyes. On the left side, the mantle is modified into a **siphon** that allows water to circulate over the fan-shaped gill. Most of the spire of the visceral mass is liver (bluish gray) and gonad (yellow, on the surface of the liver near the tip of the spire). Observe an empty *Busycon* shell and visualize how the structures fit into the shell. Look at demonstrations of the egg cases of *Busycon*. ▲

Other Gastropoda

▼ Examine any available demonstration materials that show subclass representatives. ▲

Pulmonates
    Garden slugs
    Other terrestrial snails
Prosobranchs
    Limpets
    Abalone
    Conchs
Opisthobranchs
    Sea slugs
    Sea hares

STOP AND ASK YOURSELF

1. What are three regions of the body of a mollusc?
   _____

2. What is an open circulatory system? _____
   _____
   _____

3. What is a hydraulic skeleton? _____
   _____
   _____

4. How would you characterize members of the class Gastropoda? _____
   _____

5. What is the function of the following:
   a. an operculum? _____
      _____
   b. a siphon? _____
      _____

Class Bivalvia—Clams, Oysters, Mussels

Characteristics: Body enclosed in a shell consisting of two valves, hinged dorsally; no head or radula; wedge-shaped foot: Examples: *Anodonta*, *Mytilus*, and *Venus*.

External Structure

▼ Examine a preserved freshwater mussel. Note the two **valves** and their point of attachment on the dorsal surface. The two valves are held together by an elastic **hinge ligament.** A swollen area **(umbo)** is near the anterior end of the hinge and is the oldest part of the shell. On opening the shell and studying the internal structure of your bivalve, you will find that the mouth opening is at this end of the bivalve and the anus is at the opposite end. Note the **lines of growth** that diverge from the umbo toward the edge of the shell. Using the position of the hinge and umbo, orient yourself as to dorsal, ventral, anterior, and posterior.

Internal Structure

Bivalves usually come from suppliers with the valves "pegged" open. Locate the position of **anterior** and **posterior adductor muscles** that are holding the valves closed (fig. 13.3). Slip a scalpel between the mantle and the left valve. Cut each adductor muscle at its point of attachment to the shell. When both muscles are cut, loosen the mantle over the entire area of the left valve and open the valves.

Examine the inner surface of the empty valve and the **mantle** covering the body of the bivalve (see fig. 13.3). Just below the hinge, the mantle covers the **pericardial sac** and the **heart.** These structures are very delicate: be careful not to disturb them until directed to do so. Examine the adductor muscles that you cut on opening the bivalve and their points of attachment on the empty shell. The adductor muscles have a "catch" mechanism that allows them to remain contracted for long periods with little energy expenditure and without fatigue. When the adductor muscles relax, elasticity of the **hinge ligament** forces the valves open. Next to each adductor muscle is a smaller **retractor muscle** that brings the foot into the shell. Find the retractor muscles and their points of attachment on the shell. Below the anterior adductor muscle is the **anterior protractor muscle** that helps extend the foot.

A thickening at the margin of the mantle, the **pallial muscle,** attaches the mantle to the shell. The **pallial line** on the shell marks a line of attachment for this muscle. The pallial line arches between the points of attachment of the adductor muscles. The mantle of the left and right valves comes together posteriorly to form **incurrent** and **excurrent apertures** that allow water to enter and leave the mantle cavity (Plate 1). Water enters through the ventral incurrent aperture and leaves through the dorsal excurrent aperture. In many bivalves these apertures are modified into long **siphons.** After a bivalve has burrowed into the substrate, the siphons can be extended to the water/substrate interface for gas exchange and filter feeding.

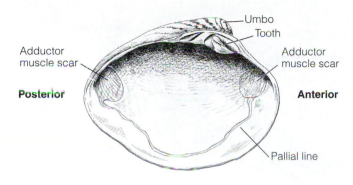

Posterior   Anterior

Figure 13.3   Internal structure of a bivalve shell.

Return to the empty valve. Just below the hinge are tongue-and-groove modifications of the shell, called **teeth,** that prevent twisting. Detach the empty valve and break it with a hammer. Examine a cross section of the shell. There are three layers in all mollusc shells. Outside is a layer of polymerized protein called **periostracum.** The middle **prismatic layer** is the thickest and is made of calcium carbonate. The inner layer is also calcium carbonate and is called the **nacreous layer.**

The space between the mantle and the body is the mantle cavity. Lift the mantle to expose the visceral mass, foot, and associated structures (fig. 13.4). The muscular, wedge-shaped **foot** is at the ventral aspect of the body. The soft tissue making up the bulk of the body is the **visceral mass.** At the anterior margin of the visceral mass, note the flaplike **labial palps.** Labial palps surround, and direct food toward, the mouth. Water coming in from the incurrent aperture reaches the ventral aspect of the gills and passes dorsally through the gills into a space above the gills formed by the attachment of the gills to the mantle on one side and the visceral mass on the other. Once in this **suprabranchial chamber,** water is directed posteriorly and out of the mantle cavity through the excurrent aperture (fig. 13.5 and Plate 2). In the process, suspended food particles are filtered and gas exchange occurs. Food particles are transported by cilia to **food grooves** along the ventral margin of the gills. Cilia in the food grooves transport food to the labial palps (see fig. 13.4a).

Digestion begins as food is entangled in a mucoid string, which enters the stomach. A consolidated mucoid mass, the crystalline style, projects into the stomach from a diverticulum, called the style sac. Rotation of the crystalline style against a chitinized gastric shield dislodges enzymes from the style (see fig. 13.4b). Undigestible materials are sent on to the intestine, and partially digested food enters a digestive gland where intracellular digestion occurs. The intestine and digestive gland will be observed later.

The circulatory system of bivalves is an open system. A heart is enclosed by the **pericardium** (homologous to a reduced coelomic peritoneum) and located dorsal to the visceral mass. The pericardium is a thin membrane covering the top of the visceral mass. Carefully remove the pericardium to see the heart. The **heart** wraps around the intestine where the intestine emerges from the visceral mass. The **intestine** is running posteriorly to empty at the excurrent aperture. The heart consists of two parts, a thick-walled **ventricle** surrounding the intestine and two thin-walled **auricles** attached at either side of the ventricle. If you were careful in removing the pericardium, you should see both.

Blood leaves the ventricle through the anterior and posterior aortae and goes to tissue sinuses for exchange of nutrients, gases, and wastes. Blood in the posterior aorta goes to the mantle and then returns to the auricles. Blood in the anterior aorta goes to the viscera and the foot. This blood returns to the auricles via the nephridia and the gills.

The excretory system consists of a pair of dark **nephridia** lying below the pericardial sac. Nephridia remove metabolic waste products from the blood and release the waste into the mantle cavity near the excurrent aperture.

Use your scalpel to make a sagittal section of the foot and visceral mass (Plate 3). Within the visceral mass, note cut sections of intestine. The yellowish tissue surrounding the intestine is gonad. Anteriorly you will have cut through a greenish digestive gland (fig. 13.6).

Sexes in bivalves are separate, but difficult to distinguish. Gonads are within the visceral mass, and gametes are released in the vicinity of the excurrent aperture. Fertilization is usually external, with swimming larval stages. In marine bivalves, the first larval stage, the **trochophore** (fig. 13.7a), is followed by the **veliger** (fig. 13.7b) and the adult. In freshwater bivalves, fertilization occurs within the mantle cavity of the female, and zygotes develop into **glochidia** (fig. 13.7c) within the gill chambers. Glochidia have tiny valves with teeth that are used in attaching to the gill filaments, fins, or skin of a fish. Each glochidium lives as a parasite for several weeks, then drops off to take up adult life. Examine slides of any larval stage available. ▲

Other Bivalves

▼ Examine any demonstration materials available. ▲

Class Cephalopoda—Squid, Octopuses, Nautili

Characteristics: Foot modified into a circle of tentacles and a siphon; shell reduced or absent; head in line with the elongate visceral mass.

The cephalopods are the most highly evolved molluscs and in many ways the most advanced of all nonchordates. They are all marine, dioecious, and active predators. Experiments have shown them to be capable of degrees of learning unparalleled by other nonvertebrates. Cephalopods use jet propulsion in locomotion and are, therefore, fast swimmers. Ecologically, cephalopods are less important. They

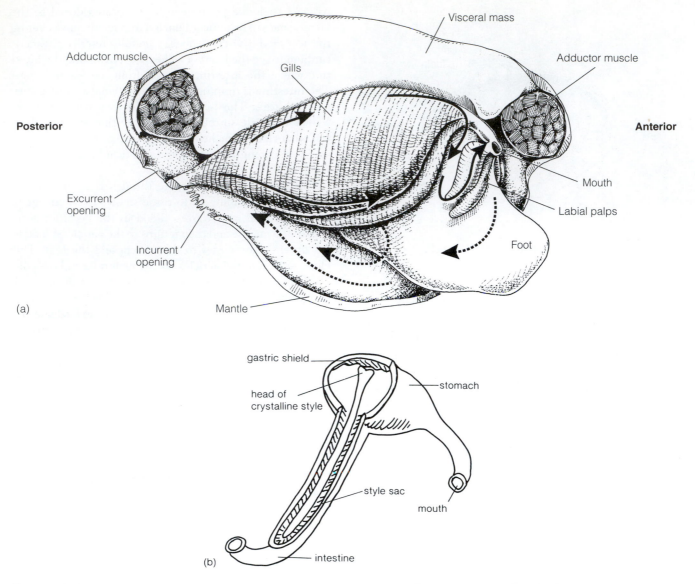

Figure 13.4 (a) Internal structure of a bivalve that was removed from its shell. Arrows show path of food particles after being filtered by gills. Dashed arrows show path of the particles rejected by gills and labial pulps. (b) The stomach and associated structures of a bivalve. Food particles entering the stomach are digested as enzymes and are released from the rotating crystalline style.

make up less than 0.5% of all molluscan genera. The reasons for this are not completely known.

Loligo—The Squid, External Structure

▼ Examine the external structure of *Loligo* (fig. 13.8). Note the head, which bears eight **arms** and two **tentacles.** Eyes on the head are structurally much like the eyes of vertebrates. They can apparently form clear images and discriminate some colors. Note that the head is lined up with the elongate **mantle.** The mantle bears **lateral fins** and encloses the **visceral mass.** Near the region where the mantle and the head meet, note the conical **siphon** protruding from below the mantle. Water drawn into the mantle cavity can be forcefully expelled through the siphon when muscles of the mantle contract. Jetlike propulsion results. Muscles attached to the siphon can direct the jet of water in different directions.

Look closely at the eight arms and two longer tentacles that encircle the mouth (fig. 13.9). In males, one arm (the **hectocotylus**) is modified for sperm transfer. The suckers on it are smaller and attached to the arm by a longer stalk. Using the hectocotylus, the male reaches into his mantle cavity, removes a spermatophore (package of sperm), and transfers it to the mantle cavity of the female. At the base of the arms and tentacles is the mouth opening. Spread the tentacles and note the chitinous **beak** used in tearing prey. Below the beak is the radula (not seen without dissection). The shell of *Loligo* is embedded in the mantle and is referred to as the **pen.** The tip of the chitinous pen can be seen as a point in the mantle opposite the siphon. The shell of cephalopods varies from being well developed in the nautili to being completely gone in the octopuses. ▲

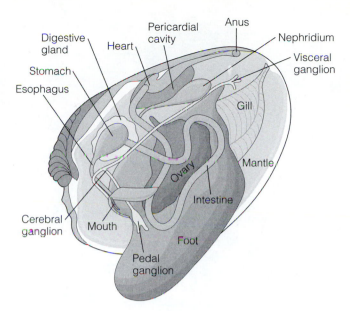

Figure 13.6 Internal structure of a bivalve. Sagittal section of the head-foot and visceral mass.

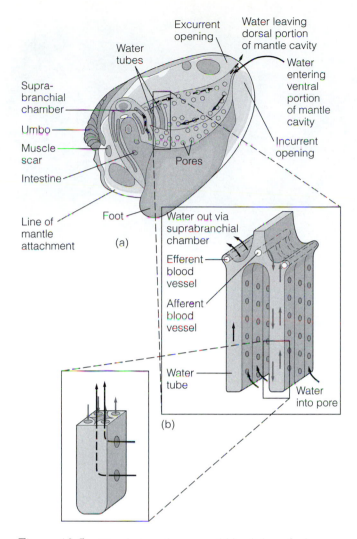

Figure 13.5 Circulation of water and blood through the gills of a bivalve. Arrows show path of water. (*a*) Incurrent (solid arrows) and excurrent (dashed arrows) water currents circulate through the bivalve gills. Food is filtered as water enters water tubes through pores in the gills. (*b*) Cross section through a portion of a gill. Water (light arrows) passing through a water tube passes in close proximity to blood (dark arrows). Gas exchange occurs between water and blood in the water tubes.

Other Cephalopods

▼ Examine any demonstrations available including

octopuses
cuttlefish
Nautilus shells
cuttlebone of *Sepia*
pen of *Loligo* ▲

Class Polyplacophora—Chitons

Characteristics: Elongate, dorsoventrally flattened; head reduced in size; shell consisting of eight dorsal plates. Example: *Chiton.*

▼ Examine demonstrations of chitons. Note the dorsal plates and the broad, flat, muscular foot. This foot allows chitons to attach to the rocky substrate common to intertidal regions. The mantle cavity is reduced to a groove running on either side of the body between the foot and the margin of the animal. ▲

Class Scaphopoda—Tooth Shells or Elephant's-Tusk Shells

Characteristics: Body enclosed in a tubular shell that is open at both ends; tentacles; no head. Example: *Dentalium.*

▼ Examine demonstration specimens. Note especially the structure of the shell. The foot protrudes from the larger end of the shell. The small posterior end is open to facilitate the circulation of water. As the scaphopod burrows through sand and mud, the posterior end protrudes from the substrate to allow water to circulate through the mantle cavity. ▲

Class Monoplacophora

Characteristics: Single-arched shell; foot broad and flat; certain structures serially repeated. Example: *Neopilina.*

This is a group of molluscs that, prior to 1952, was known only from fossils. In 1952, this limpetlike animal was dredged up from a depth of 3,520 meters off the Pacific coast of Costa Rica. For many years *Neopilina* was thought to represent the ancestral form of the molluscs and, because of a suggestion of metamerism (serial repetition of body parts), might have linked the molluscs closely with the arthropods and annelids. Most biologists have now discarded this idea; they view the Monoplacophora as an interesting group, but not representative of ancestral molluscs. The suggested metamerism is secondarily derived.

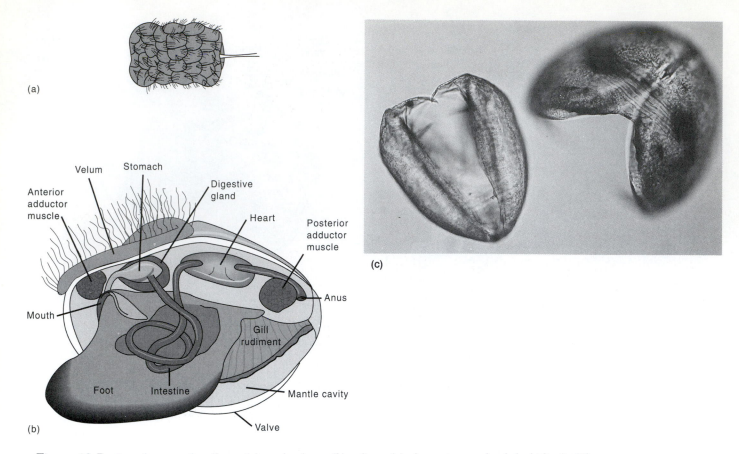

(a)

Velum   Stomach
Anterior
adductor   Digestive
muscle   gland

                            Heart
                                      Posterior
                                      adductor
                                      muscle

Mouth   Anus

                            Gill
                            rudiment

Foot   Intestine   Mantle cavity

(b)   Valve

(c)

Figure 13.7 Larval stages of molluscs: (*a*) trochophore; (*b*) veliger; (*c*) photomicrograph of glochidia (×50).

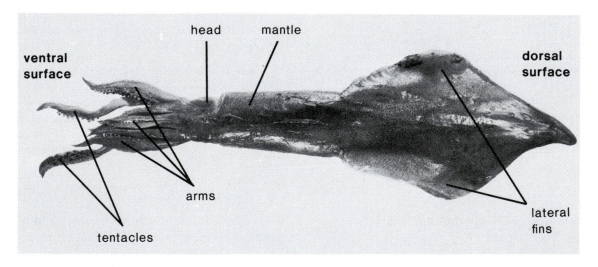

head   mantle

ventral   dorsal
surface   surface

arms   lateral
                                                 fins

tentacles

Figure 13.8 *Loligo*, anterior view.

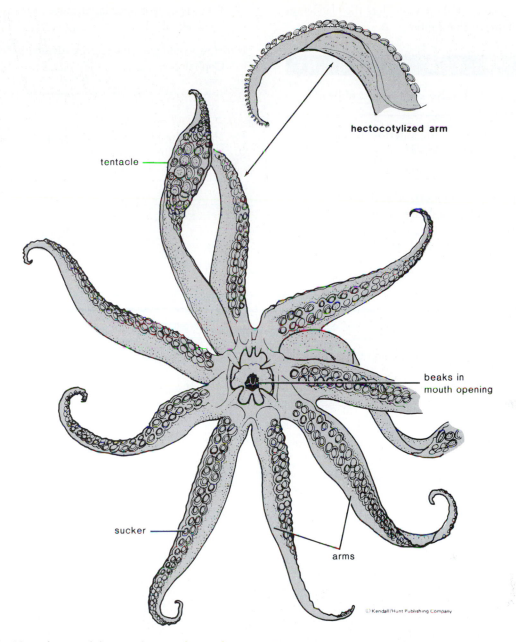

tentacle

hectocotylized arm

beaks in mouth opening

sucker

arms

© Kendall/Hunt Publishing Company

Figure 13.9 Ventral view of the mouth area of a squid. © Kendall/Hunt Publishing Company.

Class Aplacophora—Solenogastres

Characteristics: Shell, mantle, and foot lacking; wormlike; head poorly developed; grooved ventral surface. Example: *Neomenia*.

This is a small group of moderately deep-water marine molluscs.

▼ Examine any specimens available. ▲

Class Caudofoveata

Characteristics: Wormlike with scalelike spicules on the body wall; lack shell, foot, and nephridia.

Members of the class Caudofoveata range in size from 2 mm to 14 cm and live in vertical burrows on the deep-sea floor. Approximately 70 species have been described. Little is known of their natural history.

EVOLUTIONARY RELATIONSHIPS

▼ A cladogram showing a traditional interpretation of the evolutionary relationships among the molluscan classes is shown in worksheet 13.1 (fig. 13.10). Examine the synapomorphies depicted on the cladogram. Based upon your studies in this week's laboratory, you should be able to determine the position of the classes studied. Complete the cladogram and answer the questions that follow it. ▲

6. The oldest part of a bivalve shell is marked by a swollen area called the _____ .

7. How does one determine the anterior end of a bivalve when its shell is closed? _____

   _____

8. Just before it enters the mouth of a bivalve, food is passed by cilia to what flaplike structures on either side of the mouth? _____

   _____

9. What is the parasitic larval stage of many freshwater bivalves called? _____

10. How would you characterize members of the class Cephalopoda? _____

    _____

    _____

11. How is the shell of a chiton modified from that of other molluscs? _____

    _____

    _____

KEY TERMS

glochidia   165
head-foot   162
hydraulic skeleton   163
mantle   162

mantle cavity   162
open circulatory system   162
protostomes   162
radula   162

trochophore   165
veliger   165
visceral mass   162

WORKSHEET 13.1 Mollusca

Evolutionary Perspective

1. What animal phyla are included in the protostome grouping?

Mollusca

2. What functions are carried out by
   a. the mantle in bivalves?

   b. the mantle in cephalopods?

   c. the head-foot in gastropods?

   d. the radula in gastropods?

   e. the visceral mass in all molluscs?

3. Members of which large molluscan class lack a radula?

4. Describe the path of water through the gills of a bivalve.

5. Describe the path of food particles after being filtered from the water while passing through a bivalve gill.

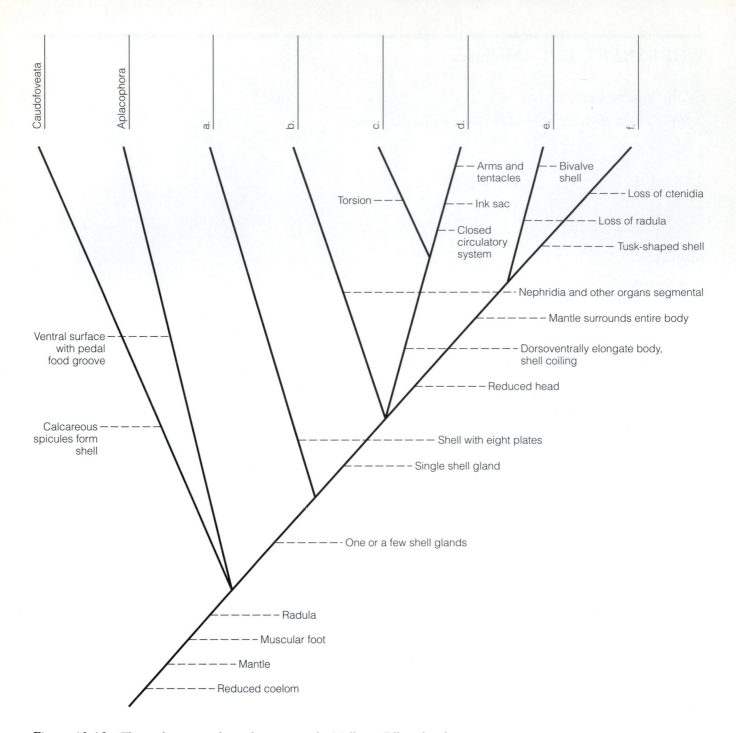

Caudofoveata

Aplacophora

a.

b.

c.

d.

e.

f.

— Arms and tentacles

— Bivalve shell

Torsion — —

— — Ink sac

— — Loss of ctenidia

— — Loss of radula

— Closed circulatory system

— — Tusk-shaped shell

- Nephridia and other organs segmental

— Mantle surrounds entire body

Ventral surface with pedal food groove — — — — —

— Dorsoventrally elongate body, shell coiling

— Reduced head

Calcareous — — — — spicules form shell

— Shell with eight plates

— Single shell gland

- One or a few shell glands

— Radula

— Muscular foot

— Mantle

— Reduced coelom

Figure 13.10 The evolutionary relationships among the Mollusca. Fill in the class names.

6. Members of what class of molluscs were once thought to display segmentation similar to that seen in annelids and arthropods?

7. What is the function of the coelom of molluscs?

8. Name three symplesiomorphic characters of the phylum Mollusca. _____ ,
_____ , _____

9. Is the presence of a shell an ancestral character for all molluscs? Explain your answer.

10. What derived character distinguishes the classes Gastropoda and Cephalopoda from other molluscs? _____
_____

11. The presence of a bivalve shell and the absence of a radula are synapomorphies for members of the class
_____. These characters reflect important aspects of the life-style of these animals.
Explain why this statement is true.

12. Is the protective structure of a bivalve a single shell of two pieces or two shells that work together? Explain your
answer based on your examination of figure 13.10.

Exercise 14

Annelida

Learning Objectives

After completing this exercise, you should be able to do the following:

1. Describe the evolutionary significance of the Annelida.
2. Describe the evolutionary relationships between the annelid classes.
3. Describe metamerism and tagmatization and explain why they are important in annelids.
4. Describe the annelid body form.
5. Characterize each of the annelid classes studied, including habitat, feeding and reproductive habits, and distinctive features of body organization.
6. Recognize class representatives.
7. Describe the structure and function of organ systems studied in each class.

Prelaboratory Quiz

Study this week's laboratory exercise and then complete the following quiz to assess your preparation for the laboratory.

1. The annelids are most closely related to members of which of the following phyla?
   a. Arthropoda
   b. Mollusca
   c. Nematoda
   d. Echinodermata

2. The serial repetition of body parts is called
   a. tagmatization.
   b. metamorphosis.
   c. metamerism.
   d. paedomorphosis.

3. All of the following are consequences of metamerism EXCEPT
   a. lessened impact of bodily injury.
   b. the ability to specialize certain regions of the body to specialized functions.
   c. efficient locomotion using localized changes in body shape and hydrostatic pressure.
   d. All of the above are consequences of metamerism.

4. The largest of the three annelid classes is
   a. Polychaeta.
   b. Oligochaeta.
   c. Hirudinea.

5. All of the following are annelid features EXCEPT
   a. an open circulatory system.
   b. nephridia used in excretion.
   c. a complete digestive tract.
   d. a ventral nervous system.
   e. a body wall with circular and longitudinal muscle layers.

6. True/False   Polychaetes have paired, lateral extensions of the body wall in each segment called parapodia.

7. True/False   A clitellum is a swollen area in the anterior portion of oligochaetes and polychaetes that secretes mucus during sperm exchange.

8. True/False   Members of the classes Oligochaeta and Hirudinea are monoecious. Most members of the class Polychaeta are dioecious.

9. True/False   Members of the class Oligochaeta have free-swimming trochophore larval stages.

10. True/False   Seminal receptacles of the earthworm store sperm prior to sperm exchange with another worm.

EVOLUTIONARY PERSPECTIVE

Members of the phylum Annelida are protostomates and have a complete digestive tract, a solid ventral nerve cord, and segmentally arranged body parts. One hypothesis of their origin is that they were derived from ancient triploblastic acoelomate animals, which would have involved a segmental splitting of mesoderm to form segmentally arranged coelomic compartments. A second hypothesis describes the coelom as being derived from outpockets of the gut and that mesodermal tissues were secondarily derived. If this is true, annelids would have probably originated from early diploblastic animals.

Virtually all zoologists agree that, because of similarities in the development of metamerism between the arthropods and annelids and their shared protostomate characteristics, the two groups are closely related. A common ancestor of both groups was probably a marine, wormlike, bilateral ancestor with segmentally arranged body parts. Evolutionary divergence from this common annelid-arthropod ancestor may have occurred with the evolution and specialization of paired appendages and an external skeleton in the arthropod lineage.

Within the Annelida, the Polychaeta (mostly marine worms) is the ancestral class. Primitive polychaetes diverged to give rise to modern polychaetes and ancestral freshwater oligochaetes. Modern oligochaetes (e.g., earthworms) and the Hirudinea (i.e., leeches) diverged from ancestral oligochaetes.

THE TRIPLOBLASTIC COELOMATE ANIMALS—ANNELIDA

The Annelida, or segmented worms, is a relatively large phylum containing the earthworms, leeches, and marine worms. The most striking and significant evolutionary feature of the annelids is metamerism. **Metamerism** refers to the serial repetition of body parts. In their primitive form, metameric animals consist of identical repeating segments, each capable of operating relatively independently (i.e., they each contain their own excretory, nervous, reproductive, and circulatory structures). No animal shows this degree of metameric development, but the annelids come the closest. Advantages that result from metamerism include the following:

1. more efficient locomotion, utilizing localized changes in body shape and hydrostatic pressure (used in efficient crawling, burrowing, and swimming);
2. lessened impact of bodily injury; and
3. **tagmatization,** the ability to devote certain segments of the body to specialized functions (e.g., specialization for locomotion, sensory function, and feeding).

Annelids are triploblastic, bilaterally symmetrical, and coelomate. All of these features have been described for previous phyla. The coelom is especially important in the Annelida in its role as a hydrostatic skeleton. The metameric organization of the body allows hydrostatic compartments to be used independently in locomotion. For example, in anterior segments, circular muscles might be contracting, causing elongation and forward extension during crawling. At the same time, posterior segments can have longitudinal muscles contracted to increase the cross-sectional area of the worm and thus maintain contact with the substrate.

In addition, annelids have

1. a body wall with circular and longitudinal muscle layers,
2. a complete digestive tract,
3. a ventral nervous system showing some cephalization,
4. a closed circulatory system,
5. nephridia used in excretion, and
6. both monoecious and dioecious taxa.

CLASSIFICATION

Class Polychaeta
Class Oligochaeta
Class Hirudinea

Class Polychaeta

Characteristics: Mostly marine; head with eyes and tentacles; parapodia bearing numerous setae; trochophore larvae. Examples: *Nereis, Arenicola, Sabella.*

The class Polychaeta is the largest of the three annelid classes. Polychaetes are mostly marine, living in muddy and sandy offshore areas. The prototype might be visualized as having many undifferentiated segments, each bearing lateral extensions (parapodia) containing many setae (hence the name). Evolution has resulted in variations on this theme. Many modern polychaetes show various degrees of tagmatization and cephalization. They may be free-swimming, burrowing, or tube-dwelling. They may feed as predators or on detritus or suspended organic particles. In the latter instances, they may develop intricate filtering devices. Some polychaetes obtain a substantial portion of their nutrients from organic matter dissolved in seawater. While the uptake of dissolved organic matter has been suggested for many aquatic organisms, it has been demonstrated to be significant in relatively few.

Nereis—The Clamworm

▼ Examine a preserved specimen with the aid of a hand lens or dissecting microscope. Examine the head (figs. 14.1

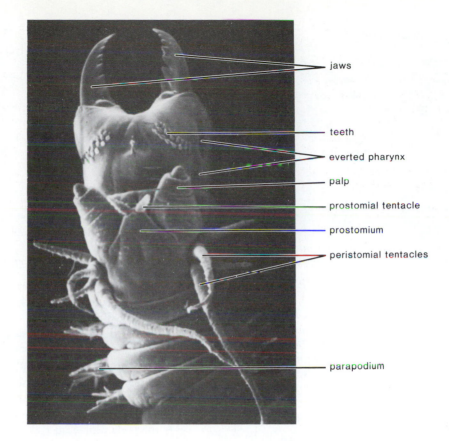

jaws

teeth

everted pharynx

palp

prostomial tentacle

prostomium

peristomial tentacles

parapodium

Figure 14.1 *Nereis*, dorsal view of the head (×48). Scanning electron micrograph courtesy of Betsy Brown.

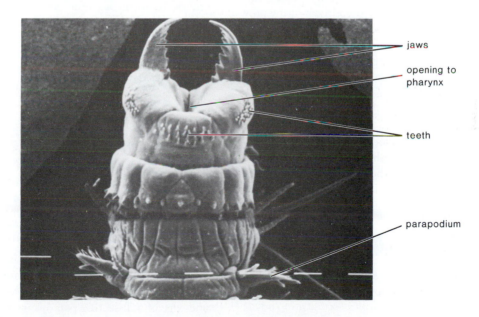

jaws

opening to pharynx

teeth

parapodium

Figure 14.2 *Nereis*, ventral view of the head (×53). Scanning electron micrograph courtesy of Betsy Brown.

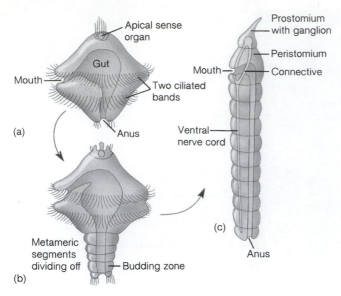

Mouth

Apical sense organ

Gut

Two ciliated bands

Anus

(a)

Metameric segments dividing off

Budding zone

(b)

Mouth

Prostomium with ganglion

Peristomium

Connective

Ventral nerve cord

(c)

Anus

Figure 14.3 The development of a polychaete trochophore larva. (*a*) A trochopore buds segments (*b*) near the anus and (*c*) settles to the substrate as an immature worm. From *A Life of Invertebrates* by W. D. Russell-Hunter. Copyright © 1979 W. D. Russell-Hunter. Reprinted by permission.

and 14.2). Note the **prostomium,** which bears four dark eyes, a pair of small **tentacles,** and fleshy **palps.** In back of the prostomium is the **peristomium** bearing four pairs of **peristomial tentacles.** The pharynx, bearing pincerlike **jaws,** is probably inverted into the head. In feeding, the pharynx is everted, and the jaws are used in capturing prey. Insert fine-tipped forceps or a teasing needle into the pharyngeal opening and evert the pharynx to see the jaws. Posteriorly, each segment bears a pair of lateral **parapodia.** Each parapodium is divided into a dorsal lobe (notopodium) and a ventral lobe (neuropodium). The internal structure of *Nereis* will not be studied, since it is similar to the earthworm to be studied next. *Nereis*, like most polychaetes, is dioecious. Fertilization and development are external. A ciliated larval stage called a **trochophore** develops, which will eventually metamorphose into the adult. Examine a slide showing a trochophore larva (fig. 14.3). Examine any available demonstrations of other polychaetes. ▲

STOP AND ASK YOURSELF

1. What characteristics link the annelids and arthropods to the same evolutionary lineage? _____

   _____

2. What is metamerism, and why is it advantageous for annelids? _____

   _____

3. How would you characterize members of the class Polychaeta? _____

---

4. What are parapodia? _____

5. The ciliated larvae of polychaetes are called

   _____ .

Class Oligochaeta

Characteristics: Few setae; no parapodia; distinct head absent; monoecious; direct development. Examples: *Lumbricus* and *Tubifex*.

The class Oligochaeta is the second largest annelid class. It is primarily terrestrial and freshwater. These annelids have few, short setae (hence the name) and never have parapodia. The tendency toward cephalization is reduced.

Lumbricus terrestris—External Structure

▼ Examine a preserved earthworm using a hand lens or dissecting microscope (fig. 14.4). Note the obvious metamerism. The slightly swollen area in the anterior third of the worm is the **clitellum.** This region secretes mucus that holds two worms together for sperm exchange. The clitellum also forms a cocoon around developing embryos. Along the lateral and ventrolateral margins of the worm are small **setae** (four pair per metamere). Run your fingers along the sides of the worm and feel the rough texture that the setae give the body wall. Can you distinguish dorsal and ventral on the worm? Setae provide secure contact with the substrate while burrowing and crawling. Setae are manipulated by muscles attached at the base of each seta. The anterior end of the worm has a mouth opening. The lobe over the mouth is the **prostomium.** The first segment, which has the mouth and is just posterior to the prostomium, is the **peristomium.** The posterior segment bears the anus. Other external features of interest are openings from various internal systems. Most are difficult to see. The openings of the sperm ducts are on the ventral surface of metamere 15 and are conspicuous. Find them. Smaller openings for ova are on the ventral surface of segment 14 but are difficult to see in preserved worms. Other openings include openings from seminal receptacles (between 9 and 10, and 10 and 11) and nephridial openings (on the ventrolateral surface of each segment except 1–3 and the last). ▲

Lumbricus—Internal Structure

▼ Recall that setae are ventral and lateral. Run your fingers along the surface of the worm. Locate the setae and distinguish dorsal and ventral. Pin your worm dorsal side up near one edge of a dissecting pan. Using a sharp scalpel, make a shallow longitudinal cut along the dorsal surface of the worm. (Be sure not to cut into the dorsal vessel and the intestinal tract, which are located just below the body wall.) Note that the body wall is held in place by **septa** (internal divisions between metameres). Using a sharp scalpel, cut the septa along the length of the worm on both sides of the intestine. These septa divide the coelom into separate cavities. Pin

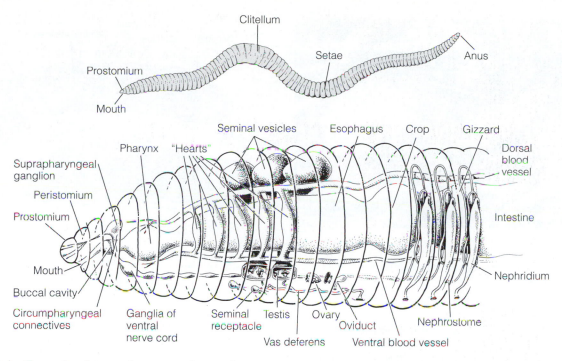

Figure 14.4 External and internal structure of an earthworm.

the body wall to the bottom of the pan, inclining the pins away from the worm. Keep your preparation moist at all times.

Digestive System Observe the digestive tract running from mouth to anus. Anteriorly there are a number of modifications of the tract. Just posterior to the mouth opening is the muscular **pharynx** (Plate 4). The torn muscles associated with the pharynx were attached to the body wall. When these muscles contract, food particles are sucked into the mouth. The **esophagus** is posterior to the pharynx and is surrounded by cream-colored bodies that will be studied later. The esophagus expands into a thin-walled storage structure, the **crop.** Probe the wall of the crop gently and note its texture. Just posterior to the crop is the muscular **gizzard.** This is a grinding structure with thick, muscular walls (obvious on gentle probing). Food is passed from the gizzard to the **intestine,** where further digestion and absorption occur. The intestine ends at the anus.

Circulatory System Circulation in the earthworm is through a series of closed vessels. The two main vessels that can be seen in your dissection are the **dorsal** and **ventral blood vessels.** These vessels are the main pumping structures. In the dorsal vessel, blood moves anteriorly. The dorsal vessel is the dark line running along the dorsal surface of the digestive tract. In the posterior third of your worm, carefully cut through and remove about three centimeters of the digestive tract. The ventral blood vessel can usually be seen adhering to the segment of intestine removed or to the ventral body wall (Plate 5). In the ventral vessel, blood moves posteriorly (an exception is noted below). Segmental branches off the ventral vessel supply the intestine and body wall with

blood (polychaetes have a third set of branches going to the parapodia). These branches eventually break into capillary beds to pick up or release nutrients and oxygen. Gas exchange occurs between the capillary beds of the body surface and the environment. Oxygen is carried by the respiratory pigment hemoglobin, which is dissolved in the fluid portion of the blood. From these capillary beds, blood is collected into larger vessels that eventually unite with the dorsal vessel. At the level of the esophagus, segmental branches are expanded into five pairs of "**hearts**" or aortic arches (see Plate 4). Although these are contractile, they only function in pumping blood between dorsal and ventral vessels. As noted earlier, dorsal and ventral vessels are the primary pumping structures. Most blood entering the ventral vessel is pumped posteriorly. Some blood moves from the aortic arches anteriorly. Locate the aortic arches. They are dark, expanded structures on either side of the esophagus. They are not present in all annelids. Other vessels associated with the nerve cord will be noted later. In an anesthetized worm, waves of contraction can be observed passing along the dorsal vessel. Observe this demonstration if available.

Reproductive System Earthworms are monoecious, with cross-fertilization the rule. Gonads are too small to see, but associated structures are clearly visible. Note the large, cream-colored structures associated with segments 9–12 (see Plate 4). These are **seminal vesicles.** Testes are associated with seminal vesicles. Sperm are passed from the testes to the seminal vesicles for storage prior to copulation. During copulation, sperm exit through a duct system opening at segment 15. The ovaries are located in segment 13, and their duct system opens to the outside in segment 14. Two

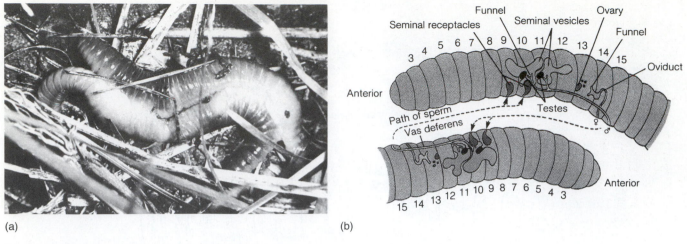

(a)

(b)

Figure 14.5 Earthworm copulation: (*a*) two worms within a mucous sheath; (*b*) path of sperm. (*b*) From *A Life of Invertebrates* by W. D. Russell-Hunter. Copyright © 1979 W. D. Russell-Hunter. Reprinted by permission.

Figure 14.6 Earthworm cocoons.

pairs of small, round, cream-colored structures are present on the ventrolateral body wall in segments 9 and 10. These **seminal receptacles** receive sperm during copulation.

During copulation, two worms line up facing in opposite directions with the ventral surfaces of their anterior ends in contact with each other. This lines up the clitellum of one worm with the genital segments of the other worm. Worms are held in place by a mucous sheath secreted by the clitellum that envelopes the anterior halves of both worms (fig. 14.5a). Sperm are released from the sperm duct and travel along the external, ventral body wall in sperm grooves formed by the contraction of special muscles (fig. 14.5b). Muscular contractions along this groove help propel sperm toward the openings of the seminal receptacles.

Following copulation, the clitellum forms a **cocoon** for the deposition of eggs and sperm (fig. 14.6). The cocoon consists of mucoid and chitinous materials that encircle the clitellum. A food reserve, albumin, is secreted into the cocoon by the clitellum, and the worm begins to back out of the cocoon. Eggs are deposited into the cocoon as the cocoon passes the oviductal openings, and sperm are released as the cocoon passes the openings of the seminal receptacles. Fertilization occurs within the cocoon and, as the

worm continues backing out, the ends of the cocoon are sealed and the cocoon is deposited in moist soil. Development is direct (without a larval stage) inside the cocoon.

Excretory System **Nephridia** are organs of excretion in the annelids. The form of the nephridium varies depending on the particular annelid. Those found in the earthworm are highly evolved metanephridia. Flood your worm with water. Observe the body wall under a dissecting microscope. Note the coiled tubule attached to the body wall of each segment.

The funnel-shaped **nephrostome** is attached to the septum dividing two segments and opens into the anterior segment. The **tubule** is within the posterior segment and connects to the body wall in that segment (see Plate 5). The tubule then opens to the outside. Body fluids, ammonia, and urea are drawn into the nephrostome and excreted.

Filtration of the blood across the tubule wall can occur because of the close association between capillaries and the nephridium. These physiological processes are not clearly understood.

Nervous System The nervous system consists of a **ventral nerve cord** with **segmental ganglia** and a pair of large ganglia anterior and dorsal to the pharynx (see Plates 4 and 5). Nerve fibers arise from segmental ganglia and innervate structures in each segment. In the region of your worm where the intestine was removed, note the ventral nerve cord with its segmental ganglia (seen as slight swellings). If your middorsal incision is not all the way to the prostomium, extend it at this time and note the **suprapharyngeal ganglion** (loosely called the "brain"). This ganglion is connected by **circumpharyngeal connectives** to the ventral nerve cord. ▲

Cross Section ▼ Using a dissecting microscope or the lowest power of a compound microscope, examine a cross section of *Lumbricus*. Locate those structures studied earlier using figure 14.7. In addition, note the noncellular, protective

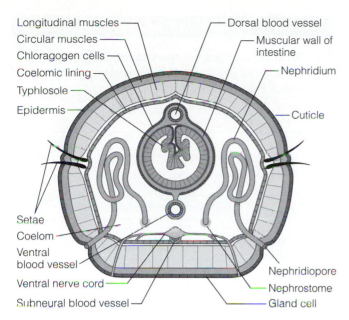

Figure 14.7 Earthworm cross section. Nephridial funnels (nephrostomes) open into the next anterior segment of the worm.

cuticle covering the epidermis (hypodermis). Just below the epidermis, note the circular muscle and below that, note the longitudinal muscle. The coelomic lining, or peritoneum, lies below the longitudinal muscle. Covering the dorsal vessel and the intestine are cells called chloragogen cells. These are involved with glycogen and fat synthesis and urea formation. Folding to the interior of the intestine is the typhlosole, which increases the surface area for secretion and absorption. Two lateral neural blood vessels and one subneural blood vessel are associated with the ventral nerve cord. Depending on the section studied you may also see a nephridium and a setum. Examine any available demonstrations of other oligochaetes. ▲

Class Hirudinea—Leeches

Characteristics: Body with 34 segments and many annuli; direct development; anterior and posterior suckers; setae absent; parapodia absent. Example: *Hirudo*.

Leeches are the most highly specialized annelids. They are predominantly freshwater, but a few are marine and terrestrial. They are predators with mouth parts usually adapted for fluid feeding, preying mostly on other invertebrates such as earthworms, slugs, and insect larvae. Those few leeches that feed on vertebrate blood have salivary secretions that prevent the clotting of blood.

▼ Examine any demonstration specimens available for study. Note that leeches lack parapodia and head appendages. They are dorsoventrally flattened and taper anteriorly. Anteriorly there is an anterior sucker, and posteriorly there is a posterior sucker. Leeches have 34 segments, but segments are difficult to distinguish externally because they have become secondarily divided. Several secondary divisions, called annuli, are in each true segment. Locate annuli on the specimen you are observing (fig. 14.8). ▲

(a)

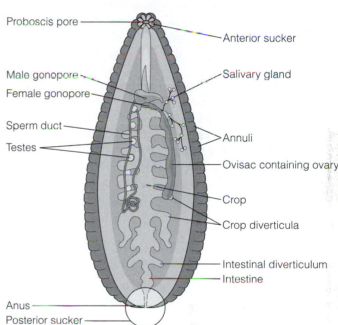

(b)

Figure 14.8 Leech structure. (a) *Hirudo medicinalis*, the medicinal leech, has been used for bloodletting. (b) Internal structure of a leech.

Leeches move by an undulating, swimming motion or by a looping action. These motions are made possible by more complex musculature than is present in other annelids. In addition, the coelom has lost its metameric partitioning; rather than using independent coelomic compartments in locomotion, the leech has a single hydrostatic cavity. After attaching the larger, posterior sucker to the substrate, circular muscles of the body wall contract to extend the leech anteriorly. The smaller anterior sucker that surrounds the mouth is attached to the substrate, and the posterior sucker is released. Longitudinal muscles then bring the posterior end of the leech anteriorly to repeat the sequence.

Leeches are monoecious, and sperm transfer and egg deposition usually occur in the same manner as described for *Lumbricus*. They possess a clitellum, which is not obvious during nonbreeding periods. A penis may aid in sperm transfer in some species. Other leeches may transfer sperm by expelling a packet of sperm (a spermatophore) from one leech through the integument and into the body cavity of another. Development is direct.

EVOLUTIONARY RELATIONSHIPS

▼ A cladogram showing a traditional interpretation of the evolutionary relationships among the annelid classes is shown in worksheet 14.1 (fig. 14.9). Examine the synapomorphies depicted on the cladogram. Based upon your studies in this week's laboratory, you should be able to determine the position of each of the classes studied. Complete the cladogram and answer the questions that follow it. ▲

Recent cladistic analysis of the phylum Annelida has caused some zoologists to doubt whether the interpretation shown in figure 14.9 is correct. Many zoologists now believe that the polychaetes were derived from a metameric ancestor independently of the oligochaetes and leeches. As suggested in worksheet 14.1, question 12, many zoologists also believe that the oligochaetes and leeches form a single clade. If these conclusions are true, the phylum Annelida is not monophyletic, and the name should be abandoned. Revisions in this taxonomy may appear in subsequent editions of this laboratory manual.

6. Why is a clitellum necessary in members of the classes Oligochaeta and Hirudinea but not in members of the class Polychaeta? _____

7. How does one determine dorsal and ventral surfaces of an earthworm? _____

8. What is the function of the pharynx of an earthworm? _____

9. Where are the seminal vesicles of an earthworm and what is their function? _____

10. What are the annuli of a leech? _____

KEY TERMS

clitellum   178

metamerism   176

nephridia   180

parapodia   178

septa   178

WORKSHEET 14.1   Annelida

Evolutionary Perspective

1. What are two hypotheses regarding the origin of the annelid-arthropod lineage?

2. Ancestral annelids were probably most similar to members of which class of modern annelids?

Annelida

3. How is metamerism useful in earthworm crawling?

4. Characterize members of the class Hirudinea.

5. Describe the path of blood flow through the earthworm's circulatory system.

6. How are sperm transferred during earthworm copulation?

7. Number from 5 to 1 the following layers of the earthworm body wall to reflect an outside-to-inside sequence.
   _____ circular muscle _____ longitudinal muscle _____ epidermis _____ peritoneum
   _____ cuticle

8. An earthworm nephridium that has its nephrostome in segment 20 would have its tubule in segment _____ .

9. What modifications of the body of a leech permit a looping type of locomotion?

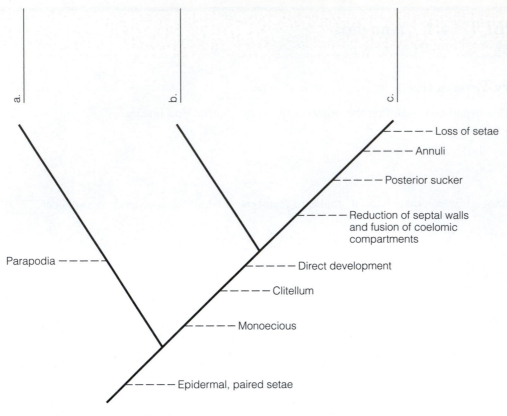

a.

b.

c.

Loss of setae

Annuli

Posterior sucker

Reduction of septal walls and fusion of coelomic compartments

Parapodia

Direct development

Clitellum

Monoecious

Epidermal, paired setae

Figure 14.9 Evolutionary relationships among the Annelida. Fill in the class names.

10. Name two synapomorphies that distinguish the oligochaetes and the leeches from the polychaetes.

11. Are there any synapormorphies that can be used to describe the oligochaetes? _____

12. The absence of oligochaete synapomorphies has suggested to some systematists that the classes Oligochaeta and Hirudinea should be combined into a single class. The name "Clitellata" has been proposed for this combined taxon. What synapomorphy of the clade does the name describe?

13. Name two synapomorphies that are used to distinguish leeches from all other annelids. _____ _____

Exercise 15

Arthropoda

Learning Objectives

After completing this exercise, you should be able to do the following:

1. Describe the evolutionary and ecological significance of the phylum Arthropoda.
2. Describe the evolutionary relationships among the three major arthropod subphyla.
3. Describe the significance of tagmatization in the arthropods.
4. Describe the significance of the arthropod exoskeleton.
5. Characterize the phylum Arthropoda.
6. Recognize class representatives.
7. Characterize the four major subphyla regarding body form and habitat.
8. Describe the structure and function of the organ systems studied in each class.

Prelaboratory Quiz

Study this week's laboratory exercise and then complete the following quiz to assess your preparation for the laboratory.

List the three arthropod subphyla containing living species and give an example of each.

Subphylum   **Example**
1. _____   _____
2. _____   _____
3. _____   _____

4. Which of the following word or phrase pairs is incorrectly associated?
   a. thorax/locomotion
   b. head/sensory and feeding functions
   c. abdomen/visceral functions
   d. chelicerae/sensory functions

5. Which of the following subphyla has members that are entirely extinct?
   a. Trilobitomorpha
   b. Chelicerata
   c. Crustacea
   d. Uniramia

6. All of the following classes belong within the subphylum Chelicerata EXCEPT
   a. Arachnida.
   b. Malacostraca.
   c. Pycnogonida.
   d. Merostomata.

7. Biramous appendages are found in the following subphylum or subphyla.
   a. Crustacea only
   b. Uniramia only
   c. Chelicerata and Crustacea
   d. Crustacea and Trilobitomorpha

8. Members of the subphylum Uniramia include all of the following EXCEPT
   a. insects.
   b. scorpions.
   c. millipedes.
   d. centipedes.

9. True/False   Millipedes have two pairs of appendages per apparent segment, while centipedes have one pair of appendages per segment.

10. True/False   In paurometabolous development, the number of molts between hatching and death is variable within a species.

11. True/False   In holometabolous development, a pupal stage precedes the last molt into an adult insect.

12. True/False   Flies, like the fruit fly, undergo paurometabolous development.

13. True/False   The appendages of a crayfish are said to be serially homologous because they are evolved from a similar basic structure.

14. True/False   Book lungs, book gills, and tracheal systems are all modifications of the exoskeleton of arthropods to facilitate gas exchange.

15. True/False   Compound eyes are found only in insects.

EVOLUTIONARY PERSPECTIVE

The evolutionary relationships of arthropods to other phyla were covered in exercise 14. Controversy regarding evolutionary relationships within the arthropods has existed for years. Traditionally, the arthropods have been considered a single phylum, which implies a common (monophyletic) origin for all of its members. This, however, is not universally agreed upon. Many zoologists believe that living arthropods are polyphyletic and should be divided into three phyla: Chelicerata, Crustacea, and Uniramia. The main point of contention between these groups of zoologists is whether or not the characteristics common to all arthropods could have originated independently three times in the annelid-arthropod lineage. Presently, this question cannot be answered for most arthropod characteristics. The living arthropods are treated in this laboratory manual as a single phylum containing three subphyla: Chelicerata, Crustacea, and Uniramia. One subphylum of extinct arthropods, Trilobitomorpha, is also briefly described.

THE TRIPLOBLASTIC COELOMATE ANIMALS—ARTHROPODA

The phylum Arthropoda is the largest phylum in the animal kingdom. More than three-fourths of all known species are members. Considering how many remain to be described, the true figure is probably even higher.

A characterization of the phylum revolves around two features of their organization: **metamerism** and the **exoskeleton.** Like the annelids, the arthropods are metameric, though they show a much higher degree of tagmatization. The specialization of body regions is usually associated with three functions: sensory and feeding functions **(head)**, locomotion **(thorax)**, and visceral functions **(abdomen)**. As we will see, there is considerable variation in degrees of tagmatization within the arthropod classes.

The exoskeleton is often cited as the major reason for arthropod success. It provides

1. structural support,
2. impermeable surfaces for the prevention of water loss,
3. a system of levers for muscle attachment and movement, and
4. protection from mechanical injury and environmental chemicals.

The exoskeleton necessitated a variety of adaptations that would allow the arthropod to live and grow within the confines of this rigid skeleton. These adaptations include

1. modifying the exoskeleton to provide for flexibility at joints;
2. modification of parts of the exoskeleton into highly developed sensory structures;
3. ecdysis, or molting, to allow for increased size; and
4. modification of the exoskeleton for gas exchange (tracheal system, book lungs, book gills).

A final characteristic of arthropods that has contributed to their success is reduced competition through **metamorphosis.** Larval forms are often very different in morphology and habits from the adults. This prevents larvae and adults from competing with one another for limited resources.

These adaptations have allowed the arthropods to fill virtually every conceivable habitat on the earth. Their impact on human history has been tremendous. Arthropods spread diseases and compete with humans for food. At the same time, they serve as food, produce useful products, and are essential for the pollination of about 65% of all plant species.

CLASSIFICATION

Subphylum Trilobitomorpha
Subphylum Chelicerata
    Class Merostomata
    Class Pycnogonida
    Class Arachnida
Subphylum Crustacea*
    Class Malacostraca
    Class Branchiopoda
    Class Cirrepedia
    Class Copepoda

*Classes listed in these subphyla are limited to those usually studied in introductory zoology laboratories.

Subphylum Uniramia*
    Class Diplopoda
    Class Chilopoda
    Class Hexapoda

*Classes listed in these subphyla are limited to those usually studied in introductory zoology laboratories.

Subphylum Trilobitomorpha

Characteristics: All extinct, from Cambrian to Carboniferous periods; body divided into three longitudinal lobes; head, thorax, and abdomen; biramous appendages; one pair of antennae.

Since trilobites are all extinct (they were probably ancestral to the arachnids and crustaceans), the only specimens available for study are fossilized.

▼ Examine any fossils available. Note the three lobes, head, (cephalon) compound eyes, thorax or trunk (the attached biramous appendages are ventral and will not be visible), and the tail-like pygidium (fig. 15.1). ▲

Subphylum Chelicerata

The chelicerate body is divided into two regions. A feeding, sensory, and locomotor tagma is called the prosoma. Visceral functions are carried out in the opisthosoma. The first pair of appendages, called chelicerae, may be pincerlike, or chelate, and are used for feeding and defense. The second pair of appendages are pedipalps and are often sensory. They may also be used in feeding, locomotion, or reproduction. Pedipalps may also be chelate.

Class Merostomata—Horseshoe Crabs

Characteristics: With book gills on the opisthosoma. Horseshoe crabs (Limulus) are marine, living along the Atlantic and Gulf coasts. Their body is very well adapted for pushing

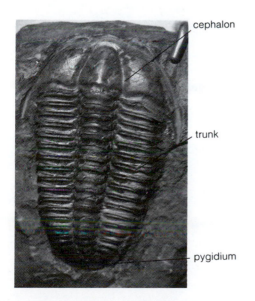

cephalon

trunk

pygidium

Figure 15.1 A trilobite fossil.

through sand and mud of coastal regions. In the process they scavenge for food. The hard exoskeleton provides excellent protection from all predators. Their body form apparently represents a nearly ideal combination of characters for the stable environment in which they live; the fossil record indicates that these arthropods have remained unchanged for over 250 million years.

▼ The head and the thorax are fused into one piece, the prosoma or cephalothorax. It is covered by a hard carapace. On the prosoma is a pair of compound eyes. Anteriorly and medially are two smaller simple eyes. The opisthosoma is located behind the prosoma. It has a series of spines around its margin and a long telson at its posterior aspect (figs. 15.2 and 15.3).

Note the appendages covering the ventral surface of the prosoma. The first pair of appendages is the chelicerae, used in food handling. Following them are the pedipalps and four pairs of walking legs. All but the last pair of appendages are chelate (pincerlike). The fourth pair of walking legs has leaflike plates at the terminal end and is used for locomotion and digging. Behind the legs is a series of plates. The anteriormost plate is the genital operculum, covering the genital pores. The series of plates posterior to the genital operculum are the book gills. In the living state, book gills are highly vascularized and serve for gas exchange. ▲

Class Pycnogonida—Sea Spiders

Characteristics: First pair of appendages modified to form chelicerae; four pairs of legs; one pair of pedipalps; lack antennae; body divided into prosoma and reduced abdomen; common in all oceans.

▼ Examine any specimens on demonstration. ▲

Class Arachnida—Spiders, Scorpions, Ticks, Mites, Harvestmen, and Others

Characteristics: First pair of appendages modified to form chelicerae; four pairs of legs; one pair of pedipalps; lack antennae; body divided into prosoma and opisthosoma. The Arachnida are primarily terrestrial and show a number of modifications for that life-style. Three modifications are particularly important. First, the book gill is modified into a book lung, an internal series of vascular lamellae used for gas exchange with the air. Second, appendages are modified for more efficient locomotion on land. Third, water conservation is enhanced by more efficient excretory structures (coxal glands and Malpighian tubules). The impermeable exoskeleton went a long way in preadapting the arthropods for life in the terrestrial environment.

The Garden Spider

▼ Examine a preserved garden spider. Note that the body is divided into two regions, the prosoma and opisthosoma. The two body regions are joined by a slender pedicel. Using a hand lens or dissecting microscope, note eight eyes on the

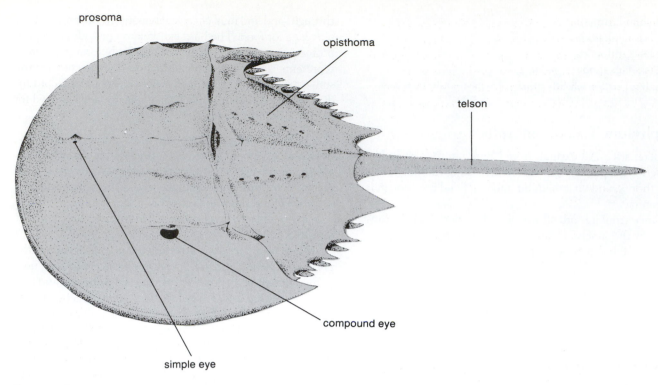

Figure 15.2 The horseshoe crab, dorsolateral view.

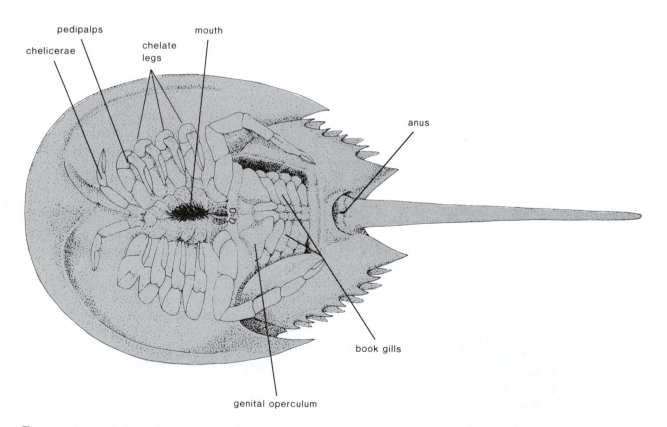

Figure 15.3 The horseshoe crab, ventral view.

anterior-dorsal surface of the prosoma. Ventrally, note six pairs of appendages. Anterior-most are the **chelicerae** that possess fangs used for injecting a paralyzing toxin into the prey. Posterior to the chelicerae are **pedipalps** that are used in manipulating food and, in the male, in sperm transfer. Posterior to the pedipalps are four pairs of **walking legs.** On the opisthosoma note the slitlike openings to the **book lungs** (anterior-ventral aspect). Book lungs are invaginations of the exoskeleton and are used in gas exchange. Spiders also possess a tracheal system similar to that subsequently described for insects. Note also three pairs of **spinnerets** used in producing silk (posterior- ventral aspect). Silk is secreted as a fluid and hardens on contact with air. ▲

Other Arachnids

▼ Examine other specimens on demonstration, including

harvestmen,
mites and ticks,
scorpions, and
other spiders. ▲

S T O P   A N D   A S K   Y O U R S E L F

1. Describe the specializations found in the three regions of the body of arthropods.

   _____

   _____

   _____

2. How has metamorphosis contributed to the success of arthropods? _____

   _____

3. How is the name "trilobite" descriptive of members of this subphylum? _____

   _____

4. What are chelicerae? _____

   _____

5. What are book lungs? _____

   _____

Subphylum Crustacea

With the exception of some members of one order (Isopoda), crustaceans are aquatic. Most are marine, although many are freshwater. They are readily recognized by their two pairs of antennae (all other arthropods have one pair or none) and their series of biramous appendages. The **biramous appendage** of a generalized crustacean consists of a basal series of segments called the **protopodite** with two rami attached distally. The medial ramus is the **endopodite,** and the lateral ramus is the **exopodite.** Originally there was one pair of appendages on each segment, with all appendages showing similar structure. With the specialization of body regions (tagmatization), appendages became modified for functions associated with the region of the body to which they were attached. Thus

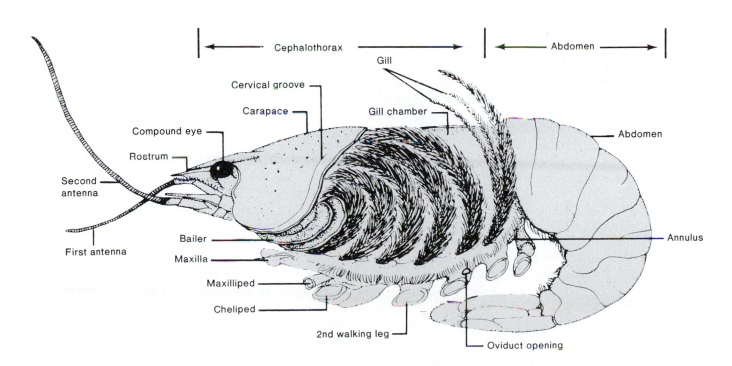

Figure 15.4 A crayfish with a portion of lateral carapace removed to show gills.

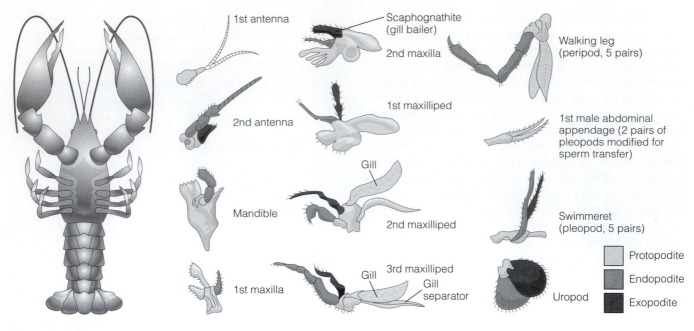

Figure 15.5 Ventral view of a crayfish showing dissected appendages.

head appendages are now involved with sensory and feeding functions, and thoracic appendages are involved with locomotion and defense. The specialization in function is reflected in specialization in structure. This will be obvious during the crayfish observations. Because all appendages are presumably evolved from a similar basic structure, they are said to be **serially homologous.**

Class Malacostraca—Crayfish, Lobsters, and Crabs

Characteristics: Appendages modified for crawling along the substrate, or the abdomen and body appendages are used for swimming.

▼ The crayfish is a common inhabitant of freshwater lakes and streams all over the world (Plate 6). Obtain a preserved specimen. The exoskeleton is made up of lipids, proteins, and chitin (a polysaccharide). Hardening results from the deposition of calcium carbonate.

Note that the body is divided into two major regions, the **cephalothorax** and the **abdomen.** The cephalothorax is covered by an arching **carapace** (see fig. 15.4). Anteriorly, note the stalked **compound eyes** attached just below a pointed extension of the carapace, the **rostrum.** Observe the surface of the compound eye under a dissecting microscope. Note the many facets that are lenses for individual photoreceptors, or **ommatidia.** The compound eye is especially adapted for detecting movement. Note also the short **first antennae** and longer **second antennae.** The sides of the carapace cover the **gill chamber.** Gently lift the carapace at the base of the legs and note the gills underneath. You will remove the portion of the carapace covering the gills on the left side of the crayfish (Plate 7). Carefully insert the tip of

your scissors just below the carapace at the dorsolateral margin where the cephalothorax and abdomen meet. Cut anteriorly, being careful not to disturb the underlying gills. When you reach the groove that represents the point of fusion of the head and thorax, turn laterally and ventrally, continuing your cut until the lateral portion of the carapace can be lifted off. Note the gills attached to the base of the legs. Medial to the gills is the body wall. Water is circulated through the gill chamber by a process on the second maxilla. This process will be observed later.

The abdomen consists of six segments. Inside, powerful **flexor muscles** are responsible for the escape response of the crayfish. When alarmed, the crayfish thrusts the tip of its abdomen ventrally and anteriorly. These powerful, rapid strokes propel the crayfish posteriorly through the water. This can be demonstrated if living specimens are available. At the tip of the abdomen is a flat process, the **telson,** which bears the **anus** ventrally.

The appendages of the crayfish are biramous (fig. 15.5). Recall the terminology associated with the generalized appendage. Beginning at the posterior tip of the abdomen, examine the **uropods** (see Plate 6). They arise from the last abdominal segment on either side of the telson. Notice the two leaflike segments (exopodite and endopodite) and the single basal segment (protopodite) to which they attach. The remaining abdominal segments have **pleopods** (swimmerets). These are typical biramous appendages. Note the three regions. In the male, the two pairs of pleopods closest to the cephalothorax are modified for sperm transfer. They are enlarged, hardened, and lie forward between the bases of the last walking legs. In the female, the anterior two pleopods are reduced in size, and all pleopods are used in drawing water across the eggs that develop attached to the ventral surface of the abdomen.

The cephalothorax bears walking legs, mouthparts, and antennae. These are best studied if removed. To remove an appendage, grasp it with forceps near its base. Loosen the appendage by probing at its attachment to the body wall with an insect pin or teasing needle. Then, with the forceps, manipulate the base of the appendage back and forth, pulling gently.

Working on the left side of the crayfish, and beginning at the posterior end of the cephalothorax, remove the next five appendages (see Plate 7). These are **pereiopods** (walking legs). In all of them the exopodite is lost. They are, therefore, secondarily uniramous. Note that the anterior three appendages are **chelate** (equipped with an opposable segment used in grasping). The anterior-most pair is greatly enlarged into defensive and offensive **chelae.** Note that, with some of the pereiopods, gills are attached at the basal segment and were removed with the leg. With others, the gills remain attached to the body wall. Observe the **genital pores** located at the basal segment of the fifth (male) or third (female) walking legs.

Five pairs of appendages are associated with the mouth. They have sensory, food-handling, and grinding functions. The posterior three pairs of mouth appendages are **maxillipeds.** Remove them posterior to anterior. Be sure to get the entire appendage. Note that the second and third bear a gill, but the first does not. Next are two pairs of **maxillae.** On the second maxilla, note that the exopodite bears a long, flat blade: the **gill bailer.** It functions in drawing water currents over the gills. The most anterior pair of mouth appendages are the **mandibles.** These are the primary organs of mastication. They are heavy, triangular structures bearing teeth along the inner margins. Probe around the base and carefully remove the mandible.

Look again at the first and second antennae. Both flagella of the first antenna derive from the exopodite. The long flagellum of the second antenna is the endopodite of that appendage, and a triangular plate just below the eye is the exopodite. At the base of the first antenna is the **statocyst,** an organ of equilibrium and balance. Remove the first antenna and slit open its base. Note the fine setae (hairs). A sand grain, the **statolith,** within the cavity of the statocyst, stimulates these hairs in response to the pull of gravity. At the base of the second antenna is a ventral opening called the **nephridiopore.** It is the opening of the excretory system.

Internal Structure

Remove the carapace covering the dorsal cephalothorax by cutting one centimeter to the right of midsagittal. Extend the cut up to the rostum, then to the left, ending under the left compound eye. Keep the inner blade of your scissors close to the body wall at all times. Lift the carapace carefully, teasing it free from underlying tissue. As you gently lift, look underneath the carapace. About one centimeter posterior to the left compound eye you will see a large **mandibular muscle** that will need to be teased free of the carapace. Remove the portion of the body wall medial to the gills by cutting with scissors near the points of attachment of the legs (fig. 15.6).

Muscular System The mandibular muscles noted previously insert on and operate the mandibles. Just medial to the mandibular muscles, note the **gastric muscles** associated with stomach movements. Laterally, a band of muscles originates that runs posteriorly and dorsally, and eventually disappears under the dorsal aspect of the abdomen. These are **extensor muscles** of the abdomen. When they contract, they bring the abdomen back into line with the cephalothorax. They oppose the powerful **flexor muscles** of the abdomen. To see the remaining portion of the extensor and flexor muscles, remove the dorsal portion of the abdomen by cutting along each side of the abdomen to the base of each uropod. Lift the dorsal portion of the abdomen carefully, freeing it from underlying tissue. The extensor muscles attach to each abdominal segment and will tend to come up with the exoskeleton. Free them with a teasing needle as you proceed. After locating extensor muscles, note the large flexor muscles nearly filling the remainder of the abdomen. Alternate contraction of flexor muscles and extensor muscles cause the abdomen to be alternately flexed and extended, which propels a crayfish posteriorly in its escape response.

Circulatory System The **heart** is located in the posterior, dorsal third of the cephalothorax. It may be injected with red latex. The circulatory system of arthropods is open. What other phylum have we studied in which some members have an open circulatory system? _____
Blood is pumped from the heart into vessels and then released into large tissue spaces called sinuses. After leaving these sinuses, blood flows to the gills, where it is oxygenated. Blood combines with the copper-containing pigment hemocyanin. Hemocyanin is dissolved in the plasma rather than being contained in cells. Carbon dioxide diffuses out of the blood at the gills. Surrounding the heart is a sac, the **pericardium.** It encloses the heart in a **pericardial sinus.** Blood from the gills is released into the pericardial sinus and then enters the heart through small openings, **ostia.** Remove the pericardium and any latex covering the heart. We will not attempt to trace the delicate vessels associated with the heart, although they can be seen in figure 15.6.

Reproductive System Paired **gonads** are just ventral to, and extend posterior to, the heart. Gonads are cream-colored and have a somewhat more uniform texture than the **digestive gland,** which is just anterior to them. In the male, the sperm duct from each testis extends back to the genital opening at the base of the fifth pereiopod. In the female, the ovarian duct from each ovary leads to the genital opening at the base of the third pereiopod.

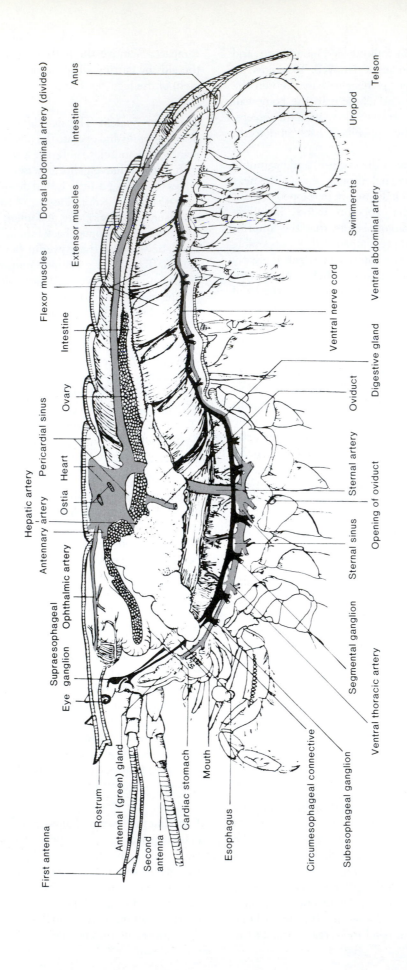

First antenna

Rostrum

Antennal (green) gland

Second antenna

Cardiac stomach

Mouth

Esophagus

Circumesophageal connective

Subesophageal ganglion

Eye ganglion

Supraesophageal ganglion

Ophthalmic artery

Antennary artery

Hepatic artery

Ostia

Heart

Pericardial sinus

Ovary

Intestine

Flexor muscles

Extensor muscles

Dorsal abdominal artery (divides)

Intestine

Anus

Telson

Uropod

Swimmerets

Ventral abdominal artery

Ventral nerve cord

Digestive gland

Oviduct

Sternal artery

Opening of oviduct

Sternal sinus

Segmental ganglion

Ventral thoracic artery

Figure 15.6 Crayfish, internal anatomy of a female.

192

Digestive System The mouth leads to a short **esophagus** and then to the **stomach.** The stomach attaches just behind the rostrum anteriorly and by the gastric muscles posteriorly. By pushing gently on the lateral margin of the stomach, one can see the esophagus extending down to the mouth. Cut open the stomach middorsally. Inside, note the **gastric mill.** It consists of a set of teeth, two lateral and one dorsal, used in grinding food. These are modifications of the exoskeleton, which lines the anterior portion of the digestive tract. On the stomach wall you may also find calcareous **gastroliths.** These are stores of calcium carbonate. Prior to ecdysis, calcium carbonate is withdrawn into the blood from the old exoskeleton and stored at the stomach walls. After ecdysis, the calcium carbonate is reused in hardening the new exoskeleton. The large **digestive gland** noted earlier secretes digestive enzymes into the stomach. The **intestine** passes under the heart, then posteriorly just under the extensor muscles. Trace it to the telson. Locate the anus ventrally on the telson.

Excretory System Excretory organs of the crayfish are called **antennal glands,** or **green glands.** They are the dark-colored glands lying against the inner body wall at the base of the second antenna. Recall your observations of the nephridiopore on the ventral surface of the basal segment of the second antenna.

Nervous System In removing the pereiopods at their base, you have exposed the ventral nerve cord in the thoracic region. Looking from the left side, follow the nerve cord posteriorly. Follow it into the abdomen by freeing flexor muscles from the lateral and ventral body wall. Separate the nerve cord carefully from the muscle. Note the **segmental ganglia** (swellings) and the lateral branches that come off each ganglion. Note that you can now see the **ventral abdominal artery** running below the nerve cord. Return to the thoracic region and follow the ventral nerve cord anteriorly. Just posterior to the esophagus is a larger ganglion, the **subesophageal ganglion.** From it, two circumesophageal connectives go around either side of the esophagus and end at the **supraesophageal ganglion** (loosely called the brain). The supraesophageal ganglion can be seen inside the anterior head wall. The segmental ganglia are integrative areas for sensory and motor functions in their respective regions. The supraesophageal and subesophageal ganglia result from the fusion of multiple-head ganglia and are important in the integration of sensory and motor functions of the mouthparts, antennae, compound eyes, statocysts, and so forth. Recall your earlier observations of these structures.
Observe demonstrations of other malacostracans. ▲

Other Crustaceans

▼ Observe demonstrations of other crustaceans, including the following:

Class Branchiopoda—fairy shrimp, water fleas, and brine shrimp. Note flattened, leaflike appendages used in filter feeding, respiration, and often locomotion.

Class Cirrepedia—barnacles. In barnacles, a planktonic larval stage is followed by settling to the substrate and metamorphosis to a sessile adult. Observe any living or preserved specimens. Can you see why they were thought to be molluscs by some early zoologists?

Class Copepoda—copepods. Copepods are among the most abundant animals in the world. They comprise an important component of marine food webs. ▲

STOP AND ASK YOURSELF

6. What are biramous appendages? _____
   _____

7. The first pair of biramous appendages of the crayfish are the _____.

8. Where are the gill chambers of a crayfish? _____
   _____

9. What sensory receptors of the crayfish are responsible for the sense of equilibrium and balance? _____

10. After the blood of the crayfish leaves the gills, it flows to the _____.
    _____

11. What is a gastrolith? _____
    _____

Subphylum Uniramia

Uniramians are distinguished from other arthropods by having one pair each of antennae, mandibles, and uniramous appendages.

Class Diplopoda—Millipedes

Millipedes are characterized by having one pair of antennae, two pairs of short legs per apparent segment, and a body that is round in cross section. Millipedes live on decaying plant and animal material and thus are very important in the decomposition of forest-floor litter. They are widely distributed terrestrial animals.

▼ Examine any demonstrations available. ▲

Class Chilopoda—Centipedes

Centipedes are active predators, feeding on a variety of other invertebrates. They have one pair of antennae, one pair of relatively long legs per segment, and their body is oval in cross section. The appendages of the first segment of the body are formed into poison claws that are used in feeding. Most are harmless to humans. Centipedes are often found beneath fallen trees on the forest floor. One group is common in human dwellings in North America.

▼ Examine any demonstrations available. ▲

Class Hexapoda—The Insects

Characteristics: Three pairs of legs; one pair of antennae; often winged (usually two pairs); body with head, thorax, and abdomen; dioecious; mouthparts variously adapted.

If the success of any animal group is measured by numbers of species and ecological adaptability, the hexapods, or insects, are the most successful eukaryotes on this planet. With the exception of the marine environment, they have successfully adapted to nearly every habitat the earth has to offer. As with all arthropods, the exoskeleton is the primary reason for their success. Adaptations to the terrestrial environment include the incorporation of waxes into the exoskeleton, which waterproofs the exoskeleton and helps prevent desiccation. Many insects can thus live with little or no external source of water. In addition, insects are the only nonvertebrates to exploit air; the exoskeleton is modified to form wings without sacrificing any appendages.

In spite of their success, the insect body plan is remarkably constant. There are three tagmata: **head, thorax,** and **abdomen.** On each of the three segments of the thorax is a pair of appendages (hence the class name, Hexapoda). Generally, two pairs of wings are carried on the second and third thoracic segments. The head bears one pair of compound eyes; zero, two, or three ocelli; one pair of antennae; and a series of mouthparts, three pairs of which derive from appendages associated with the primitive metameres of the head. In some orders, mouthparts are specialized for various feeding habits but are still built upon the same basic plan. The abdomen lacks walking legs and is associated with visceral functions (reproduction, digestion, and excretion).

The Grasshopper and Cockroach

Traditionally, in introductory courses the grasshopper is used to study insect structure. Your instructor may ask you to use a cockroach in addition to the grasshopper inasmuch as the cockroach is less specialized. The following discussion can be applied to either.

External Structure ▼ On your specimen, note the division of the body into three tagmata: head, thorax, and abdomen (fig. 15.7). The exoskeleton of the insect is very similar to that of the crustacean. Note the hardened plates (sclerites) with membranous areas between them. The hardening of the cuticle of the insect is different from that of the crustacean. Hardening is the result of polymerization of protein (it is said to be sclerotized), rather than the deposition of calcium salts.

On the head, note the single pair of **antennae,** the **compound eyes,** and the **mouthparts** (fig. 15.8). Using a dissecting microscope, examine the compound eyes closely. Note the many facets that make up each eye. Examine the

area between the compound eyes for **ocelli.** As with the crayfish, the compound eyes are adapted to detect movement. The ocelli are used to detect changes in light intensity and do not form images.

Examine the mouthparts using a dissecting microscope. We will remove them as we did with the crayfish appendages. As with the crayfish, the mouthparts derive from segmental appendages. The lower lip is the **labium.** It represents the fusion of one pair of appendages. Note the **labial palps** on each side. The labium is primarily sensory. Remove the labium. Just above the labium is a pair of **maxillae.** Note the palp on each maxilla. Maxillae are cutting and sensory structures. Remove them and note the cutting surface on the inner margin. Above the maxillae are the primary grinding structures, the **mandibles.** They are highly sclerotized and difficult to remove. You will need to probe around the base of one mandible with a teasing needle to loosen it. Note the toothed margin. The upper lip is the **labrum.** It is sensory and not derived from paired appendages. Inside the mouth cavity a tonguelike structure, the **hypopharynx,** can now be seen. It is also sensory.

Examine the thorax (see fig. 15.7). It consists of three segments devoted primarily to locomotion. The **prothorax** bears the first pair of legs. The second segment, the **mesothorax,** bears the second pair of legs and the first pair of wings. In both the grasshopper and the cockroach, the first pair of wings is narrow and leathery. The third segment, the **metathorax,** bears the third pair of legs and the second pair of wings. These wings are broad and membranous. The legs of an insect are divided into five main segments joined together by membranous joints (fig. 15.7). The **coxa** is a small segment that attaches at the ventrolateral margin of the body wall. The next three segments are the **trochanter, femur,** and **tibia.** The last segment is the **tarsus.** It is usually subdivided into smaller tarsal segments and bears terminal claws (two in the grasshopper and cockroach). Between the claws of the grasshopper is a fleshy pad, the **arolium,** which is used to help grip smooth surfaces. The legs of insects are various modified with spines, setae, and ridges that may be used for specialized functions (e.g., gripping surfaces, swimming, and making sound). In grasshoppers the metathoracic leg is enlarged and muscular and is used for jumping. In male *Romalea* grasshoppers, the inner surface of the metathoracic leg bears a row of small spines that is rubbed against the lower edge of the front wing to produce sounds used in mating. Other insects produce sounds by rubbing wings together or other legs against wings. This rubbing form of sound production is called **stridulation.**

The abdomen of insects usually consists of 10 or 11 segments. All except the terminal segments are free of appendages. Note the openings on the lateral surface of each abdominal segment. These openings are called **spiracles.** They are the external openings of the insect's gas exchange system. Spiracles lead to chitin-lined tubes called tracheae, which branch and rebranch throughout the insect's body. A tympanum is located laterally on each side of the first abdominal

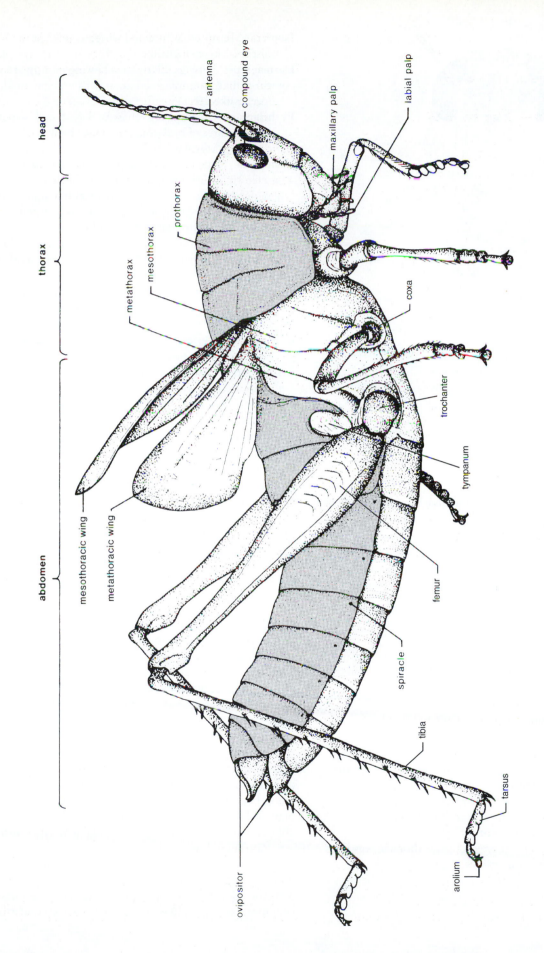

Figure 15.7 Grasshopper, lateral view of a female *Romalea*.

head

thorax

abdomen

antenna

compound eye

maxillary palp

labial palp

prothorax

mesothorax

metathorax

coxa

trochanter

tympanum

femur

spiracle

tibia

tarsus

arolium

ovipositor

mesothoracic wing

metathoracic wing

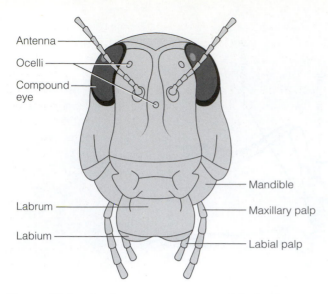

Antenna

Ocelli

Compound eye

Labrum

Labium

Mandible

Maxillary palp

Labial palp

Figure 15.8 Grasshopper, anterior view of the head.

segment. It is the auditory organ of the grasshopper. The terminal segments often bear sensory projections called **cerci** (sing. cercus). Both the cockroach and grasshopper show these structures. An ovipositor is frequently present in females. The female grasshopper has an ovipositor consisting of pairs of lateral pointed plates at the tip of the abdomen.

The Insect Orders

Examine demonstrations showing representatives of the following major insect orders. Note the characteristics that distinguish each. The mouthparts of insects have been especially important in their success. Note where mouthparts have been adapted for different feeding methods (e.g., piercing, sucking, lapping). ▲

Thysanura Bristletails. Wingless; tapering abdomen; scales on the body; terminal tail filaments; long antennae.

Ephemeroptera Mayflies. Elongate; abdomen with two or three tail filaments; wings held above the body at rest; short antennae.

Odonata Dragonflies and damselflies. Elongate; membranous wings; abdomen long and slender; compound eyes occupying most of the head.

Phasmida Walking sticks, leaf insects. Body elongate and sticklike; wings reduced or absent; some tropical forms are flattened and leaflike.

Orthoptera Grasshoppers, crickets. Forewing long, narrow, and leathery; hind wing broad and membranous; cerci present.

Mantodea Mantids. Prothorax long; prothoracic legs long and armed with strong spines for grasping prey.

Blattaria Cockroaches. Body oval and flattened; head concealed from above by a shieldlike extension of the prothorax.

Isoptera Termites. White and wingless (except in dispersal stages); antlike.

Dermaptera Earwigs. Elongate; chewing mouthparts; threadlike antennae; abdomen with unsegmented, forcepslike cerci.

Phthiraptera Sucking and chewing lice. Small, wingless ectoparasites of birds and mammals; body dorsoventrally flattened.

Hemiptera True bugs. Proximal portion of forewing sclerotized, distal portion membranous; sucking mouthparts hang ventrally off anterior margin of head.

Homoptera Cicadas, leaf hoppers, aphids, whiteflies. Wings entirely membranous; sucking mouthparts appear to hang ventrally off posterior margin of head.

Thysanoptera Thrips. Small bodied; sucking mouthparts; wings narrow and fringed with long setae; plant pests.

Neuroptera Lacewings, snakeflies, ant lions. Wings membranous and net veined; hind wings held rooflike over body at rest; antennae long and threadlike.

Coleoptera Beetles. Forewings sclerotized, forming cover over abdomen (elytra).

Trichoptera Caddisflies. Mothlike with setae-covered antennae; chewing mouthparts; wings covered with setae and held rooflike over abdomen at rest; larvae aquatic and often dwell in cases that they construct.

Lepidoptera Moths and butterflies. Wings broad and covered with scales; mouthparts formed into a sucking tube.

Diptera Flies. Mesothoracic wings well developed, metathoracic wings reduced to knoblike halteres; mouthparts various modified, never chewing.

Siphonaptera Fleas. Laterally flattened; sucking mouthparts; jumping legs; parasites of birds and mammals.

Hymenoptera Ants, bees, and wasps. Anterior portion of abdomen constricted into a threadlike "waist"; wings (when present) membranous.

Insect Development

Evolution of insects has resulted in the divergence of immature and adult body forms and habitats. If immature and adult insects do not compete for the same food resources and the same space, greater numbers of individuals can be supported in an area. Developmental patterns of insects reflect degrees of divergence between immatures and adults and are classified into three (four by some zoologists) categories.

Ametabolous development occurs in the Thysanura and other orders. The primary differences between adults and immatures are body size and degree of sexual development. Both adults and immatures are wingless. The number of molts in the ametabolous development of a species is variable, and molting continues after sexual maturity is reached.

Paurometabolous development involves a species-specific number of molts between egg and adult stages. The immature stages (**instars**) gradually take on the adult form. This includes the development of external wing pads (except in those insects, like the lice, that have secondarily lost wings), the attainment of adult body size and proportions, and the development of genitalia. No further molts normally occur once adulthood is reached. Except for their smaller size, sexual immaturity, and incompletely developed wings, immatures (called **nymphs**) usually resemble adults. Those insects listed on page 196 between, but not including, the Thysanura and the Neuroptera are paurometabolous (with the possible exception, as explained in the following, of the Ephemeroptera and Odonata). Some authors use an additional classification for those insects that have a gradual series of changes in their development, but whose immatures possess a body form different from the adult. These include some of the insects with aquatic immature stages (e.g., Ephemeroptera and Odonata). These **hemimetabolous** insects have immatures that are called **naiads.** They frequently have long developmental periods (up to two years in some Odonata) and short-lived adult stages (a few hours to days in some Ephemeroptera).

Holometabolous development occurs in the Neuroptera through the Hymenoptera in the list on page 196. Eggs hatch into **larvae** that are very different from the adult in body form, behavior, and habitat. Larvae molt through a species-specific number of larval instars. The last larval molt forms the **pupal stage.** The pupal stage is often visualized as a period of apparent inactivity, but is actually a time of radical cellular change. The pupal stage may be enclosed in a protective case (cocoon) or by the last larval exoskeleton (chrysalis, puparium). The **adult,** which is usually winged, emerges after the pupal stage.

Paurometabolous (Gradual) Metamorphosis ▼ You will observe paurometabolous development of the hemipteran *Oncopeltus fasciatus* (milkweed bug). Your instructor will provide you with some eggs in a cotton substrate. Preserve two to five eggs in a vial of 70% ethanol. Place the remaining eggs into a container; provide the container with raw, shelled sunflower seeds and water (as directed by your instructor). Cover tightly with a double layer of cheesecloth. Watch your culture over the next eight weeks. An approximate schedule of molts is outlined in table 15.1 in worksheet 15.1, and a similar sequence of molts is shown in figure 15.9. As the various instars appear, record the date in table 15.1 and remove and preserve one representative of each stage. At the end of the developmental cycle, you will have eggs, five nymphal instars, and the adult. Be prepared to hand in your vial when the exercise is completed.

Holometabolous (Complete) Metamorphosis You will use the fruit fly, *Drosophila melanogaster*, to illustrate holometabolous metamorphosis. Figure 15.10 illustrates holometabolous metamorphosis of the housefly (*Musca domestica*). Your instructor will give you a container of medium and provide you with etherized flies. Place the flies into the container and cover. Eggs and first and second instars will be found in the medium. Third instars and pupae will be found on the sides of the container. The life cycle is relatively short (two weeks from egg to adult), and larval instars proceed rapidly. You will need to check your cultures frequently (as indicated in table 15.2 in worksheet 15.1). Because eggs are small and the first two instars crawl through the medium, these stages are difficult to find. As the life cycle proceeds, you may see larvae next to the side of your vial. You will certainly see tracks the larvae make while crawling through the medium. Use a dissecting microscope to help with these observations. Because egg and larval stages are difficult to find, you will not be asked to hand in a representative of each stage. ▲

EVOLUTIONARY RELATIONSHIPS

▼ Cladograms showing two interpretations of the evolutionary relationships among the arthropod subphyla are shown in worksheet 15.2 (fig. 15.11). Examine the synapomorphies depicted on the cladograms. Based upon your studies in this week's laboratory, you should be able to determine the position of each of the subphyla studied. Complete the cladogram and answer the questions that follow it. ▲

STOP AND ASK YOURSELF

12. How does one distinguish between members of the classes Chilopoda and Diplopoda? _____
_____

13. What are the characteristics of members of the class Hexapoda? _____
_____

14. Which segments of the insect thorax may bear wings? _____
_____

15. What stages occur in the life cycle of a holometabolous insect? _____
_____

16. What are ocelli? _____
_____

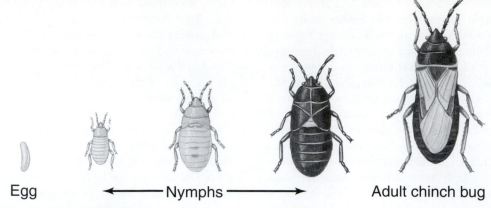

Egg ⟵ Nymphs ⟶ Adult chinch bug

Figure 15.9   Life cycle of the chinch bug, *Blissus leucopterus*.

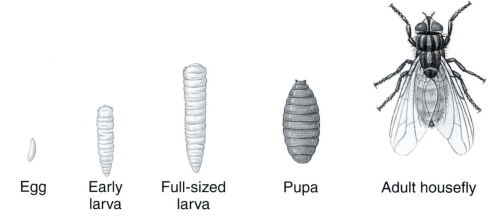

Egg   Early   Full-sized   Pupa   Adult housefly
       larva   larva

Figure 15.10   Life cycle of the house fly, *Musca domestica*.

KEY TERMS

WORKSHEET 15.1   Insect Development

TABLE 15.1   Development of *Oncopeltus fasciatus*

Week Number/ Date	Instar/Body Form	Description
1–2 _____	Egg	Light yellow early, bright orange prior to hatching.
2–3 _____	First Instar	Small, orange, no wing pads obvious.
3–4 _____	Second Instar	Larger with four wing pad spots on dorsal thorax.
4–5 _____	Third Instar	Larger with wing pads just reaching metathoracic legs.
5–6 _____	Fourth Instar	Longer with wing pads just beyond metathoracic legs. Two dorsal black spots at tip of abdomen.
6–8 _____	Fifth Instar	Wing pads reaching well beyond metathoracic legs. Three black spots on dorsal surface of posterior abdomen.
8–9 _____	Adult	Wings fully developed.

TABLE 15.2   Development of *Drosophila melanogaster*

Day/Date	Instar/Body Form	Description
1–2 _____	Egg	Small, white, with two fingerlike projections at one end.
2–4 _____	First Instar	Small maggot in media.
4–5 _____	Second Instar	Larger maggot in media.
5–7 _____	Third Instar	Larger maggot, found on sides of vial.
7–14 _____	Pupa	Attached to side of vial. Turns from light to dark as development proceeds.
14–16 _____	Adult	Wings and adult body form fully developed.

WORKSHEET 15.2 Arthropoda

Evolutionary Perspective

1. What are two ideas regarding the phylogeny within the phylum Arthropoda?

Arthropoda

2. Why is the exoskeleton often cited as the major reason for arthropod success?

3. What animals are included within the subphylum Chelicerata?

4. Explain what is meant by the concept of serial homology. Give an example of where it occurs.

5. The circulatory system of arthropods is an open system. What does that mean?

6. The last larval molt in holometabolous metamorphosis produces a stage called the _____.

7. How could one distinguish an adult paurometabolous insect from one of the nymphal instars?

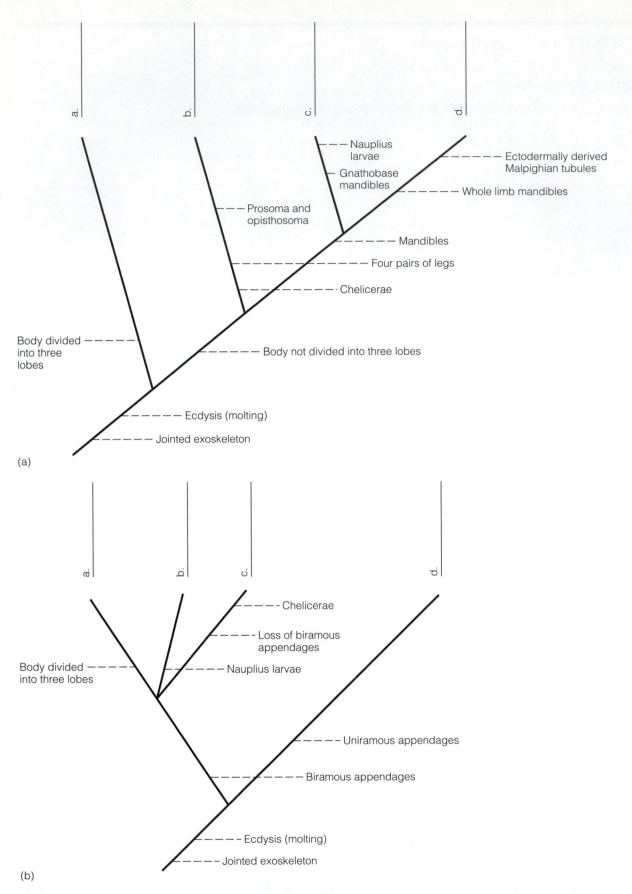

(a)

(b)

Figure 15.11 Two interpretations of evolutionary relationships among the Arthropoda. Fill in the subphylum names.

8. Name one symplesiomorphic character for the phylum Arthropoda.

9. According to interpretation (a), in figure 15.11, are the mandibles of crustaceans homologous to the mandibles of uniramians? (That is, do the mandibles in both groups share a common ancestry?)
_____ According to interpretation (b), in figure 15.11, are the mandibles of crustaceans homologous to the mandibles of uniramians? _____

10. Interpretation (b) implies that biramous appendages are a synapomorphic character of the three subphyla
_____, _____, and _____.

11. According to interpretation (a), what can be said regarding the relationship between the biramous appendages of the trilobites and the crustaceans? _____

_____

12. Summarize the importance of the trilobites in the phylogeny of the arthropods as depicted in both cladograms.

13. Summarize the basic differences in interpretation of arthropod phylogeny presented in these cladograms.

Exercise 16

Echinodermata

Learning Objectives

After completing this exercise, you should be able to do the following:

1. Describe the evolutionary significance of the Echinodermata.
2. Characterize the phylum Echinodermata.
3. Identify representatives of the echinoderm classes.
4. Describe the following:
   a. the water-vascular system,
   b. pentaradial symmetry, and
   c. the calcareous endoskeleton.
5. Recognize and describe the function of structures studied in the class representatives.

Prelaboratory Quiz

Study this week's laboratory exercise and then complete the following quiz to assess your preparation for the laboratory.

1. The echinoderms are most closely related to the
   a. Annelida.
   b. Chordata.
   c. Arthropoda.
   d. Mollusca.

2. The echinoderms possess
   a. biradial symmetry.
   b. bilateral symmetry.
   c. pentaradial symmetry.
   d. asymmetry.

3. All of the following are characteristics of the echinoderms EXCEPT
   a. water-vascular system.
   b. calcium carbonate endoskeleton.
   c. oral and aboral surfaces.
   d. cephalization.

4. Members of the class Crinoidea include the
   a. sea stars.
   b. brittle stars.
   c. sea urchins.
   d. sea lilies.

5. Members of the class Opiuroidea include the
   a. basket stars.
   b. sea stars.
   c. sea cucumbers.
   d. sea urchins.

6. True/False   Members of the class Holothuroidea include the sea cucumbers.

7. True/False   A chewing apparatus called Aristotle's lantern is found in members of the class Ophiuroidea.

8. True/False   Pyloric ceca are digestive glands found in the arms of members of the class Asteroidea.

9. True/False   In the water-vascular system of echinoderms, tube feet attach to the lateral canal by way of the ring canal.

10. True/False   The madreporite of sea stars is the external opening of the water-vascular system.

EVOLUTIONARY PERSPECTIVE

Most zoologists believe that echinoderms share a common ancestry with the chordates. Much of the evidence for these evolutionary ties comes from similarities in the early embryological development of members of these phyla. These shared developmental characteristics are the basis for grouping these phyla (as well as a few other smaller phyla) together as **deuterostomes.** Unfortunately, no fossils have been discovered that document a common ancestor for these phyla or explain how the deuterostomate lineage derived from ancestral diploblastic or triploblastic stocks.

Adult echinoderms have a form of radial symmetry called pentaradial symmetry (see the following discussion). It is generally accepted, however, that echinoderms evolved from bilaterally symmetrical ancestors. Evidence for this comes from the observation that echinoderm larval stages are bilaterally symmetrical. In addition, certain extinct echinoderms, called carpoids, were bilaterally symmetrical or asymmetrical.

THE TRIPLOBLASTIC COELOMATE ANIMALS—ECHINODERMATA

The echinoderms share many embryological features with the chordates and thus are believed to be the major non-chordate phylum most closely related to the chordates. Based on the morphology of living forms, however, they are very different from any other phylum.

The echinoderms are marine and include sea stars, sea urchins, sea cucumbers, and sea lilies. All echinoderms show **pentaradial symmetry** arranged around an oral-aboral axis. Radial symmetry was described earlier as it applied to the Cnidaria. In the echinoderms, body parts are arranged in "fives" around the oral-aboral axis. Radial symmetry is an adaptation to sedentary life-styles. Why would this be so?

_____

_____

_____

Echinoderms are dioecious with external fertilization. In development, a bilateral larva metamorphoses into the radial adult. This is evidence that the echinoderm pentaradial symmetry is a secondary adaptation and that echinoderms are descended from bilateral ancestors.

Two other functionally related characteristics of the echinoderms are a **water-vascular system** and a **calcium carbonate endoskeleton** made of numerous interlocking os-

sicles. The ossicles frequently protrude through the thin body wall, hence the name Echinodermata (spiny skin). The ossicles may form a rigid test (as in the sea urchin) or may be reduced to microscopic plates (as in the sea cucumber). The water-vascular system is a series of water-filled canals to which tube feet attach. Tube feet protrude through the calcareous skeleton and may be modified for locomotion, food gathering, and attachment, and they also function in respiration. The skeleton anchors the tube feet firmly in the body wall so they can be effective locomotor structures.

CLASSIFICATION (MAJOR LIVING FORMS)*

Class Crinoidea
Class Asteroidea
Class Ophiuroidea
Class Concentricycloidea
Class Echinoidea
Class Holothuroidea

Class Asteroidea

Characteristics: Star-shaped; rays not distinctly set off from central disk; ambulacral grooves with tube feet; suction disks on tube feet; pedicellariae present. Example: _Asterias._ The asteroids are the familiar sea stars. They are often found on hard substrates in marine environments, but some species are also found in sand or mud. They feed on snails, bivalves, crustaceans, polychaetes, corals, detritus, and a variety of other food items. Sea stars are often brightly colored in red, orange, blue, or gray. _Asterias_ is a common orange sea star found along the Atlantic coast of North America and is studied in the following exercise.

External Structure

▼ Examine a preserved specimen of _Asterias_ and note the **oral** and **aboral surfaces** (fig. 16.1). On the aboral surface, note the opening to the water-vascular system, the **madreporite.** The body is covered by a thin epidermis through which modifications of the skeleton protrude. Using a dissecting microscope, note the calcareous **spines** and, around their bases, the pincerlike **pedicellariae.** The pedicellariae are primarily used in cleaning the body surface. Fleshy projections on the surface of the body are **dermal branchiae.** These, along with the tube feet, allow for gas exchange. A small, pigmented **eyespot** is located at the tip of each ray but is difficult to see in preserved specimens.

Turn your specimen over to its oral surface. Note that each ray has a groove running down its length. Along either side of this **ambulacral groove** are rows of **tube feet.** Note

*This listing reflects a phylogenetic arrangement; however, in the exercises that follow, the echinoderms that are familiar to most students are studied first.

Figure 16.1 Sea star, oral view of *Asterias*.

the suction disks at their tips. Part the tube feet along the midline to see the groove more clearly. The central mouth is surrounded by larger, movable **oral spines.**

Internal Structure

Again, place the specimen aboral side up. With a pair of sharp scissors, pierce the body wall along the lateral surface of one ray. Carefully cut along the entire margin of that ray and then around the top of the central disk, leaving the madreporite attached to the body. Lift the body wall gently, freeing any tissue that adheres to the portion of the body wall being removed using the blade of your scalpel. Use figures 16.2 and 16.3 and Plates 8 and 9 to aid in the following observations.

Note the large **pyloric ceca** (digestive glands) in the ray. In following the pyloric ceca toward the central disk, note that they unite with a membranous sac, the **stomach.** The stomach consists of two parts. The **pyloric portion** may have been removed with the body wall covering the central disk. The larger **cardiac portion** is easily seen. A rudimentary intestine and anus are attached to the portion of the body wall removed.

When a sea star feeds, it wraps itself around the ventral margin of a bivalve. The tube feet attach to the outside of the shell, and the body-wall musculature forces the valves apart. When the valves are opened about 0.1 mm, the cardiac portion of the stomach is everted through the mouth and into the bivalve shell by increased coelomic pressure.

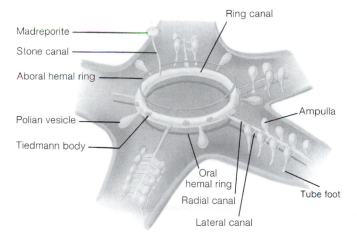

Figure 16.2 Sea star structure. Dissection showing the water-vascular system.

Digestive enzymes are released from the pyloric ceca, and partial digestion occurs within the bivalve shell. This weakens the bivalve's adductor muscles, and the shell eventually opens completely. Partially digested tissues are taken into the pyloric portion of the stomach, and into pyloric ceca, for further digestion and absorption.

Remove the pyloric cecae from the ray. Note the ridge of ossicles that represents the aboral side of the ambulacral groove. Lying along either side of the ambulacral ridge near

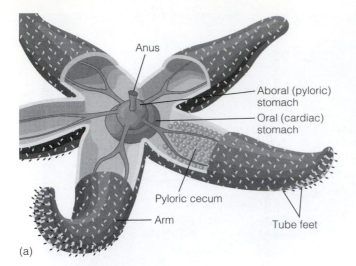

Anus

Aboral (pyloric) stomach

Oral (cardiac) stomach

Pyloric cecum

Arm

Tube feet

(a)

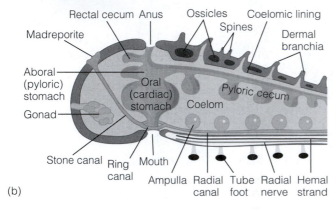

Rectal cecum Anus

Madreporite

Ossicles

Spines

Coelomic lining

Dermal branchia

Aboral (pyloric) stomach

Oral (cardiac) stomach

Pyloric cecum

Gonad

Coelom

Stone canal Ring canal

Mouth

Ampulla Radial canal Tube foot Radial nerve Hemal strand

(b)

Figure 16.3 Digestive structures in a sea star. A mouth leads to a large cardiac (oral) stomach and a pyloric (aboral) stomach. Pyloric ceca extend into each arm. (*a*) Aboral view. (*b*) Lateral view through the central disk and one arm.

the central disk is a thin strand of muscle. This muscle attaches to the ossicles of the ambulacral ridge and to the stomach wall. These **stomach retractor muscles** are used in returning the stomach to the body cavity after feeding (see Plate 8). At the base of the ray, note the **gonads.** Recall that sexes are separate. The sex of an individual cannot be determined by a general inspection of the gonads. Gonads open separately through aboral pores.

In the ray, note also the internal manifestations of the tube feet. Each tube foot consists of a tube extending to the outside, with a bulblike **ampulla** at its internal apex (see Plate 9). Contraction of the ampulla forces water into the tube and causes the tube foot's extension. Each tube foot connects via a **lateral canal** to a **radial canal** running within the ambulacral ossicles. If injected, the radial canal can be seen in the ambulacral groove (oral surface). Each radial canal connects to a **ring canal** surrounding the stomach. The madreporite connects to the ring canal via a **stone canal.**

Also present in the ambulacral groove are a **radial nerve** and a **radial blood vessel.** These can be seen in a prepared cross section. Observe a slide under low power, noting these and other structures described earlier (fig. 16.4).

Observe a slide showing sea star larvae. The development that follows external fertilization results in a free-swimming bilateral larva called the **bipinnaria.** The bipinnaria larva is followed by a second larva, the **brachiolaria.** The mature brachiolaria larva attaches to the substrate and metamorphoses into an adult. ▲

STOP AND ASK YOURSELF

1. What two well-known animal phyla are included in the deuterostomate lineage? _____
_____

2. What is pentaradial symmetry? _____
_____

3. What is the function of the following structures of a sea star?

    a. madreporite _____
_____

    b. pedicellariae _____
_____

    c. stomach retractor muscles _____
_____

    d. ampulla _____
_____

Class Ophiuroidea—Brittle Stars, Basket Stars

Characteristics: Arms sharply marked off from the central disk; body cavity limited to central disk; tube feet without

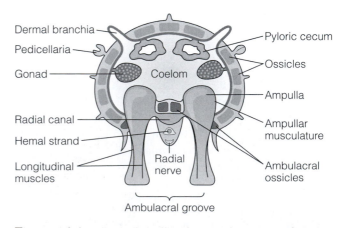

Dermal branchia

Pedicellaria

Gonad

Radial canal

Hemal strand

Longitudinal muscles

Coelom

Ambulacral groove

Radial nerve

Pyloric cecum

Ossicles

Ampulla

Ampullar musculature

Ambulacral ossicles

Figure 16.4 The body wall and internal anatomy of a sea star arm. A cross section through one arm of a sea star shows the structures of the water-vascular system and the tube feet extending through the ambulacral groove.

suckers. Examples: *Ophiothris, Ophioderma*. Brittle stars and basket stars are located in all oceans, at all depths. Brittle stars have long, snakelike, unbranched arms; basket stars have arms that branch repeatedly. The arms contain a reduced coelomic space, ossicles, and large muscles. They are used to pull the animal along the substrate and sweep the substrate to collect prey or decaying animal matter. Some ophiuroids are filter feeders and wave their arms in the water to trap plankton on mucus-covered tube feet.

▼ Observe any demonstrations that are available. ▲

Class Echinoidea—Sea Urchins, Sand Dollars

Characteristics: Globular or disk-shaped, without rays; movable spines; skeleton (test) of closely fitting plates. Example: *Arbacia*. Echinoids are widely distributed in all oceans, at all depths. Sea urchins are specialized for living on hard substrates, often wedging themselves into crevices and holes in rock or coral. Sea urchins feed on algae, coral polyps, and dead animal remains. Food is manipulated by tube feet surrounding the mouth and passed to a chewing apparatus, called **Aristotle's lantern,** which projects from the mouth. Aristotle's lantern consists of about 35 ossicles and attached muscles and cuts food into small pieces for ingestion.

▼ Observe living or preserved sea urchins and a dried sea urchin skeleton or test (fig. 16.5). Look for evidence of pentaradial symmetry. Examine an Aristotle's lantern that has been removed from a test. ▲

Sand dollars and heart urchins usually live in sand or mud near the surface of the substrate. They are flattened and their symmetry is modified to a nearly bilateral or asymmetrical form. Sand dollars and heart urchins use tube feet to catch organic matter that settles on, or over, them. Sand dollars often live in very dense beds, which favor efficient reproduction and feeding.

Figure 16.5 Class Echinoidea: Tests and live specimens of *Strongylocentrotus purpuratus*.

▼ Examine any specimens available. ▲

Class Holothuroidea—Sea Cucumbers

Characteristics: No rays; elongate along the oral-aboral axis; microscopic ossicles embedded in muscular body wall; circumoral tentacles. Example: *Stichopus*. Sea cucumbers are found at all depths in all oceans, where they crawl over hard substrates or burrow through sand or mud. They have no arms, they are elongate along the oral-aboral axis, and they lie on one side—which is usually flattened as a permanent ventral side. This results in sea cucumbers having a secondary bilateral symmetry. The oral end of a sea cucumber is surrounded by tentacles, which are really modifications of the water-vascular system. Tentacles are used in feeding by trapping food particles that settle on them or by sweeping across the substrate. The water-vascular system is also evidenced externally as five rows of tube feet that run between the oral and aboral ends of the animal.

▼ Examine preserved specimens for these aspects of external structure (fig. 16.6). Examine a demonstration slide showing cucumber ossicles. Normally these ossicles are embedded in the thick connective tissue of the body wall. A ring of larger ossicles surrounds the oral end of the digestive tract and serves as a point of attachment for body-wall muscles. ▲

Class Crinoidea—Sea Lilies, Feather Stars

Characteristics: Rays present; aborally attached to substrate through a stalk of ossicles; spines, madreporite, and pedicellariae absent. Example: *Antedon*. Sea lilies are widely distributed and abundant. They are attached permanently to their substrate by a **stalk,** which bears a flattened disk, or rootlike extensions, at the attached end. The unattached end of a sea lily is called the **crown.** The aboral end of the crown is attached to the stalk and is supported by a set of ossicles, called the **calyx.** Five arms attach at the calyx, and the branching of the arms gives the sea lily a featherlike appearance. Feather stars are similar to sea lilies except they lack a stalk and are able to swim and crawl.

Crinoids feed by using their outstretched arms. When a planktonic organism contacts a tube foot, it is trapped and carried to the mouth by cilia within ambulacral grooves of the arm. This method of feeding is thought to reflect the original function of the water-vascular system.

▼ Observe any demonstrations that are available. ▲

Class Concentricycloidea—Sea Daisies

Characteristics: Two concentric water-vascular rings encircle a disklike body. No digestive system. Internal brood pouches. No free-swimming larval stage. This class contains a single, recently discovered species: *Xyloplax medusiformis*, the sea daisy (fig. 16.7). Sea daisies live in deep oceans on wood and other debris. A thin membrane in contact with

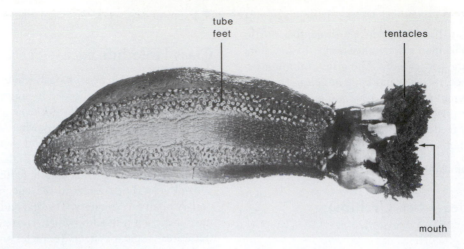

Figure 16.6 Sea cucumber, *Cucumaria*.

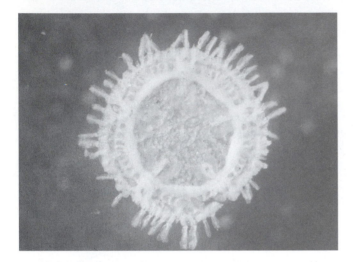

Figure 16.7 Class Concentricycloidea. Photograph of a preserved sea daisy *(Xyloplax medusiformis)*. This specimen is 3 mm in diameter.

their substrate digests and absorbs nutrients. No specimens will be available for your study.

EVOLUTIONARY RELATIONSHIPS

▼ A cladogram showing the evolutionary relationships among the echinoderm classes is shown in worksheet 16.1 (fig. 16.8). Examine the synapomorphies depicted on the cladogram. Based on your studies in this week's laboratory, you should be able to determine the position of each of the classes studied. Complete the cladogram and answer the questions that follow it. ▲

4. How can one distinguish a sea star from a brittle star?

_____

_____

5. What is the feeding apparatus of members of the class Echinoidea? _____

6. How is pentaradial symmetry manifested in sea urchins? _____

_____

7. How is the symmetry of sea cucumbers different from that of other echinoderms? _____

_____

8. How do sea lilies feed? _____

_____

_____

KEY TERMS

Aristotle's lantern   211

calcium carbonate endoskeleton   208

deuterostomes   208

pedicellariae   208

pentaradial symmetry   208

water-vascular system   208

WORKSHEET 16.1   Echinodermata

Evolutionary Perspective

1. Why are ancestral echinoderms believed to have been bilaterally symmetrical?

Echinodermata

2. Diagram and label the following in the water-vascular system of a sea star: ring canal, stone canal, madreporite, one radial canal, lateral canals, and the tube feet.

3. What are the functions of the water-vascular system you just illustrated?

4. Explain how the water-vascular system described above is different from that of the following:
   a. holothuroideans

   b. ophiuroideans

5. Characterize members of the following classes:
   a. Ophiuroidea

   b. Holothuroidea

   c. Echinoidea

6. Describe the method by which a sea cucumber feeds.

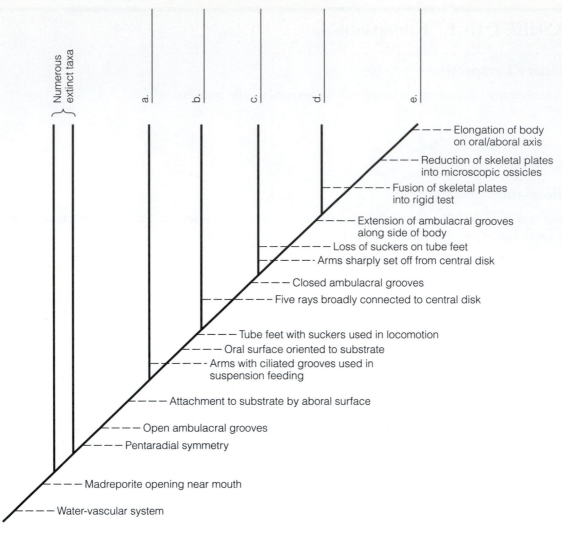

Figure 16.8 Evolutionary relationships among the Echinodermata. Fill in the class names. The Concentricycloidea have been omitted from this cladogram because their phylogenetic status is highly speculative.

7. Name one symplesiomorphic character for the phylum Echinodermata. _____

8. Name one synapomorphy that can be used to distinguish members of each of the following classes from other echinoderms.
   a. Crinoidea _____   d. Echinoidea _____

   b. Asteroidea _____   e. Holothuroidea _____

   c. Ophiuroidea _____

9. Is pentaradial symmetry symplesiomorphic for the phylum? Explain your answer.

10. Some systematists think that the Asteroidea and Ophiuroidea should be depicted as being more closely related than they are represented to be in figure 16.8. What characteristic is shared by these two groups that could be a derived character uniting these two classes into a single clade?

Exercise 17

Chordata

Learning Objectives

After completing this exercise, you should be able to do the following:

1. Describe the evolutionary significance of the Chordata.
2. Characterize the phylum Chordata.
3. Recognize subphylum representatives and describe characteristics of each chordate subphylum.
4. Recognize members of the vertebrate classes and describe evolutionary relationships among them.
5. Recognize and describe the functions of structures studied in subphylum representatives.

Prelaboratory Quiz

Study this week's laboratory exercise and then complete the following quiz to assess your preparation for the laboratory.

1. All of the following are unique features of members of the phylum Chordata EXCEPT the
   a. dorsal tubular nerve cord.
   b. notochord.
   c. closed circulatory system.
   d. pharyngeal slits (pouches).

2. Members of the subphylum Urochordata include
   a. amphioxus.
   b. hagfish.
   c. tunicates.
   d. lampreys.

3. All of the following are members of the subphylum Vertebrata EXCEPT
   a. lancelets.
   b. hagfish.
   c. lampreys.
   d. frogs and toads.

4. Adult urochordates are
   a. fishlike predators.
   b. fishlike filter feeders.
   c. sessile filter feeders.
   d. planktonic predators.

5. Which of the following is most dorsal in position in virtually all chordates?
   a. notochord
   b. nerve cord
   c. pharyngeal slits
   d. gut tract

6. True/False   The amniotes include the amphibians, reptiles, and birds.

7. True/False   Sharks, skates, and rays are members of the class Osteichthyes.

8. True/False   Members of the class Cephalaspidomorphi possess a rasping tongue and a sucking mouth with teeth.

9. True/False   The subphylum Vertebrata is divided into two superclasses based on the presence or absence of jaws and paired appendages.

10. True/False   Both the birds and the mammals are endothermic and possess amniotic eggs.

EVOLUTIONARY PERSPECTIVE

The relationship of chordates to echinoderms and the relationship of the echinoderm-chordate line to previous phyla were discussed in exercise 16. Zoologists have not yet found evidence documenting the origin of chordates from nonchordates. There are three subphyla within the phylum Chordata: Urochordata, Cephalochordata, and Vertebrata. Of these, members of the subphylum Urochordata have been used in one hypothesis of chordate origins. As you will see in this exercise, adult urochordates are sessile filter feeders and are most similar to other chordates as larvae. It is suggested that a urochordate-like ancestor gave rise to the chordate lineage when sexual maturity developed in the larval body form. (This is called "paedomorphosis" and is seen in a number of animal phyla.) These ancestral chordates then could have given rise to modern urochordates, cephalochordates, and vertebrates.

THE TRIPLOBLASTIC COELOMATE ANIMALS—CHORDATA

Since the remaining laboratory exercises will cover details of chordate structure and function, this exercise is intended to introduce the major characteristics of the phylum, subphyla, and major classes.

It is relatively easy to characterize the phylum Chordata. Even though they show remarkable diversity of form and function, all chordates at some time in their life history have four easily recognized, unique features: the **notochord, pharyngeal slits (pouches)**, a **dorsal tubular nerve cord**, and a **postanal tail** (fig. 17.1).

The notochord is a supportive rod that extends most of the length of the animal along the dorsal body wall. It consists of a connective tissue sheath packed with vacuolated cells. It supports the body along the anterior-posterior axis, yet is flexible enough to allow freedom of body movement. In many vertebrate adults, the notochord is partly or entirely replaced by bone.

Pharyngeal slits are a series of openings between the digestive tract, in the pharyngeal region, and the outside of the body. Although the term "gill slits" is often used to refer to these openings, it is misleading. The earliest chordates used the openings for filter feeding and did not have gills associated with these pharyngeal slits. In some chordates that function is retained. Other chordates, however, have developed gills within the slits and use them for gas exchange. Like the notochord, the pharyngeal slits of higher vertebrates are mainly embryonic features. The eustachian tube

of tetrapods, connecting the middle ear to the pharynx, is derived from a pharyngeal slit. The eustachian tube is used in equalizing pressure within the middle ear to prevent damage to the eardrum.

The dorsal tubular nerve cord and its associated structures are largely responsible for the success of the chordates. It runs along the dorsal axis of the body, just above the notochord, and is usually modified anteriorly into a brain. These nervous structures have allowed the development of the most complex animal systems for sensory perception, integration, and motor response.

The fourth characteristic is a tail that extends posterior to the anal opening. The tail is generally supported by the notochord or by the vertebral column, which replaces the notochord in many adult chordates. In addition, chordates are coelomate, triploblastic, and bilaterally symmetrical.

CLASSIFICATION

Subphylum Urochordata
Subphylum Cephalochordata
Subphylum Vertebrata
    Agnathans
        Class Cephalaspidomorphi
        Class Myxini
    Gnathostomes
        Class Chondrichthyes
        Class Osteichthyes
        Class Amphibia
        Class Reptilia
        Class Aves
        Class Mammalia

The vertebrates have traditionally been divided into superclasses based on whether they lack jaws and paired appendages (Agnatha) or possess these structures (Gnathostomata). Cladistic analysis shows that some members of the Agnatha (lampreys and some ostracoderms) are more closely related to jawed fishes (Gnathostomata) than to other agnathans (hagfishes). Therefore, most researchers have rejected the traditional groupings. The traditional terms are still common in zoological literature, and they are used as convenient nontaxonomic groupings in this laboratory manual.

Subphylum Urochordata—Tunicates

Characteristics: Notochord, nerve cord, and postanal tail present only in free-swimming larvae; pharyngeal slits present in larvae and adults; pharyngeal basket; adults sessile, or occasionally planktonic, and enclosed in a cellulose-containing tunic. Example: *Ciona*.

Members of the subphylum Urochordata are the tunicates or sea squirts. Most are sessile as adults and are either solitary or colonial. Others are planktonic. Sessile urochordates attach their saclike bodies to rocks, pilings, the hulls of ships, and other solid substrates.

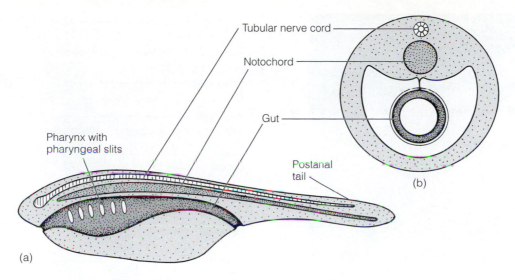

Figure 17.1 The generalized chordate body plan: (*a*) sagittal section; (*b*) cross section.

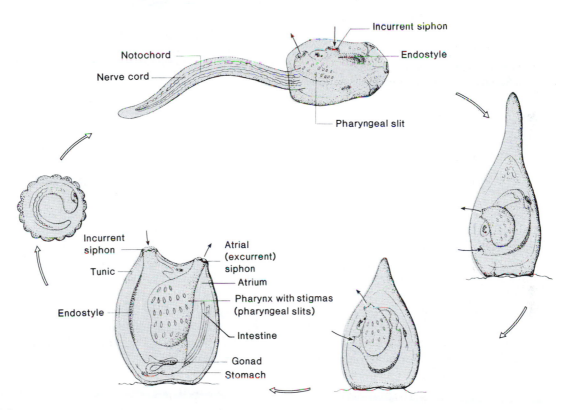

Figure 17.2 Tunicate structure and life cycle. Solid arrows show the path of water moving through the pharyngeal basket (pharynx).

▼ Examine tunicates on demonstration (fig. 17.2). Note the body wall, which encloses a cavity called the atrium. At the unattached end are two siphons. The incurrent siphon brings seawater into an expanded pharyngeal basket, which filters food from the water. A structure called the endostyle aids in filter feeding. It is a ciliated groove that produces a mucous sheet for trapping food particles and moving them to the intestine. (The endostyle is believed to be the evolutionary forerunner of the vertebrate thyroid gland.) Water leaves the tunicate via the atrial (excurrent) siphon.

Tunicates are monoecious. Gametes are shed through the atrial siphon for external fertilization, or eggs may be retained within the atrium for fertilization and early development. Cross-fertilization is the rule. Development results in the formation of a tadpole-like larva that possesses all four chordate characteristics (see fig. 17.2). After a brief free-swimming existence, the larva settles to a firm substrate and attaches by adhesive papillae located near the mouth. During metamorphosis, there is a 180° rotation of internal structures that results in the development of the adult body

(a)

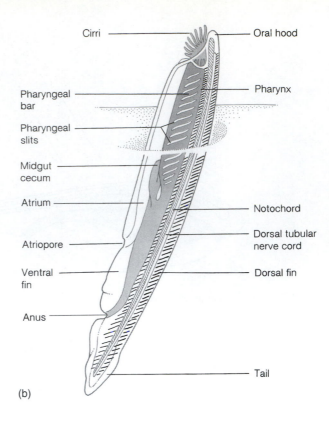

(b)

Figure 17.3 Subphylum Cephalochordata. *(a)* Amphioxus in its partially buried feeding position. *(b)* Amphioxus structure.

form (see fig. 17.2). Examine a slide showing a tunicate larva under the compound microscope. ▲

Subphylum Cephalochordata

Characteristics: Body laterally compressed and transparent, fishlike; all four chordate characteristics persist throughout life; buccal apparatus used in feeding.

The Cephalochordata (lancelets) is a relatively small group of fishlike chordates. This group is used extensively in introductory courses because it shows features characteristic of the phylum Chordata. These features are probably similar to their appearance in ancestral chordates.

Lancelets are marine, spending most of their life with their posterior half buried in sandy substrates of shallow water (fig. 17.3a). They feed by filtering suspended organic particles from water in a manner similar to that described for tunicates. We will study the genus *Branchiostoma*, commonly known as amphioxus.

Whole Mount

▼ Obtain an amphioxus whole mount. Examine it under a dissecting microscope or low power of a compound microscope. Compare your observations to figure 17.3b. Note that anteriorly there is an **oral hood** with a terminal **rostrum** and fingerlike **oral cirri** that surround the mouth. These structures form a buccal apparatus used in feeding. Oral cirri are largely sensory, with some straining functions. Cilia lining the oral hood draw water currents into the gut tract, where food particles are caught in mucus secreted by the **endostyle** (to which the vertebrate thyroid gland has been homologized). Note the **pharyngeal slits** margined by supportive **pharyngeal bars.**

Water passing into the digestive tract during filter feeding passes through pharyngeal slits to the **atrium,** the cavity surrounding the pharyngeal slits. Pharyngeal slits of cephalochordates, as with urochordates, are primarily involved with filter feeding. From the atrium, water moves to the outside through the ventral **atriopore.** Follow the digestive tract posteriorly. Just behind the pharyngeal slits, note that a diverticulum off the gut tract, called the **hepatic (gastric) cecum,** extends anteriorly. Enzymes secreted by this diverticulum partially digest food particles before absorption. Digestion is both extracellular and intracellular. The digestive tract ends at the **anus.** Locate the anus and note the postanal tail. Locate the **notochord** dorsal to the digestive tract. Dorsal to the notochord is the **nerve cord.** Dorsally, muscles of the body wall are organized into units called **myomeres.**

Cross Section

Obtain a slide showing one or more cross sections of amphioxus. The slide is likely to show sections through representative regions of the body. If so, locate a section through the pharyngeal region similar to that shown in figure 17.4a. Locate those structures described as being unique to the phylum: tubular nerve cord, notochord, and pharyngeal slits. In addition, at the ventral margin of the pharynx, note the endostyle. The coelom is restricted to the space dorsal and lateral to the pharynx and the space below the endostyle. Most of the chamber around the pharynx is the atrium. Lateral to the pharynx, you should also see **gonads.** (Cephalochordates are dioecious.) An ovary has large, nucleated cells, while a testis has much smaller germ cells that appear as darkly stained dots. Gametes are shed into the atrium and

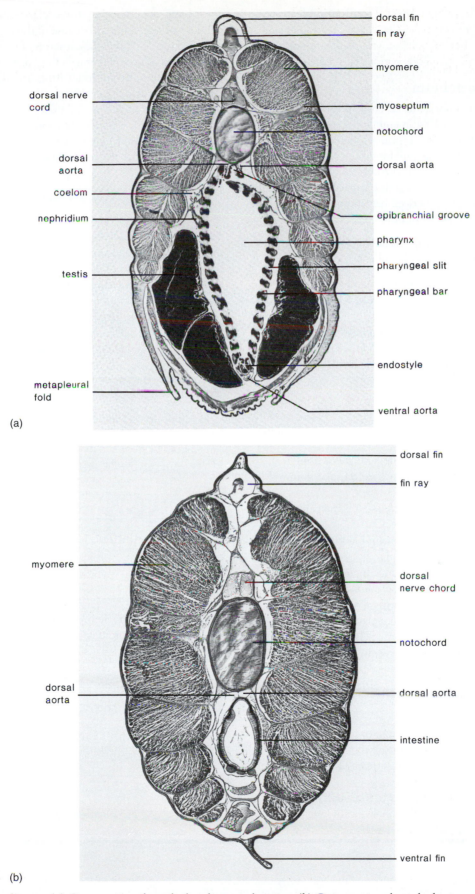

dorsal fin
fin ray
myomere
dorsal nerve cord
myoseptum
notochord
dorsal aorta
dorsal aorta
coelom
epibranchial groove
nephridium
pharynx
pharyngeal slit
testis
pharyngeal bar
endostyle
metapleural fold
ventral aorta

(a)

dorsal fin
fin ray
myomere
dorsal nerve chord
notochord
dorsal aorta
dorsal aorta
intestine
ventral fin

(b)

Figure 17.4 Amphioxus. (a) Cross section through the pharyngeal region. (b) Cross section through the intestinal region.

leave through the atriopore. External fertilization leads to a bilaterally symmetrical larva.

Examine the other sections on your slide and associate them with their respective body regions (fig. 17.4b). ▲

Subphylum Vertebrata

Characteristics: Notochord, nerve cord, postanal tail, and pharyngeal slits present at least in embryonic stages; vertebrae surround nerve cord and serve as primary axial support; skeleton modified anteriorly into a skull for protection of the brain.

There are eight classes of vertebrates. In subsequent laboratory exercises you will examine aspects of vertebrate evolution, structure, and function. It is important for you to be able to recognize major vertebrate classes.

▼ Examine specimens of the following classes, noting as many of the characteristics as possible. ▲

Agnathans—Lack jaws and paired appendages. Have cartilaginous skeleton, persistent notochord, and two semicircular canals.
  Class Cephalaspidomorphi—Lampreys
    Sucking mouth with teeth and rasping tongue. Seven pairs of pharyngeal slits. Blind olfactory sacs.
  Class Myxini—Hagfishes
    Mouth with four pairs of tentacles. Olfactory sacs open to the mouth cavity. Five to fifteen pairs of pharyngeal slits.
Gnathostomes—Jaws and paired appendages present. Notochord may be replaced by a vertebral column. Three semicircular canals.
  Class Chondrichthyes—Sharks, skates, rays, ratfishes
    Tail with large upper lobe. Cartilaginous skeleton. Most lack opercula, swim bladder, and lungs.
  Class Osteichthyes—Bony fish
    Most have a bony skeleton. Single gill opening covered by a bony operculum. Pneumatic sac(s) function as lungs or swim bladders.
  Class Amphibia—Frogs, toads, salamanders
    Skin with mucoid secretions. Usually undergo metamorphosis. Moist skin functions as a respiratory organ.
  Class Reptilia—Snakes, lizards, alligators, turtles
    Dry skin with epidermal scales. Amniotic egg.
  Class Aves—Birds
    Feathers. Skeleton modified for flight. Metabolic heat used in temperature regulation (endothermic). Amniotic egg.
  Class Mammalia—Mammals
    Have hair, mammary glands. Endothermic temperature regulation. Amniotic egg.

Members of the last three classes are relatively independent of water, partly because of the presence of amniotic eggs. Amniotic eggs possess a series of membranes that form bags for the nontoxic storage of nitrogenous wastes and prevent the embryo from desiccating. In addition, the eggs of amniotes other than most mammals possess a shell that protects the embryo from mechanical injury. This common feature, along with other adaptations to terrestrial environments, has resulted in reptiles, birds, and mammals being lumped together as the **amniotes.**

EVOLUTIONARY RELATIONSHIPS

▼ A cladogram showing the evolutionary relationships among the chordates is shown in worksheet 17.1 (fig. 17.5). Examine the synapomorphies depicted on the cladogram. Based on your studies in this week's laboratory, you should be able to determine the position of each of the taxa studied. Complete the cladogram by matching the number of the taxon or character listed in worksheet 17.1, question 7, with the appropriate blank on the cladogram. Answer the questions that follow the cladogram. ▲

STOP AND ASK YOURSELF

1. What are four unique chordate characteristics?
   a. _____
   b. _____
   c. _____
   d. _____

2. What was the function of pharyngeal slits in the earliest chordates?
   _____
   _____

3. What stage in the life cycle of a urochordate is most like other chordates? _____

4. How do adult urochordates feed? _____
   _____

5. Where is the endostyle of a cephalochordate located?
   _____

6. What is the function of the endostyle? _____
   _____

KEY TERMS

amniotes 220   notochord 216   postanal tail 216

dorsal tubular nerve cord 216   pharyngeal slits (pouches) 216

WORKSHEET 17.1   Chordata

Evolutionary Perspective

1. What is paedomorphosis? How might it have been important in chordate evolution?

Phylum Chordata

2. Describe the life cycle of a sea squirt.

3. List the functions of the following cephalochordate structures:
   a. cirri

   b. atrium

   c. hepatic cecum

   d. notochord

4. What characteristics can be used to distinguish between members of the three chordate subphyla?

5. Characterize members of the following classes:
    a. Cephalaspidomorphi

    b. Osteichthyes

    c. Reptilia

    d. Mammalia

6. What is an amniotic egg?

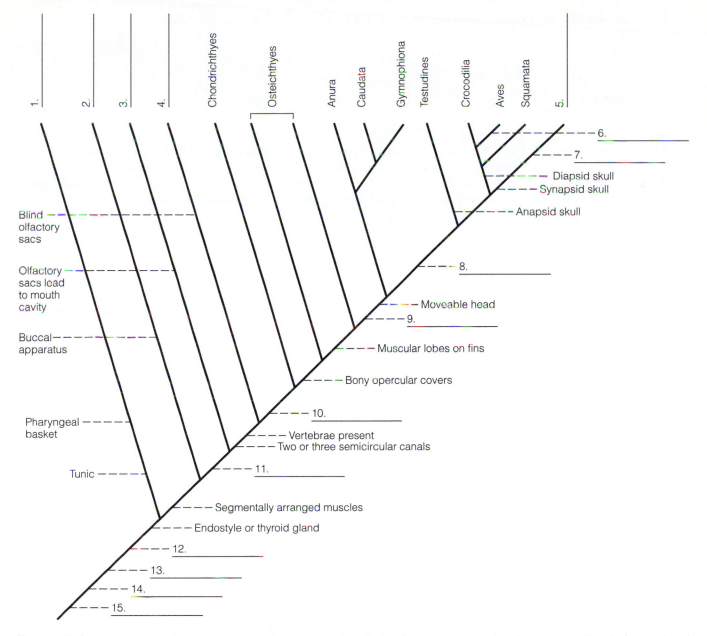

Figure 17.5 Evolutionary relationships among the Chordata. Match the character or taxon listed in question 7 with the numbered blank in the cladogram.

7. Match the following taxa or characters to the appropriate blank on the cladogram (fig. 17.5).
   a. Endoskeleton including cranium
   b. Four limbs/terrestrial locomotion
   c. Feathers and skeletal modifications for flight
   d. Pharyngeal slits
   e. Jaws and paired appendages
   f. Postanal tail
   g. Hair and mammary glands
   h. Extraembryonic membranes
   i. Notochord
   j. Dorsal tubular nerve cord
   k. Mammalia
   l. Myxini
   m. Urochordata
   n. Cephalaspidomorphi
   o. Cephalochordata

8. Name and describe four symplesiomorphic characters for members of the phylum Chordata.

9. The taxonomy of the phylum Chordata is currently unsettled. The traditional eight-class system is being reevaluated because of recent cladistic analysis. Notice in the cladogram that the reptiles are divided into three orders: Testudines (turtles), Crocodilia (crocodiles and alligators), and Squamata (snakes and lizards). How is the cladogram shown in figure 17.5 inconsistent with the traditional class designations for birds and reptiles?

10. Birds are sometimes called "glorified reptiles." Based on the cladogram, do you think that is an accurate description? Explain your answer.

11. Similarly, the bony fishes have been traditionally grouped into a single class, Osteichthyes. The two lineages depicted in the cladogram represent the ray-finned fishes, which include our modern bony fishes, and lobe-finned fishes, which include the lungfishes and the ancestors of tetrapods. Based on figure 17.5, do you think that the bony fishes form a valid clade worthy of the class rank? Explain.

12. The amphibians are divided into three orders: Anura (toads and frogs), Caudata (salamanders), and Gymnophiona (caecilians). Based on figure 17.5, do you think that they form a valid clade worthy of the class rank? Explain.

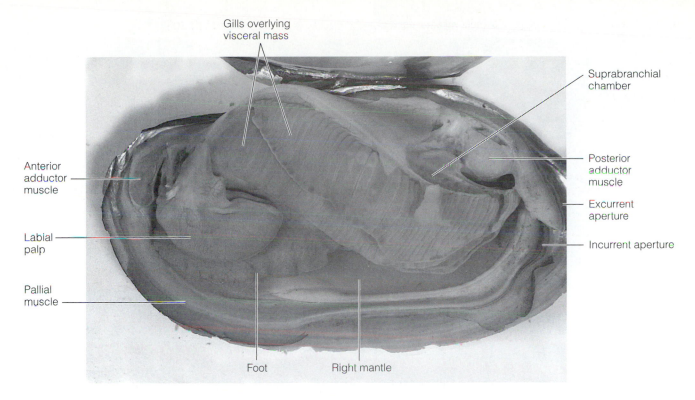

Plate 1 Internal structure of a bivalve. Left mantle cut away.

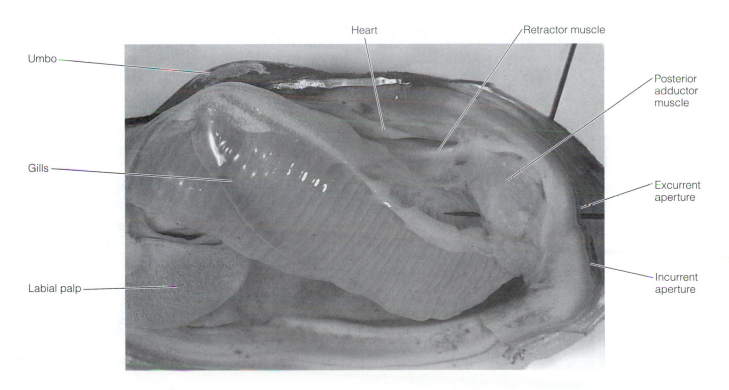

Plate 2 Internal structure of a bivalve. The mantle covering the posterior portion of the suprabranchial chamber has been removed. Note the pin inserted through the excurrent aperture with its point resting in the suprabranchial chamber. Water circulates from the incurrent aperture, through the gills, into the suprabranchial chamber, and exits the bivalve through the excurrent aperture.

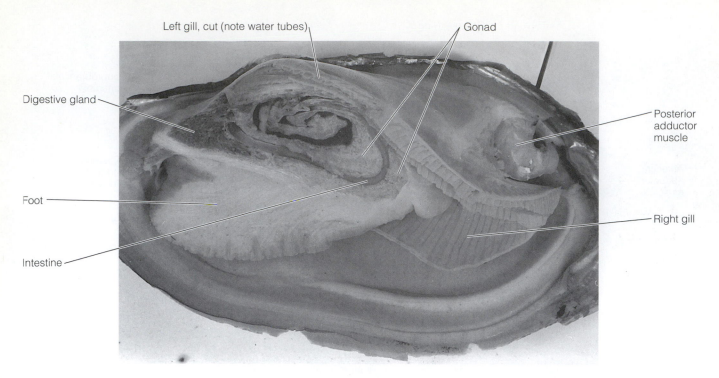

Left gill, cut (note water tubes)

Gonad

Digestive gland

Posterior
adductor
muscle

Foot

Right gill

Intestine

Plate 3   Internal structure of a bivalve. Left gill cut away. Visceral mass cut vertically to expose internal structures.

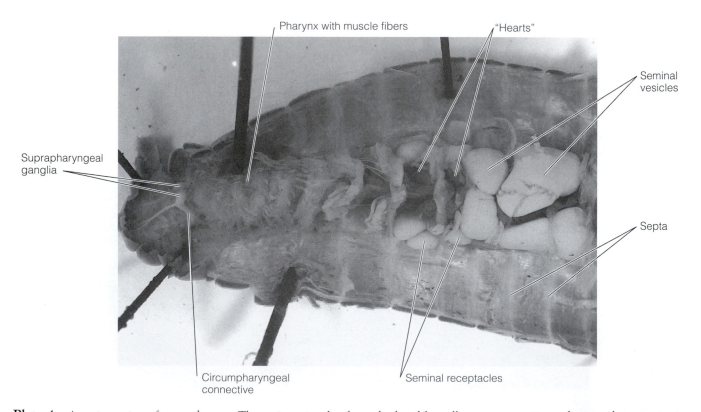

Pharynx with muscle fibers

"Hearts"

Seminal
vesicles

Suprapharyngeal
ganglia

Septa

Circumpharyngeal
connective

Seminal receptacles

Plate 4   Anterior region of an earthworm. The peristomium has been displaced laterally to expose a circumpharyngeal connective.

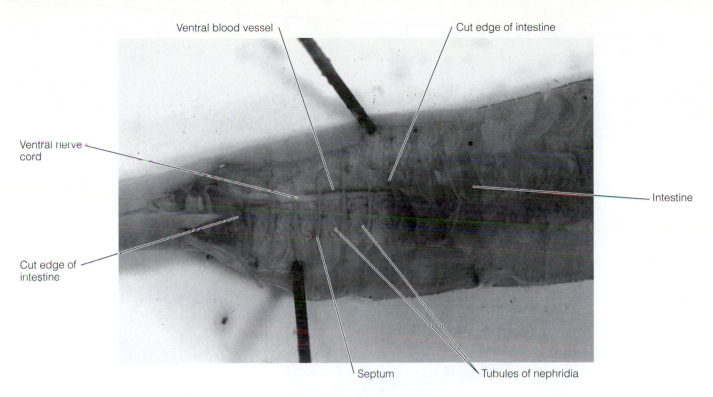

Ventral blood vessel

Cut edge of intestine

Ventral nerve cord

Intestine

Cut edge of intestine

Septum

Tubules of nephridia

Plate 5 Posterior region of an earthworm. A section of intestine has been removed to expose the ventral nerve cord and ventral blood vessel.

Plate 6 Ventral view of male (top) and female (bottom) crayfish.

228

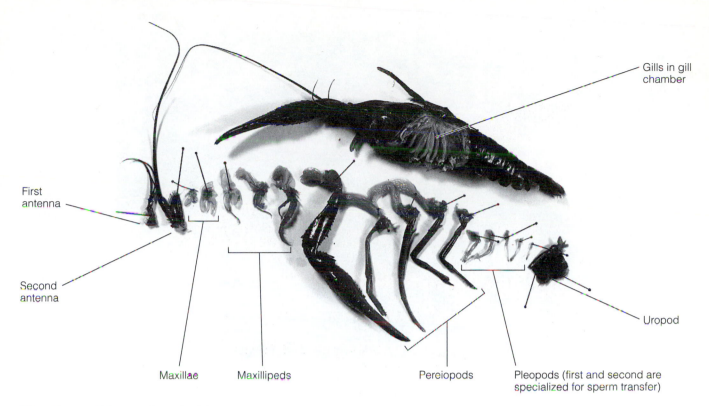

First
antenna

Second
antenna

Gills in gill
chamber

Maxillae

Maxillipeds

Pereiopods

Pleopods (first and second are
specialized for sperm transfer)

Uropod

Plate 7   Crayfish appendages (male).

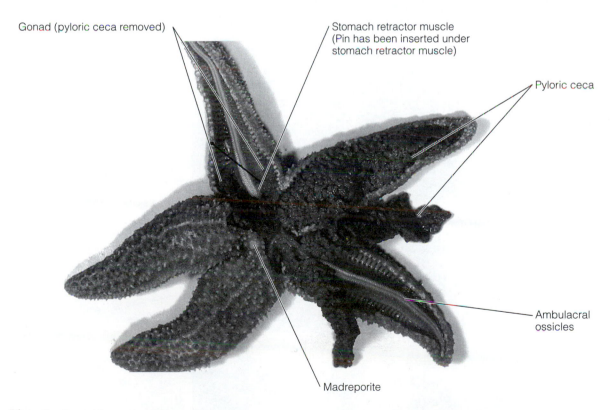

Gonad (pyloric ceca removed)

Stomach retractor muscle
(Pin has been inserted under
stomach retractor muscle)

Pyloric ceca

Ambulacral
ossicles

Madreporite

Plate 8   Internal structure of a sea star.

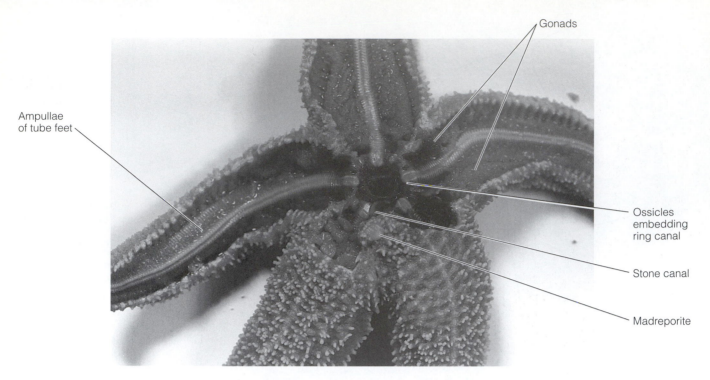

Gonads

Ampullae
of tube feet

Ossicles
embedding
ring canal

Stone canal

Madreporite

Plate 9 Internal structure of a sea star. Stomachs (oral and aboral) and gastric ceca from three arms are removed.

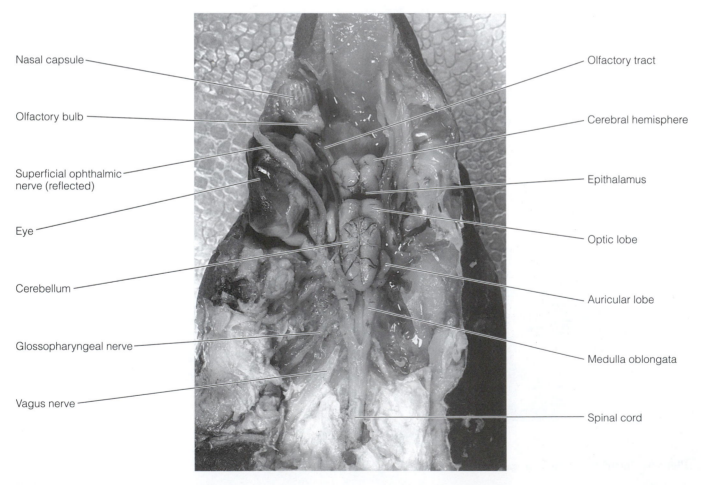

Nasal capsule

Olfactory bulb

Superficial ophthalmic
nerve (reflected)

Eye

Cerebellum

Glossopharyngeal nerve

Vagus nerve

Olfactory tract

Cerebral hemisphere

Epithalamus

Optic lobe

Auricular lobe

Medulla oblongata

Spinal cord

Plate 10 Dorsal view of a shark brain.

230

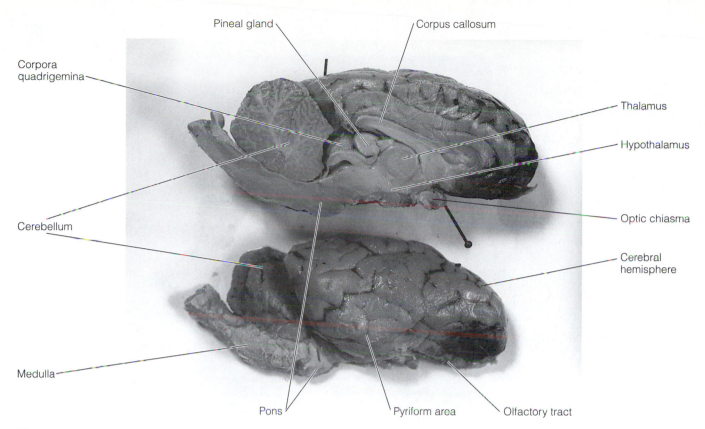

Plate 11 Midsagittal and lateral views of the sheep brain.

Pineal gland

Corpus callosum

Corpora quadrigemina

Thalamus

Hypothalamus

Cerebellum

Optic chiasma

Cerebral hemisphere

Medulla

Pons

Pyriform area

Olfactory tract

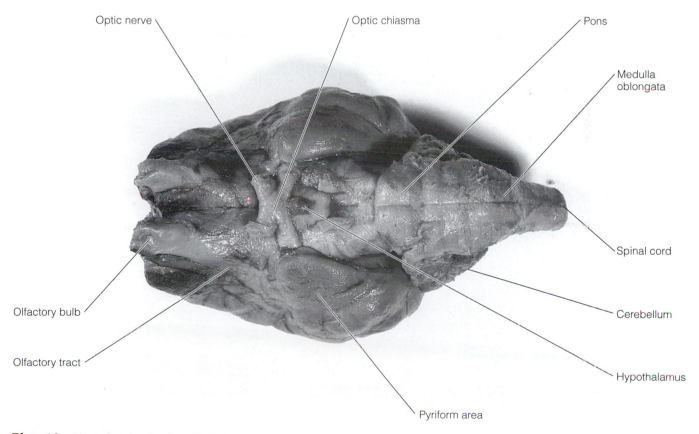

Plate 12 Ventral view of a sheep brain.

Optic nerve

Optic chiasma

Pons

Medulla oblongata

Spinal cord

Olfactory bulb

Cerebellum

Olfactory tract

Hypothalamus

Pyriform area

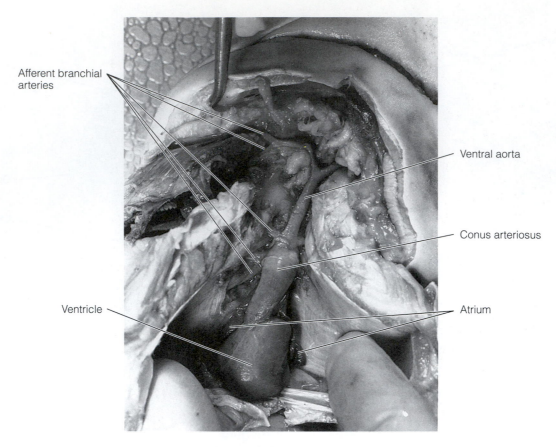

Afferent branchial arteries

Ventral aorta

Conus arteriosus

Ventricle

Atrium

Plate 13   Heart, ventral aorta, and afferent branchial arteries of a shark.

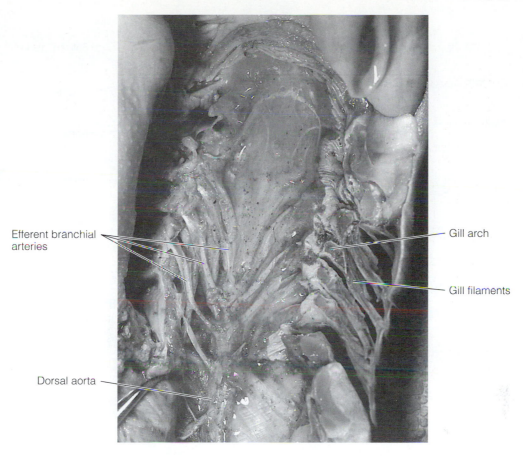

Efferent branchial arteries

Gill arch

Gill filaments

Dorsal aorta

Plate 14   Efferent branchial arteries of the shark.

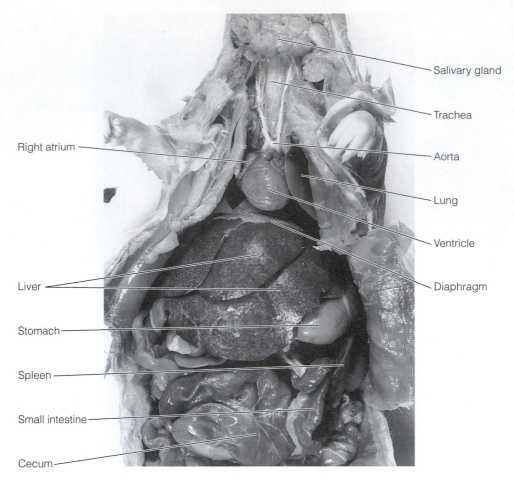

Right atrium

Liver

Stomach

Spleen

Small intestine

Cecum

Salivary gland

Trachea

Aorta

Lung

Ventricle

Diaphragm

Plate 15 Internal organs of the rat.

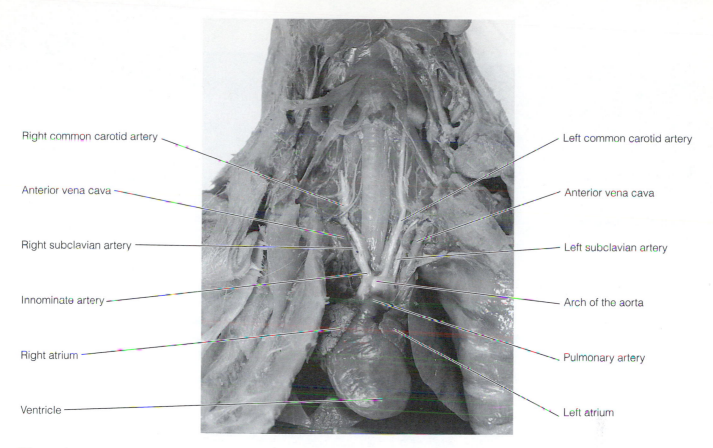

Right common carotid artery

Anterior vena cava

Right subclavian artery

Innominate artery

Right atrium

Ventricle

Left common carotid artery

Anterior vena cava

Left subclavian artery

Arch of the aorta

Pulmonary artery

Left atrium

Plate 16 The heart and major vessels of a rat.

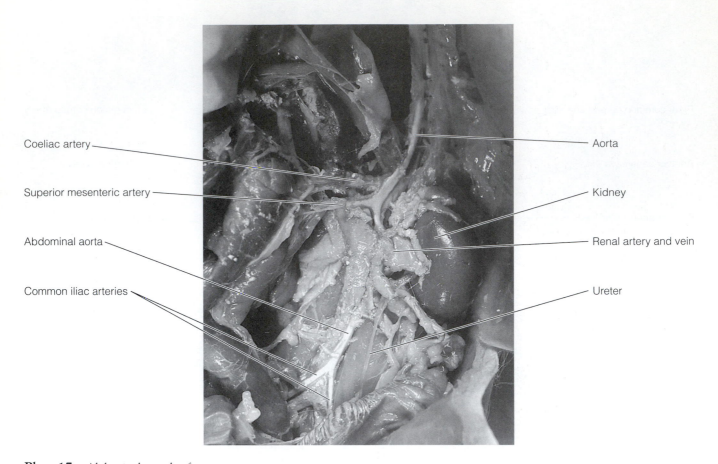

Coeliac artery

Superior mesenteric artery

Abdominal aorta

Common iliac arteries

Aorta

Kidney

Renal artery and vein

Ureter

Plate 17 Abdominal vessels of a rat.

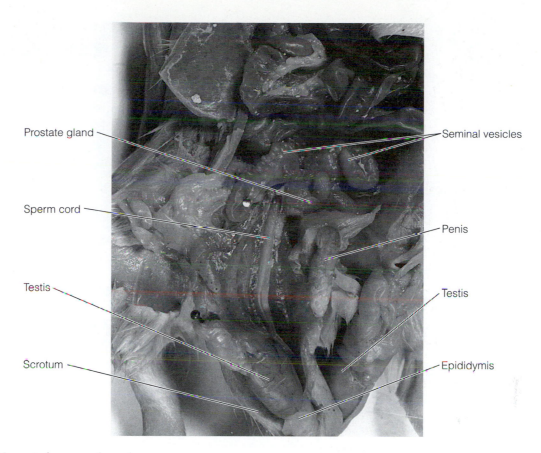

Prostate gland

Seminal vesicles

Sperm cord

Penis

Testis

Testis

Scrotum

Epididymis

Plate 18 Urogenital system of a male rat.

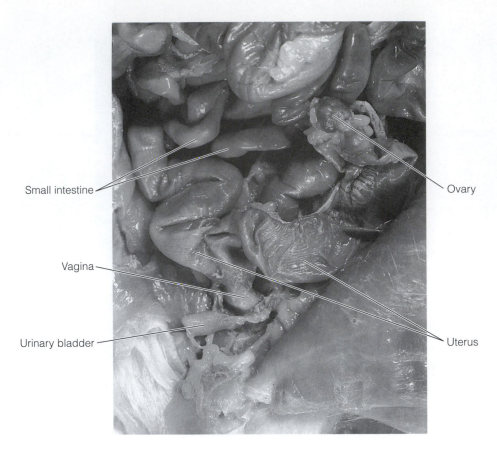

Small intestine

Vagina

Urinary bladder

Ovary

Uterus

Plate 19   The urogenital system of a female rat.

UNIT

III

An Evolutionary Approach to Vertebrate Structure and Function

Unit III is a study of the structure and function of vertebrate organ systems. The "organ system–by–organ system" format of this unit is intended to help you understand vertebrate structure and function and some aspects of vertebrate evolution. While time and space do not allow the examination of evolutionary changes in detail, you will gain an appreciation of the diversity of structure and function within the vertebrates. The rat is used as the primary animal for student dissection. Demonstration dissections of the shark and other vertebrates are used for comparative purposes. In addition, selected physiological exercises are designed to introduce you to the functions of some systems.

Exercise 18

Vertebrate Musculoskeletal Systems

Learning Objectives

After completing this exercise, you should be able to do the following:

1. Describe the importance of muscular and skeletal systems in vertebrate biology and describe their functional interdependence.

2. Describe the microscopic structure of bone tissue and identify parts of a long bone.

3. Describe the evolutionary changes in vertebrate skeletal systems and the adaptive significance of those changes.

4. Identify specific bones studied in the laboratory.

5. Identify smooth, skeletal, and cardiac muscle under the microscope.

6. Describe the cellular mechanism of muscle contraction and characteristics of a muscle's contraction.

7. Deduce a muscle's action based on a knowledge of the muscle's origin and insertion.

Prelaboratory Quiz

Study this week's laboratory exercise and then complete the following quiz to assess your preparation for the laboratory.

1. At each end of a long bone is a region called the
   a. osteon.
   b. diaphysis.
   c. epiphysis.
   d. matrix.

2. Bone cells, or osteocytes, are found within
   a. canaliculi.
   b. lacunae.
   c. osteonic canals.
   d. the matrix.

3. Which of the following vertebrate classes has members whose vertebral column consists of cervical, trunk, and caudal regions?
   a. Osteichthyes
   b. Amphibia
   c. Reptilia
   d. Aves
   e. Mammalia

4. The axial skeleton consists of all of the following EXCEPT
   a. the skull.
   b. the ribs.
   c. the girdles.
   d. the vertebral column.

5. Skeletal support for the gills of fishes is provided through
   a. visceral arches.
   b. dorsal ribs.
   c. the sternum.
   d. clavicles.

6. True/False   The vertebral column of reptiles, birds, and mammals consists of cervical, thoracic, lumbar, sacral, and caudal regions.

7. True/False   The dark bands in skeletal muscle are called I bands and are regions of overlap between actin and myosin.

8. True/False   Histologically, cardiac muscle differs from skeletal muscle in that the former has intercalated disks and is highly branched. The latter lacks intercalated disks and is unbranched.

9. True/False   A muscle twitch is divided into two periods, the contraction period and the relaxation period.

10. True/False   An action that increases the size of the angle between the anterior surfaces of the articulated bones is an extension.

In the following laboratories, we will be studying the major vertebrate organ systems. Of these, it is appropriate to study skeletal and muscular systems first, since they are the fundamental systems around which the body is built. We will deal with muscular and skeletal systems as a unit because they are functionally interdependent.

BONE STRUCTURE

We will begin our study of the skeletal system by examining a typical long bone that has been sectioned longitudinally (fig. 18.1a).

▼ Examine the outer, uncut surface. The bone is divided into three regions. At each end is an **epiphysis,** which is separated from the **diaphysis** (shaft) by **epiphyseal lines.** In life, each epiphysis is covered by **articular cartilage.** The articular cartilage represents the region of contact with the adjacent bone. The epiphyseal lines indicate the position of **epiphyseal plates.** During growth, the epiphyseal plate is cartilaginous. New bone is laid down in these regions during elongation. If available, examine a demonstration showing the epiphyseal plates of a fetus. In this preparation, all body tissues have been cleared and the bones stained. In life, the outer and inner surfaces of the bone are covered with connective tissue (periosteum and endosteum, respectively). These tissues are involved with bone growth, maintenance, and repair.

Examine the inner cut surface of the bone. Within the diaphysis note the **marrow cavity.** In life, the marrow cavity is filled with yellow marrow that functions in fat storage. The bone of the diaphysis is very compact and hard. It is called **compact bone.** In the epiphysis there is a network of bone called **cancellous bone.** Individual bony filaments are **trabeculae** (fig. 18.1b). Red marrow fills the spaces between trabeculae and functions in blood cell production. ▲

Bone and Cartilage Histology

Bone and cartilage are special connective tissues that were not covered in the histology exercises. They will be covered at this time.

▼ Examine a slide of ground, compact bone under low power of a compound microscope (fig. 18.2). Note that the tissue is organized into a series of circular **osteon systems.** In the center of each is an **osteonic canal** that contains a blood vessel in the living bone. Turn to high power and note the layers of circular **lamellae** that surround the osteonic canal. **Lacunae** are the small cavities in bone and cartilage that contain bone and cartilage cells. In bone, they are the dark spaces between lamellae and contain **osteocytes** that function in maintenance and repair. The small lines radiating from the lacunae are **canaliculi.** Cell processes of osteocytes communicate with adjacent osteocytes through canaliculi. Lamellae are made of collagenous fibers impregnated with calcium phosphate. This nonliving matrix gives the bone its strength. ▲

Cartilage takes on a variety of forms depending on the type of fibers present and the nature of the matrix. It is typically found on the ends of bones (articular cartilage), between bones (intervertebral discs), connecting bone to bone (costal cartilage), and giving support to structures like the nose and pinna of the ear. It is made of cells and fibers embedded in a nonliving matrix (fig. 18.3).

▼ Observe a slide of hyaline cartilage. Examine it under low and then high power. Note the semitransparent **matrix.** Fibers embedded in the matrix cannot be seen. **Chondrocytes** (cartilage cells) are contained within clear lacunae. ▲

THE SKELETON

The vertebrate skeleton can be divided into axial and appendicular components. The **axial skeleton** includes those parts of the skeleton located along the longitudinal axis of the body (skull, vertebrae, ribs, and sternum). The **appendicular skeleton** consists of the more laterally located skeletal elements (pelvic and pectoral girdles and limbs). In the following study, you will examine the skeletons of representative vertebrates. The goal is not to learn every bone in what might be represented as a "typical" vertebrate, but to look at evolutionary changes in major components of the vertebrate skeleton. During this exercise, you will be asked to stop and think about how aspects of what you are seeing are adaptive for the animals in question. Unless your instructor tells you otherwise, feel free to "brainstorm" with neighboring students.

The Axial Skeleton—Vertebral Column
Fishes

▼ Examine the skeleton of a fish (fig. 18.4). If you are looking at a shark (*Squalus*), remember that the skeleton is cartilaginous. This is atypical for most adult vertebrates and represents the retention of an embryonic condition rather than being a primitive situation.

The vertebral column of fishes consists of **trunk** and **caudal (tail)** vertebrae. These two regions can be distinguished

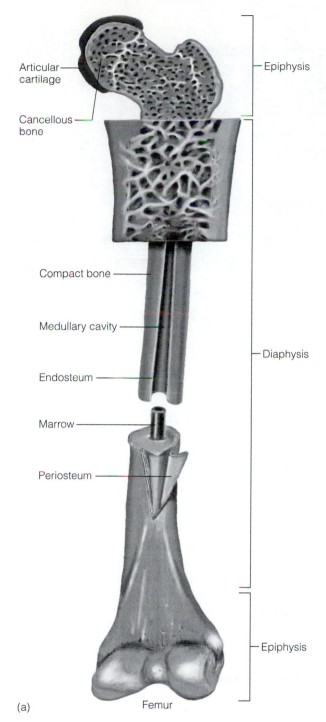

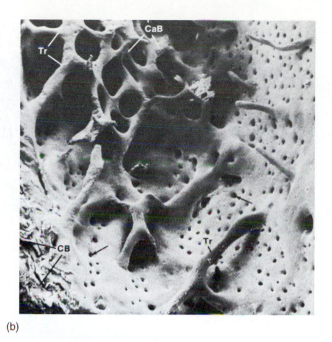

(b)

(a)　　　　Femur

Figure 18.1 Bone structure. (*a*) The structure of a long bone. (*b*) Diaphyseal region of the rat femur. Cancellous bone (CaB), trabeculae (Tr), compact bone (CB). Arrows indicate blood vessel pores (×105). (*b*) From R. G. Kessel and R. H. Kardon. *Tissues and Organs: A Text-Atlas of Scanning Electron Microscopy.* © 1979 by W. H. Freeman and Company.

by the absence of ribs in the tail region. Note that the vertebral column consists of a series of similar bones called **vertebrae.** Each vertebra contains a dorsal (neural) arch and, in the caudal region, a ventral (hemal) arch. The dorsal arches protect the spinal cord, and the ventral arches protect blood vessels and nerves of the tail. ▲

Amphibians

▼ Examine the skeleton of *Necturus* (the mud puppy). Its axial skeleton is divided into four regions. Amphibians possess a neck, and the skull attaches to the axial skeleton via a single **cervical (neck) vertebra.** This permits a vertical nodding movement of the head. Posterior to the neck is the trunk region with its attached ribs. In the evolution of the amphibians, the last trunk vertebra developed into a **sacral vertebra.** This vertebra serves as the point of attachment for the hind limbs via a modified rib. Caudal vertebrae follow the sacral vertebra. ▲

Amniotes: Reptiles, Birds, and Mammals

All amniotes have a similar organization of the vertebral column (fig. 18.5). Use specimens from each group, in the following observations.

▼ Note the proliferation of cervical vertebrae. The first two are specialized into an **atlas** (articulates with the skull and the axis) and an **axis** (articulates with the atlas and the next cervical vertebra). Examine an atlas and an axis that have been disarticulated. Fit them together and then articulate them with the condyles of a skull. How does the combination of skull, atlas, and axis move in nodding the head? ____

_____

In turning the head? _____

_____

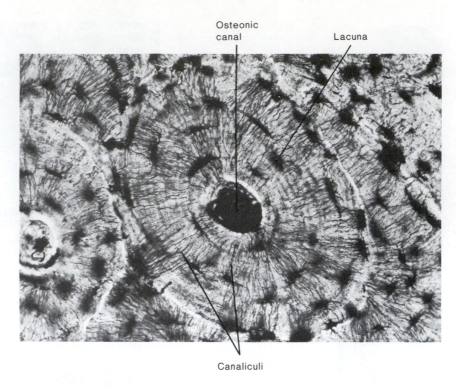

Osteonic canal   Lacuna

Canaliculi

Figure 18.2 Compact bone, the osteon.

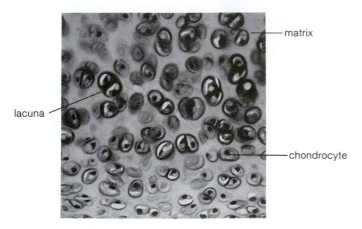

matrix

lacuna

chondrocyte

Figure 18.3 Hyaline cartilage.

Examine disarticulated cervical vertebrae from a goose or turkey. How are they different from those of a mammal?

_____

The trunk region is subdivided into thoracic and lumbar regions in amniotes. The **thoracic vertebrae** have ribs attached that protect the heart and lungs. Locate the thoracic vertebrae and ribs on the skeletons available. The large dorsal (neural) spines of anterior thoracic vertebrae are points of origin of muscles that insert on the back of the skull.

Lumbar vertebrae are present in the abdominal region of amniotes. Note that there are no ribs associated with this region. The large transverse processes are formed from the fusion of ribs and vertebrae.

There is an increased number of sacral vertebrae in amniotes (two to four) as compared to the single sacral vertebra

in amphibians. Examine a sacrum and try to determine how many vertebrae are fused into this structure. In a bird skeleton, note that parts of the thoracic, lumbar, and sacral vertebrae are fused into a single unit, called the **synsacrum.** This fusion keeps the body rigid during flight and helps form an aerodynamic surface.

The caudal vertebrae of vertebrates are variable in number, size, and shape. Compare the caudal vertebrae of a rat, a bird, and a human. The fused caudal vertebrae of birds are called the **pygostyle,** which is where the tail feathers insert. Tail feathers and the pygostyle are used in maneuvering during flight. In humans caudal vertebrae are reduced to form the **coccyx.**

During the evolution of the amphibians, ribs attached ventrally to a plate of bone called the **sternum.** This arrangement is most easily seen in the amniotes. Do all ribs of a mammal's skeleton attach to both the vertebral column and the sternum? _____

Do ribs that attach to the sternum all attach in the same manner? _____

Explain _____

_____

_____ ▲

1. Fishes have eyes on the sides of their heads and, thus, have excellent peripheral vision. Why is this advantageous for animals that lack a neck? _____

_____

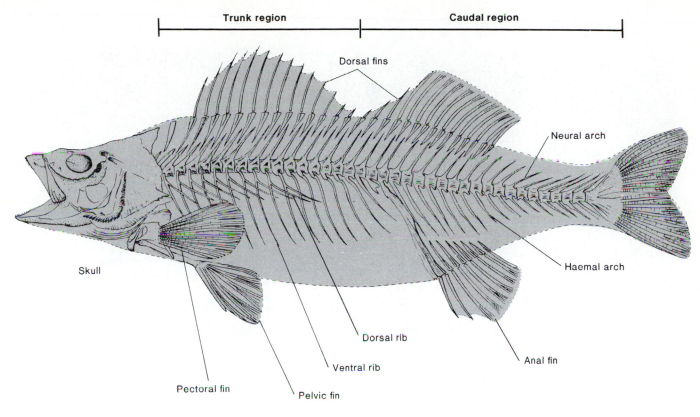

Figure 18.4 Lateral view of a perch skeleton.

2. What must a fish do to "turn its head"? _____
_____

3. What advantage resulted from the development of a neck in terrestrial vertebrates? _____
_____

4. How is the vertebral column of terrestrial vertebrates adapted for locomotion on land? _____
_____
_____

5. Birds are sometimes said to have a fifth appendage (in addition to legs and wings). What is that fifth appendage, and why is it needed in addition to the other four? _____
_____

6. Based on previous observations of live birds, what do you think are three functions of that fifth appendage?

   a. _____

   b. _____

   c. _____

7. How is the axial skeleton of a bird modified to promote these functions? _____
_____

8. Name two modifications of the axial skeleton of birds for flight.

   a. _____

   b. _____

9. What are the functions of these modifications? _____
_____

10. Why is the "hump" on the back of a bison so much larger than that on most other grazing animals?
_____

The Axial Skeleton—Skull

The skull has a very complex evolutionary history and consists of many bones. In primitive vertebrates, the skull consisted of a cartilaginous trough, which housed the brain. (The skull of a shark is cartilaginous; however, this is the result of the retention of an embryonic, not a primitive, condition.) In later vertebrates, bone is found covering the top and sides of the skull, and the cartilaginous trough is replaced by bone. This resulted in additional support and protection

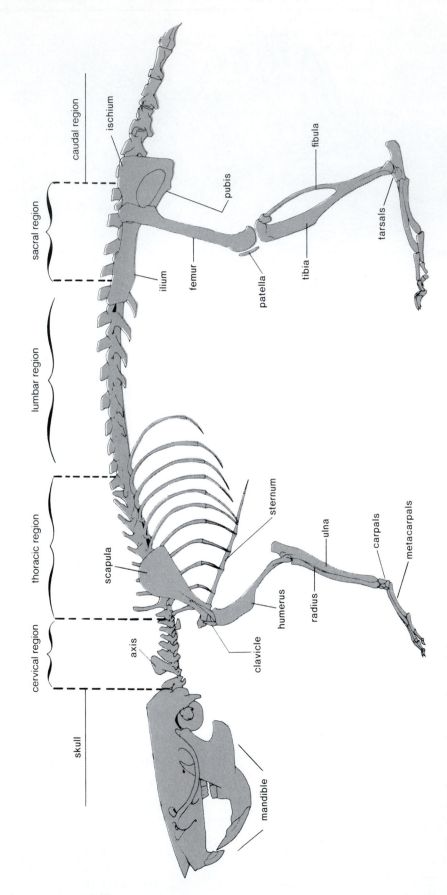

Figure 18.5 Lateral view of a rat skeleton.

for the brain. The vertebrate skull also includes gill supports in fishes and bones derived from gill supports in terrestrial vertebrates. Gill supports are called **visceral arches** and are thought to have been used originally for filter feeding and later supported gas exchange surfaces (gills). In the evolution of jawed fishes, bones of the first visceral arch were incorporated into bones of the upper and lower jaw. In terrestrial vertebrates, the visceral arches are incorporated into the hyoid bone for tongue support, the bones of the middle ear, and the cartilage supporting the larynx (voice box).

▼ Examine a shark skull and visceral arches. Note the similarity in the structure of the jaw and the visceral arches. Examine a human skeleton to see a hyoid bone and models and embedded mounts of the middle ear ossicles. ▲

The Appendicular Skeleton

The appendicular skeleton consists of pelvic and pectoral girdles as well as bones of the paired appendages. The organization of these bones is remarkably consistent throughout the vertebrates.

The Pectoral Girdle

The evolution of the pectoral girdle is complex. We will not describe it in detail. In fishes, the pectoral girdle consists of a number of bones and is usually not attached to the axial skeleton (see shark skeleton). In terrestrial vertebrates, the general trend is to reduce the number of bones and to attach these bones to the axial skeleton. In mammals, the result is a **scapula** that articulates with the arm at the **glenoid cavity.** The scapula serves as a point of attachment for muscles of the arm. The scapula is anchored through the **clavicle** and **sternum** to the axial skeleton. In the bird, the clavicles are fused into a single **furcula,** or wishbone, which serves for the attachment of flight muscles.

▼ Examine skeletal material provided. ▲

The Pelvic Girdle

The evolution of the pelvic girdle is less complex. All terrestrial vertebrates have three pairs of bones that unite at the point of articulation with the hind limb, the **acetabulum.** Forming the anterior ventral border of each half of the pelvic girdle is the **pubis.** Along the posterior border and extending laterally to the acetabulum is the **ischium.** Extending dorsally from the acetabulum and attaching to the sacrum is the **ilium.** These three bones are fused in adults, and their individuality is obscured. The medial border of the pubis and ischium is joined with the same bone from the other side of the body through fibrocartilage. The immovable joint between the pubic bones is referred to as the **pubic symphysis.** The origin of the pelvic girdle is speculative.

▼ Examine skeletal materials provided. Note again the synsacrum of the bird, which shows fusion of the bones of the vertebral column and pelvic girdle. ▲

The Paired Appendages

The appendages of terrestrial vertebrates are all similarly organized. Homologies with bones of fishes have been suggested, but they are speculative. The organization of fore- and hind limbs is similar. In the discussion that follows, the forelimb will be described, and the analogous structure of the hind limb will be given parenthetically.

▼ The bone that articulates with the **glenoid cavity (acetabulum)** is the **humerus (femur).** Distally, the humerus (femur) articulates with two bones, the **ulna (tibia)** and **radius (fibula).** Note on the human skeleton the joint between the radius and humerus. This joint allows a rotation of the lower arm so that the radius crosses over the ulna distally. This allows one to turn palms up or down. Distal to these bones are the bones of the wrist, the **carpals** (the ankle, the **tarsals**). We will not learn the names of individual wrist (ankle) bones. Following these are the bones of the hand, the **metacarpals** (foot, **metatarsals**) and the bones of the fingers, the **phalanges** (toes, **phalanges**). Examine a variety of terrestrial skeletons for an appreciation of the consistent organization of the arm and leg bones of vertebrates. These similarities are a striking example of how homology is used in demonstrating evolutionary relationships. Examine a bird skeleton. Note the fusion and loss of appendage bones that occurred in the evolution of flight. In addition to the previously mentioned adaptations, the bones of birds are porous. ▲

STOP AND ASK YOURSELF

11. What function does the clavicle play in the appendicular skeleton of most vertebrates? _____

    _____

12. Examine the skeleton of a cat. Find its clavicle. How is it different from that of a human and a rat?

    _____

    _____

13. What are the implications of this arrangement for a cat that is walking? (Hint: What do you notice about the scapulae when a cat walks?) _____

    _____

14. What are the implications of this arrangement for a cat that jumps from a tree to the ground and lands on its front legs? (Hint: What happens to humans when we fall from a tree and land on our outstretched arms?) _____

    _____

    _____

15. What problem does the structure of the mammalian pelvic girdle present during the birth process? _____

    _____

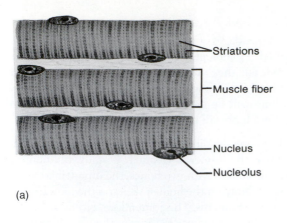

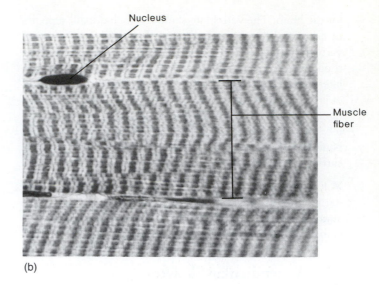

(a)

(b)

Figure 18.6 Skeletal muscle as viewed in the light microscope: (*a*) diagram; (*b*) photomicrograph. A small portion of these cells is shown in both (*a*) and (*b*).

THE MUSCULAR SYSTEM—HISTOLOGY

Because the structure of muscle is closely tied to its function, the study of muscle histology has been delayed until now. Muscles are classified by structure, location, and contractile characteristics.

Skeletal Muscle

Skeletal muscles, as the name implies, are associated with the skeleton. They can be consciously controlled, and they appear striated under the microscope.

▼ Observe a prepared slide of skeletal muscle under low and high powers of a compound microscope (figs. 18.6 and 18.7). Skeletal muscle fibers are long and multinucleate (syncytial). Within each muscle cell are thousands of regularly arranged **myofilaments (actin** and **myosin)** that give the tissue a striated appearance. Note the banding.

Associate the banding seen in the compound microscope with the diagrammatic representation shown in figure 18.8. ▲

A muscle cell or muscle fiber consists of bundles of myofilaments called **myofibrils.** Surrounding myofibrils is a membrane system, actually modified endoplasmic reticulum, called **sarcoplasmic reticulum.** Associated with the sarcoplasmic reticulum are invaginations of the sarcolemma (muscle plasma membrane) called **transverse tubules (t-tubules).** The regular arrangement of the myofilaments, actin and myosin, results in the banding pattern observed in the microscope. The dark bands running across each muscle fiber are called **A bands.** They represent regions of overlap between actin and myosin. The light bands are called **I bands** and are regions of actin filaments. Down the middle of each I band is a **Z line** (not easily seen in the light microscope). The Z line is made of protein that runs across the muscle fiber and anchors longitudinally oriented actin filaments. The distance from one Z line to another is a single

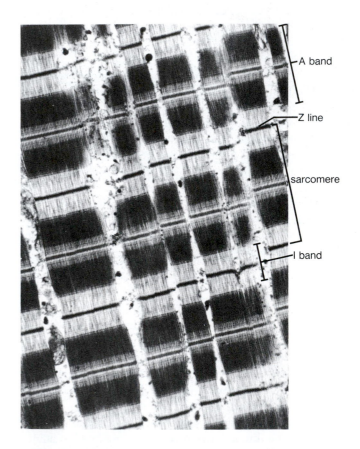

Figure 18.7 Skeletal muscle as viewed in a transmission electron micrograph. A small portion of a single, vertically oriented cell is shown here.

contractile unit called a **sarcomere.** As a muscle action potential (impulse) passes along the sarcolemma, it turns into the transverse tubules (t-tubules) at a Z line. As the action potential passes along the t-tubule, calcium ions are released from the sarcoplasmic reticulum. Calcium ions diffuse to the actin and myosin filaments, initiating contraction. During

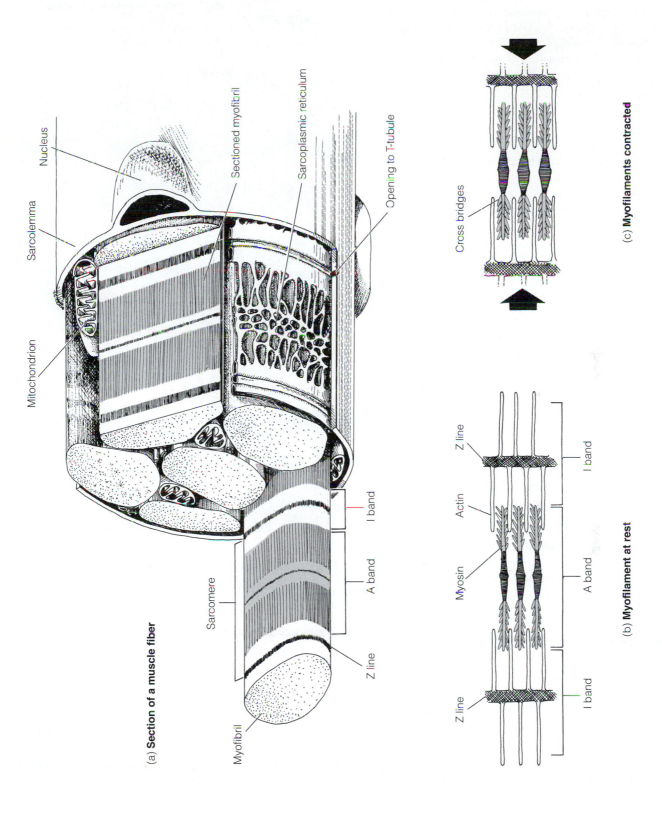

(a) **Section of a muscle fiber**

Nucleus

Sarcolemma

Mitochondrion

Sectioned myofibril

Sarcoplasmic reticulum

Opening to T-tubule

Sarcomere

Myofibril

Z line

I band

A band

Cross bridges

(c) **Myofilaments contracted**

Z line

Actin

Myosin

Z line

I band

A band

I band

(b) **Myofilament at rest**

Figure 18.8 Skeletal muscle ultrastructure. (*a*) Section of a muscle fiber (cell). A muscle fiber contains bundles of myofilaments, called myofibrils. Note the banding pattern of striated muscle shown in the myofibril. (*b*) Myofilaments (actin and myosin) at rest. (*c*) Contracted myofilaments.

contraction, cross bridges on the myosin molecules interact with the actin to cause a sliding of actin relative to myosin, thus shortening the sarcomere. When multiplied over the length of the muscle fiber, the sliding produces a pronounced shortening of the muscle. Essential for the process is the enzymatic breakdown of ATP to ADP and Pi. The energy released facilitates the interaction of cross bridges and actin.

Smooth Muscle

Smooth muscles are associated with internal organs, cannot be consciously controlled, and are nonstriated.

▼ Examine a slide of smooth muscle (fig. 18.9). Myofilaments are not as regularly arranged in smooth muscle as in striated muscle, and so no banding is seen. Each muscle cell is spindle shaped and has a single nucleus. Try to find a region in your slide where muscle cells have been separated so individual cells can be seen. This is most likely to happen near one edge of the section. Observe under high power. ▲

Cardiac Muscle

Cardiac muscle is associated with the heart. Cardiac muscle shares characteristics of both smooth and striated muscle, being involuntary, striated, and uninucleate. Its cell boundaries between the ends of the cells are referred to as **intercalated disks.** These are plasma membranes specialized for conducting electrical impulses between cells. Cardiac muscle fibers are highly branched (fig. 18.10).

▼ Examine a slide of cardiac muscle. Under high power and reduced light, note the striations, the branched fibers, and the intercalated disks. ▲

MUSCLE PHYSIOLOGY

The following exercise illustrates properties of skeletal muscle contraction and will be done as a demonstration. Your instructor has either destroyed the brain of a frog (by pithing) or decapitated it. The skin covering one leg is removed (fig. 18.11a), and the sciatic nerve and gastrocnemius muscle are exposed (fig. 18.11b). The nerve-muscle preparation is then attached to a muscle transducer, and the nerve is placed across stimulating electrodes. (The transducer is attached to a recorder, which can be used to measure the strength and duration of muscle contraction.) The muscle and nerve must be kept moist throughout the following exercises.

▼ With the paper of the recorder moving slowly, apply single electrical stimuli directly on the gastrocnemius in one-volt increments from zero until contraction is observed. The voltage at which the contraction is first seen is called the **threshold.** Once threshold is determined, increase stimulus strength in five-volt intervals until a **maximal response** is attained. Record the voltage of each stimulus on the paper where the corresponding muscle response occurred. Repeat,

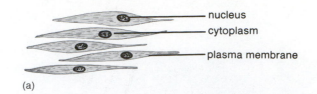

(a)

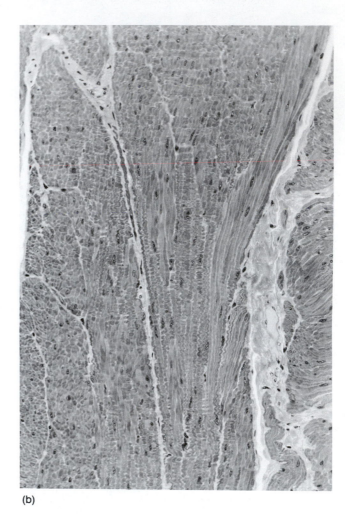

(b)

Figure 18.9 Smooth muscle: (*a*) diagram of smooth muscle with cells teased apart; (*b*) photomicrograph of smooth muscle in cross section (left) and longitudinal section (right).

applying stimuli to the nerve. Record the voltage that resulted in threshold and maximal responses for the nerve and muscle in worksheet 18.1. ▲

The Muscle Twitch

The response of a muscle to a single effective stimulus is called a twitch. A twitch has three components. The **latent period** is the time it takes for the stimulus to cause Ca^{++} release from the sarcoplasmic reticulum and for Ca^{++} to activate actin-myosin interaction. This period lasts only about 10 milliseconds and may not show up on a recording. The **contraction period** is the time during which actin and myosin are interacting and the muscle is shortening. The **relaxation period** is the time after contraction when the muscle passively returns to its original shape.

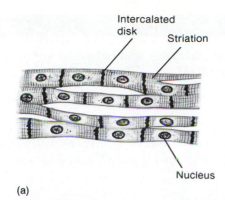

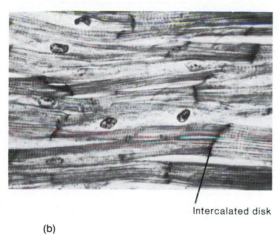

Figure 18.10 Cardiac muscle: (*a*) diagram; (*b*) photomicrograph.

▼ To calculate latent, contraction, and relaxation periods, one must know the rate at which paper is moving through the recorder. Set the paper speed at a high setting and record that value. A muscle twitch is demonstrated by giving the muscle a single stimulus at a voltage above threshold. Reproduce the tracing of a muscle twitch in figure 18.12 of worksheet 18.1. Be sure to accurately record the time on the X-axis so you can calculate latent, contraction, and relaxation periods. Also record stimulus duration and stimulus strength. ▲

Multiple Motor-Unit Summation

A **motor unit** consists of a group of muscle fibers innervated by a single nerve cell. Different motor units of a muscle have different individual thresholds. Therefore, as stimulus strength increases, motor units are added to the response, and the muscle contracts more forcefully. This is called **multiple motor-unit summation.** Multiple motor-unit summation can be demonstrated by gradually increasing the stimulus strength from threshold to maximal response strength, at which point all motor units will contract. When applying the stimulus, the paper should be stopped but advanced a small distance before a subsequent stimulus.

▼ Record the stimulus strength for each contraction. Reproduce the tracing in figure 18.13 in worksheet 18.1. ▲

Temporal (Wave) Summation

If stimuli for submaximal responses are given to a muscle such that a second stimulus occurs before relaxation is

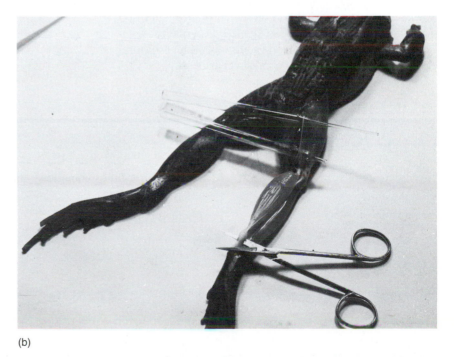

Figure 18.11 Procedure for exposing the gastrocnemius muscle and the sciatic nerve. (*a*) The skin is cut around the base of the thigh and stripped past the ankle. (*b*) Muscles of the thigh are separated with glass probes, and the sciatic nerve is exposed. When attaching the muscle to a transducer, the Achilles tendon is cut as shown.

completed, the second contraction will be more forceful than the first. This is **temporal summation.** If the stimulus frequency increases to 100 to 200 stimuli per second, the muscle cramps. This is **tetanization.**

▼ Reproduce the tracing in figure 18.14 of worksheet 18.1. Label where summation and tetanization occurred. ▲

MUSCULOSKELETAL INTERACTIONS

The interdependence of skeletal muscle and the skeletal system is obvious. Neither could function without the other. A muscle functions by being attached at an immovable point at one end (the muscle's **origin**) and a movable point at the other end (the muscle's **insertion**). A muscle attaches on the skeleton at one or more of the tuberosities, trochanters, ridges, and processes. The action of a muscle is determined by the muscle's origin, insertion, and the nature of the joint over which it acts. The following terms are used to describe muscle actions:

Flexion   Decreases the size of the angle between the anterior surfaces of the articulated bones. Flexing movements are bending or folding movements.

Extension   Increases the size of the angle between the anterior surfaces of the articulated bones. Restores bones to the original position after flexing.

Abduction   Moves a bone away from the median plane of the body.

Adduction   Moves a bone toward the median plane of the body.

Rotation   The pivoting of a bone on its own axis.

▼ Examine a human skeleton on which origins and insertions of five muscles have been labeled. Describe the location of the origins, insertion, and actions of each of the muscles by filling in table 18.1 in worksheet 18.1. ▲

16. What are bundles of myofilaments within a muscle fiber called? _____

17. What is a sarcomere? _____

_____

18. How can you distinguish between skeletal muscle and cardiac muscle under a light microscope?

_____

_____

19. How does one determine the threshold of a muscle?

_____

_____

20. Why does increasing stimulus strength to a nerve increase the strength of contraction of an associated muscle? _____

_____

_____

21. Which is shorter: the latent period or the contraction period? _____

22. While lifting a fork from your plate to your mouth, what action occurs at the elbow joint? _____

_____

KEY TERMS

appendicular skeleton   242

axial skeleton   242

cardiac muscle   250

multiple motor-unit summation   251

osteon system   242

sarcomere   248

skeletal muscle   248

smooth muscle   250

temporal summation   252

WORKSHEET 18.1   Vertebrate Musculoskeletal Systems

Muscle Physiology

stimulus for threshold response _____

stimulus for maximal response _____

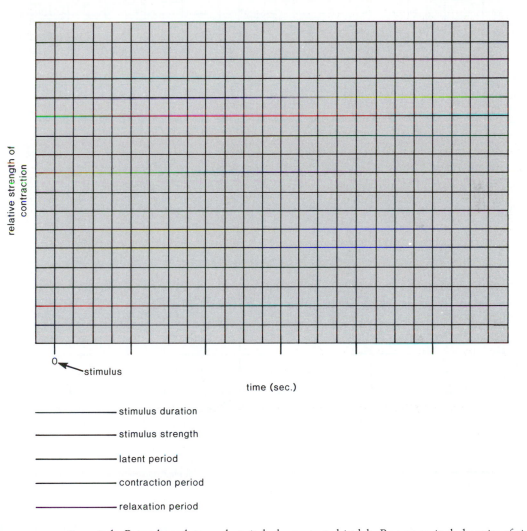

Figure 18.12   The gastrocnemius twitch. Reproduce the muscle twitch demonstrated in lab. Be sure to include units of time on the X-axis. Label and calculate the duration of the latent period, the contraction period, and the relaxation period.

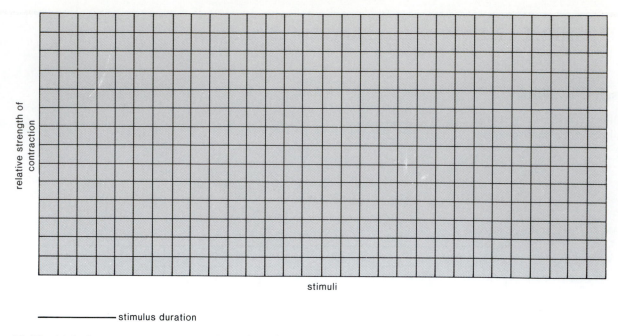

stimuli

——————————— stimulus duration

Figure 18.13 Multiple motor-unit summation. Reproduce the tracing. Label each contraction with the strength of stimulus delivered.

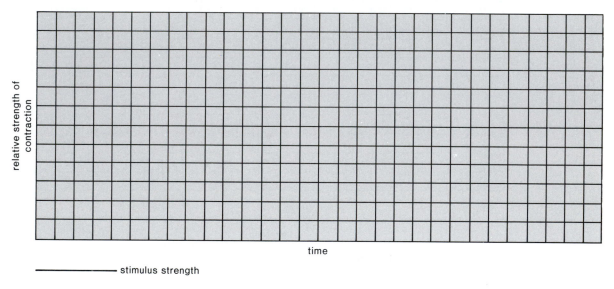

time

——————————— stimulus strength

Figure 18.14 Temporal summation. Reproduce the tracing. Label the X-axis with the frequency of stimuli.

Musculoskeletal Interactions

TABLE 18.1	The Origins, Insertions, and Actions of Human Muscles		
Muscle	**Origin**	**Insertion**	**Action (Including Which Bone Moved)**
1. Sternocleidomastoid			
2. Deltoid			
3. Biceps brachii			
4. Rectus femoris			
5. Adductor magnus			

WORKSHEET 18.2   Vertebrate Musculoskeletal Systems

Bone Structure

1. What is the significance of the epiphyseal plate of a long bone?

2. In a long bone, as in the humerus, where would you find compact bone? Cancellous bone?

3. Where do osteocytes occur in compact bone?

The Skeleton

4. What changes occurred in the vertebrate axial skeleton during the evolution of terrestrial vertebrates?

5. How does the combination of skull, atlas, and axis move in nodding the head? In rotating the head?

6. What skeletal feature helps a bird keep a rigid skeleton during flight?

The Muscular System

7. Dark bands of skeletal muscle are the result of what aspect of muscle cell ultrastructure?

8. What is tetanization, and how was it produced in the laboratory?

9. What are two ways in which the body can alter the strength of contraction of a muscle?

10. While you are using a screwdriver, what action occurs at the wrist joint?

Exercise 19

Vertebrate Nervous Regulation

Learning Objectives

After completing this exercise, you should be able to do the following:

1. Describe the function of the nervous system.
2. Describe the components of a responding system.
3. Describe the structure and function of a neuron, and identify its parts observed in a microscope.
4. Describe the structure of the spinal cord.
5. Identify and describe the functions of selected parts of shark and mammalian brains.
6. Describe the major trend in the evolution of the vertebrate brain.
7. Describe the divisions and functions of the peripheral nervous system.

Prelaboratory Quiz

Study this week's laboratory exercise and then complete the following quiz to assess your preparation for the laboratory.

1. A nerve impulse is carried toward the central nervous system through a(n)
   a. motor neuron.
   b. sensory neuron.
   c. effector organ.
   d. sensory receptor.

2. A(n) _____ initiates an impulse in a sensory neuron.
   a. effector organ
   b. motor neuron
   c. sensory receptor
   d. neurolemmocyte

3. The cell body of a multipolar (motor) neuron is associated with
   a. a sensory receptor.
   b. the central nervous system or a ganglion.
   c. an effector organ.
   d. Both "a" and "c" are correct.

4. In this laboratory you will be examining a prepared microscope slide of a neuromuscular junction. In this slide you will be seeing
   a. a synapse between a nerve and a muscle.
   b. a synapse between two nerves.
   c. the dendritic region of a motor neuron.
   d. the dendritic region of a sensory neuron.

5. In this exercise, you will be studying the brain structure of
   a. a fish and a bird.
   b. a bird and a mammal.
   c. an amphibian and a mammal.
   d. a fish and a mammal.

6. True/False   The embryonic region of the vertebrate brain, called the metencephalon, develops into the cerebral hemispheres of the adult vertebrate.

7. True/False   The region of the vertebrate brain that is continuous with the spinal cord is the medulla oblongata.

8. True/False   The pituitary gland connects to the olfactory tract of the vertebrate brain.

9. True/False   The embryonic diencephalon develops into the epithalamus, hypothalamus, and optic lobes of the adult vertebrate brain.

10. True/False   The peripheral nervous system has two subdivisions: the somatic nervous system and the autonomic nervous system.

The regulation and coordination of physiological processes in the vertebrate body are the roles of two closely linked systems: the **nervous system** and the **endocrine system.** These coordinating systems function by allowing the body to respond to changes in the environment, resulting in the maintenance of constant internal conditions.

The nervous system functions through very rapid **nerve impulses** that travel along nerve cells (neurons). A response always involves a **sensory receptor** receiving a stimulus and initiating an impulse in a **sensory neuron** (fig. 19.1). The sensory neuron carries the impulse to the **central nervous system** (brain or spinal cord). After processing this sensory information, the central nervous system then initiates an impulse that travels via **motor neurons** to an **effector organ** (the responding structure).

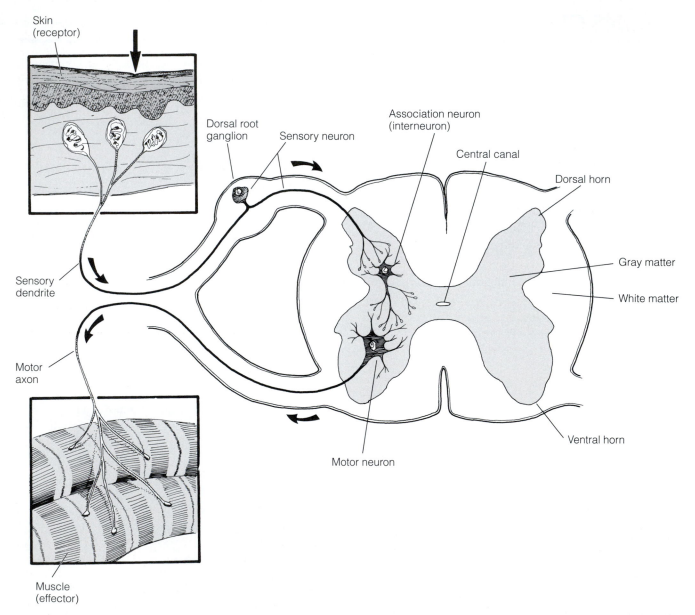

Figure 19.1   Cross section of the spinal cord showing components of a response. A response involves a sensory receptor receiving a stimulus and initiating an impulse in a sensory neuron, which conducts the impulse to the central nervous system. A motor neuron then conducts an impulse to the responding structure.

The endocrine system has a similar basic organization. Replacing the nerve impulse, however, is a chemical called a **hormone,** and replacing the nerve cell is the bloodstream. The nervous system often acts as an intermediary between the environment and the endocrine system. That is, the nervous system receives information regarding environmental changes and conveys that information to the endocrine system. An endocrine gland then responds by secreting a hormone that initiates changes in the animal, which compensate for the environmental change.

THE NEURON

The basic functional unit of the nervous system is the nerve cell, or **neuron.** Two of its variations are shown diagrammatically in figure 19.2. The most common of these will be used to illustrate neuron structure. The **multipolar neuron** is a motor neuron whose nerve cell body is usually associated with the central nervous system or a ganglion (see fig. 19.2a). It ends at an effector structure. The enlarged **nerve cell body** houses the nucleus and other cytoplasmic organelles involved with cell maintenance. A **dendritic region** has small processes radiating out from the nerve cell body. These fibers are where the impulse is generated.

▼ Examine a slide of an ox spinal cord smear. Under low power locate star-shaped nerve cell bodies of multipolar neurons. Switch to high power and locate a relatively clear nucleus containing a darkly stained nucleolus. The processes radiating out from the nerve cell body are mostly **dendrites** (fig. 19.3). Although it is difficult to distinguish, one of the processes radiating out from the nerve cell body is the axon. ▲

The **axon** propagates the impulse from the nerve cell body to the effector structure. The axon is covered with a fatty insulating material called myelin. **Myelin** is produced by and contained in **neurolemmocytes,** which wrap around the axon. Figure 19.4b shows a transmission electron micrograph of a myelinated axon. Myelin is a lipid (fat) found in the plasma membrane of the neurolemmocyte. The figure also shows diagrammatically how degrees of myelination can be achieved by the neurolemmocyte wrapping around the axon. Axons are said to be white, or myelinated, if they contain relatively large quantities of myelin. Axons are said to be gray, or nonmyelinated, if they contain less myelin. The significance of neurolemmocytes is that they increase nerve impulse velocity. This is called **saltatory conduction** and can be visualized as the impulse "jumping" from node (region between neurolemmocytes) to node.

▼ Examine a cross section of a medullated nerve under low power. Note that a nerve is a bundle of many neurons. To see individual neuron cross sections, focus on the interior of the nerve and switch to high power. Each neuron can be seen as a light dot surrounded by layers of darkly stained membranes. These membranes are the myelin-containing neurolemmocyte membranes. ▲

The axon of a multipolar neuron ends at the effector structure, where fine branches called terminal fibers end at a **synapse.** The synapse is the region where an impulse is transferred to an effector structure or another nerve cell. A motor neuron ending on a muscle cell has a synapse referred to as a **neuromuscular junction.** Here, the impulse reaching the synapse causes the release of a **neurotransmitter** from **synaptic vesicles.** The neurotransmitter diffuses across the **synaptic cleft** to initiate a muscle action potential and contraction of the muscle.

▼ Examine a slide showing the neuromuscular junction. Figures 19.5 and 19.6 show nerve cells synapsing with other cells. ▲

STOP AND ASK YOURSELF

1. How are each of the following involved in a response?

   a. sensory neuron _____

   _____

   b. central nervous system _____

   _____

   c. motor neuron _____

   _____

2. What are three parts of a neuron that were visible in the spinal cord smear that you studied?

   a. _____

   b. _____

   c. _____

3. What is saltatory conduction? _____

   _____

   _____

4. What are synaptic vesicles? What is their function in the neuromuscular junction? _____

   _____

   _____

THE CENTRAL NERVOUS SYSTEM

The central nervous system, consisting of the **brain** and the **spinal cord,** is hollow and contains **cerebrospinal fluid.** Recall that a tubular hollow nervous system is one of the unique characteristics of the Chordata. In the central nervous system, sensory information from a variety of sources is processed. This processing is called **integration** and occurs within the gray matter. The central nervous system is organized into regions of gray and white matter. **Gray matter** (usually inside) consists of nerve cell bodies and nonmyelinated axons. **White matter** is made up of myelinated fiber tracts carrying impulses from one part of the central nervous system to another.

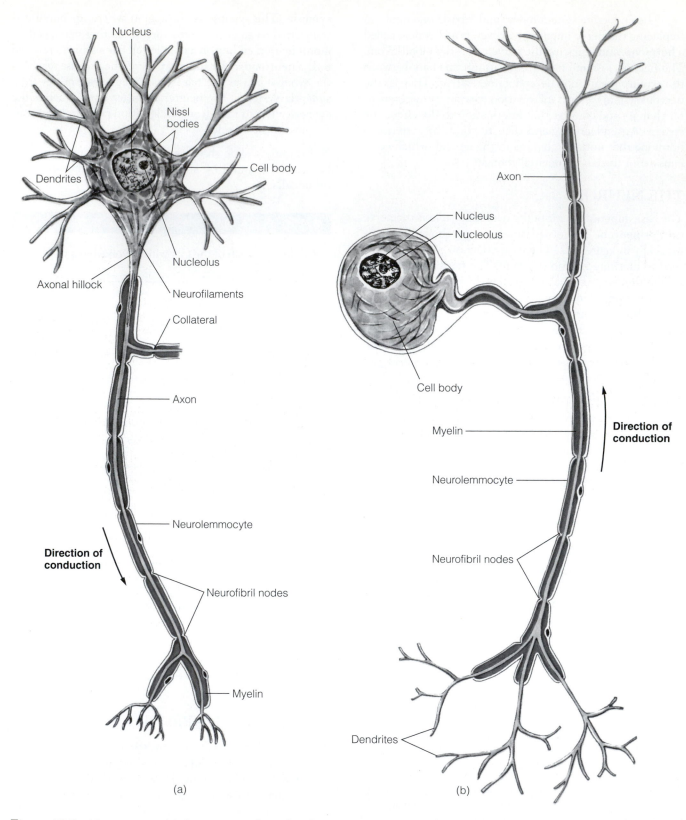

Labels for figure (a):
- Nucleus
- Nissl bodies
- Cell body
- Dendrites
- Nucleolus
- Axonal hillock
- Neurofilaments
- Collateral
- Axon
- Neurolemmocyte
- Direction of conduction
- Neurofibril nodes
- Myelin

Labels for figure (b):
- Axon
- Nucleus
- Nucleolus
- Cell body
- Myelin
- Direction of conduction
- Neurolemmocyte
- Neurofibril nodes
- Dendrites

(a)   (b)

Figure 19.2 Neuron types: (*a*) the structure of a multipolar, or motor neuron, and (*b*) a modified bipolar, or sensory neuron.

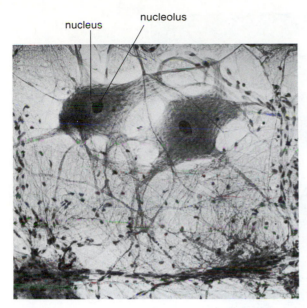

Figure 19.3 Photomicrograph of motor neurons from the vertebrate spinal cord.

nucleus nucleolus

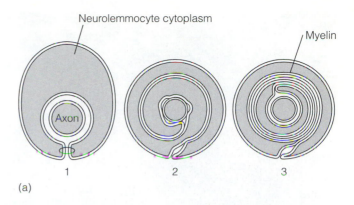

Neurolemmocyte cytoplasm

Myelin

Axon

(a)

TABLE 19.1	The Fate of the Three Embryonic Brain Regions of Vertebrates	
Embryonic Structures		**Adult Structures**
Prosencephalon (Forebrain)	Telencephalon	Cerebral hemispheres
	Diencephalon	Epithalamus, thalamus, hypothalamus
Mesencephalon (Midbrain)	Mesencephalon	Optic lobes (tectum, corpora quadrigemina)
Rhombencephalon (Hindbrain)	Metencephalon	Cerebellum
	Myelencephalon	Medulla oblongata

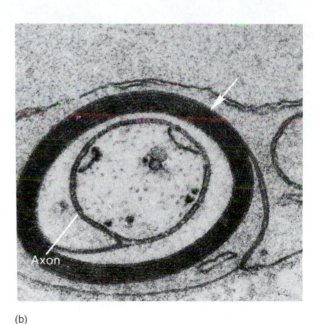

Axon

(b)

Figure 19.4 Cross section of an axon and neurolemmocyte. (a) Diagrammatic representation showing how degrees of myelination can be achieved. Successive wraps of a neurolemmocyte provide increasing myelination. (b) Transmission electron micrograph. The arrow points to the myelin sheath (×73,000).

The Spinal Cord

▼ Use lower power of a compound or dissecting microscope to examine a spinal cord sectioned through the dorsal root ganglion (see fig. 19.1). Distinguish between the white matter and gray matter. The gray matter is organized into an H-shaped central region. White matter lies outside the gray. Note the small central canal in the center of the gray matter. The **central canal** extends up to the brain, where it is continuous with the ventricles (canal system) of the brain. Note the roots of the spinal nerves associated with the spinal cord. In the mammal, the **dorsal root** has sensory neurons entering the spinal cord, and the **ventral root** has motor neurons leaving the spinal cord. The enlarged **dorsal root ganglion** is the location of nerve cell bodies of sensory neurons. ▲

The Brain

The primary area for integration within the central nervous system is the brain. Embryologically, the brain develops as three anterior swellings of the spinal cord. The **prosencephalon (forebrain)** develops as an anterior swelling associated with olfaction. The **mesencephalon (midbrain),** which develops just posterior to the prosencephalon, functions in visual integration. The posterior-most swelling is the **rhombencephalon (hindbrain).** It develops in association with the sense of equilibrium and balance and, later, hearing. Each of these regions then further differentiates as indicated in table 19.1.

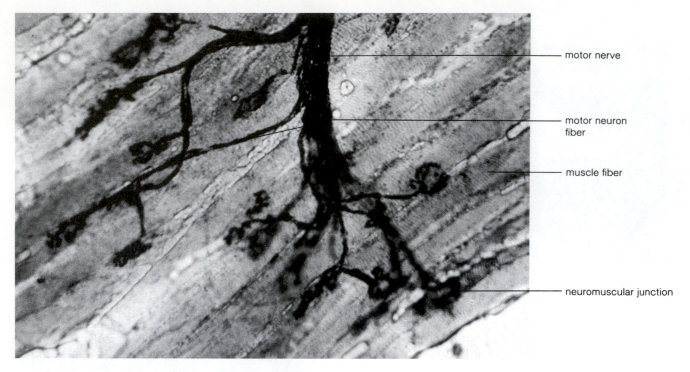

motor nerve

motor neuron fiber

muscle fiber

neuromuscular junction

Figure 19.5 Photomicrograph of the neuromuscular junction (×175).

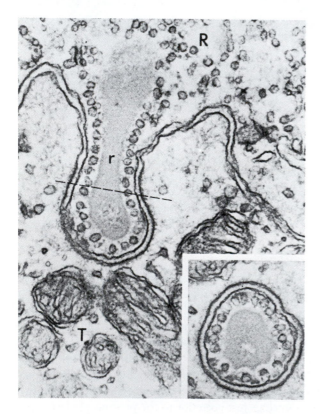

Figure 19.6 Transmission electron micrograph of a synapse in the ampullary organ of a catfish (×85,000). Presynaptic cell showing synaptic vesicles (r). Postsynaptic cell (T). Inset shows cross section through region at dashed line (×97,000).

The Shark Brain

The brain of *Squalus* is typical of many lower vertebrates (figs. 19.7 and 19.8). In it we can see the general form for the five primary regions of the brain. A demonstration dissection has been provided for you (Plate 10).

▼ In the anterior-most region, note the large **nasal capsules,** which are the locations for sensory receptors associated with the sense of smell. Attached to the nasal capsules are the **olfactory bulbs,** which narrow posteriorly to form the **olfactory tract.** This tract represents the pathway for olfactory information coming back to the **cerebral hemispheres** (swellings located just posterior and slightly medial to the olfactory bulbs and olfactory tracts). In fishes the cerebral hemispheres process olfactory information.

Just posterior to the cerebral hemispheres is the **thalamus.** Its very thin roof is the **epithalamus.** The **epiphysis** is attached to the epithalamus. The epiphysis is a thin stalk of tissue that hormonally regulates functions influenced by photoperiod. This is very delicate and is probably missing from your specimen. The bulk of the thalamus is an area of synapse for sensory and motor information traveling between the cerebrum and the mesencephalon. The floor of the diencephalon is the **hypothalamus.** If the brain you are looking at is still embedded in the chondrocranium, the hypothalamus will be impossible to see. The hypothalamus is the primary visceral control center. Attached to it ventrally is the **pituitary gland.** The hypothalamus is the area that coordinates nervous and endocrine systems.

Posterior to the thalamus is the midbrain. In lower vertebrates this region is also referred to as the **optic lobes.** Note the two large dorsal swellings. The roof of the mid-

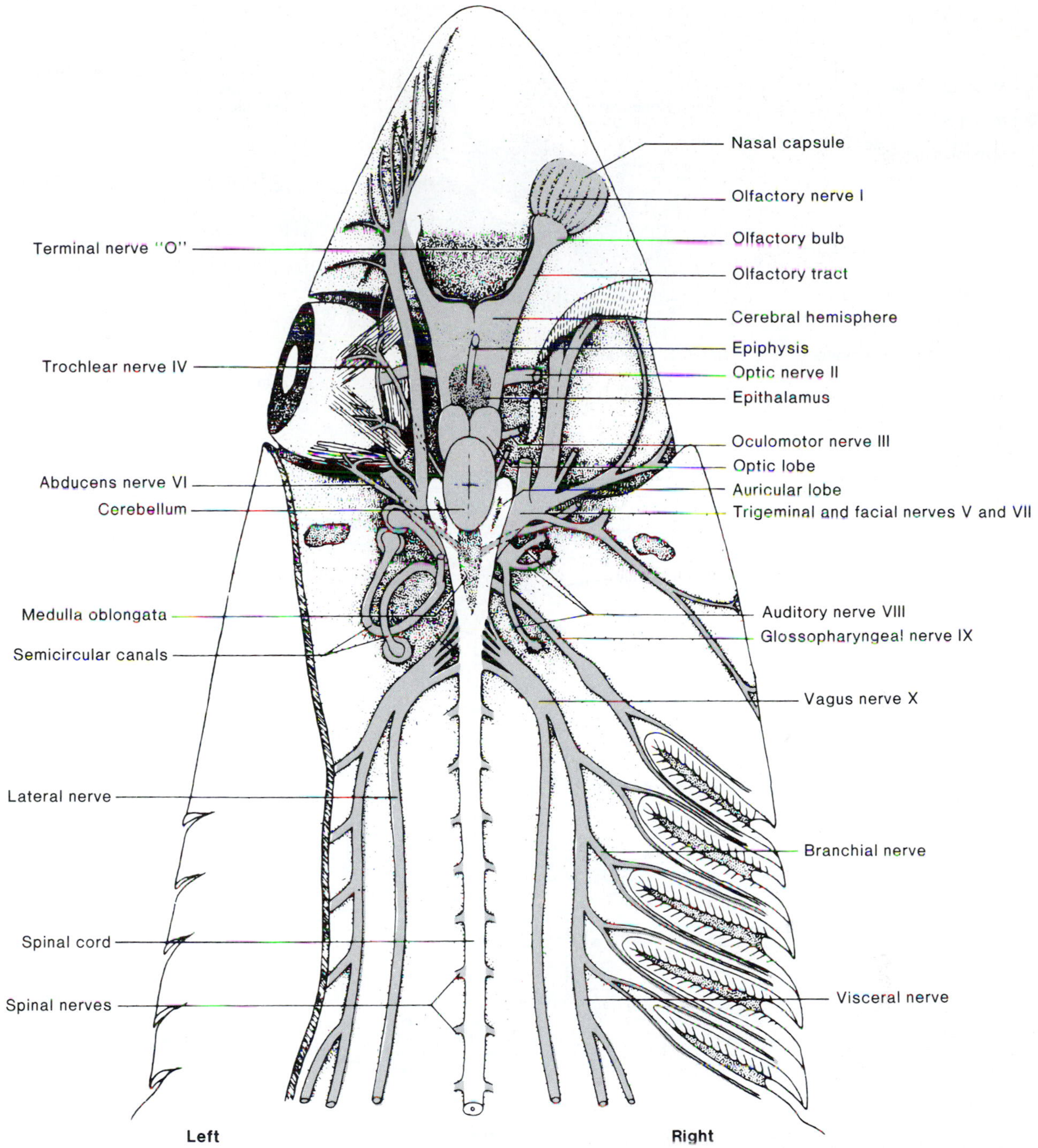
Nasal capsule
Olfactory nerve I
Olfactory bulb
Olfactory tract
Cerebral hemisphere
Epiphysis
Optic nerve II
Epithalamus
Oculomotor nerve III
Optic lobe
Auricular lobe
Trigeminal and facial nerves V and VII
Auditory nerve VIII
Glossopharyngeal nerve IX
Vagus nerve X
Branchial nerve
Visceral nerve
Terminal nerve "O"
Trochlear nerve IV
Abducens nerve VI
Cerebellum
Medulla oblongata
Semicircular canals
Lateral nerve
Spinal cord
Spinal nerves
Left
Right

Figure 19.7 Shark head showing brain and cranial nerves (dorsal view).

brain is referred to as the **tectum.** In fishes it is the primary area for the integration of sensory information and coordination of motor responses.

Posterior to the midbrain is the **cerebellum.** The **body of the cerebellum** is the large, median, oval mass lying over the top of the brain stem. Extending laterally are the earlike **auricular lobes.** The body of the cerebellum serves as a center for muscle coordination, and the auricular lobes function in equilibrium and balance.

Posterior and ventral to the cerebellum is the **medulla oblongata.** It tapers into the spinal cord and functions as a conducting pathway between the spinal cord and higher brain centers. It is also the point of origin for a number of cranial nerves. ▲

The Mammalian (Sheep) Brain

▼ If your specimen has the **meninges** removed, observe a demonstration specimen that still has at least some of the meninges intact. The outer tough layer of connective tissue is the **dura mater.** Below the dura is a loose, weblike **arachnoid layer,** and tightly adhering to the surface of the brain is

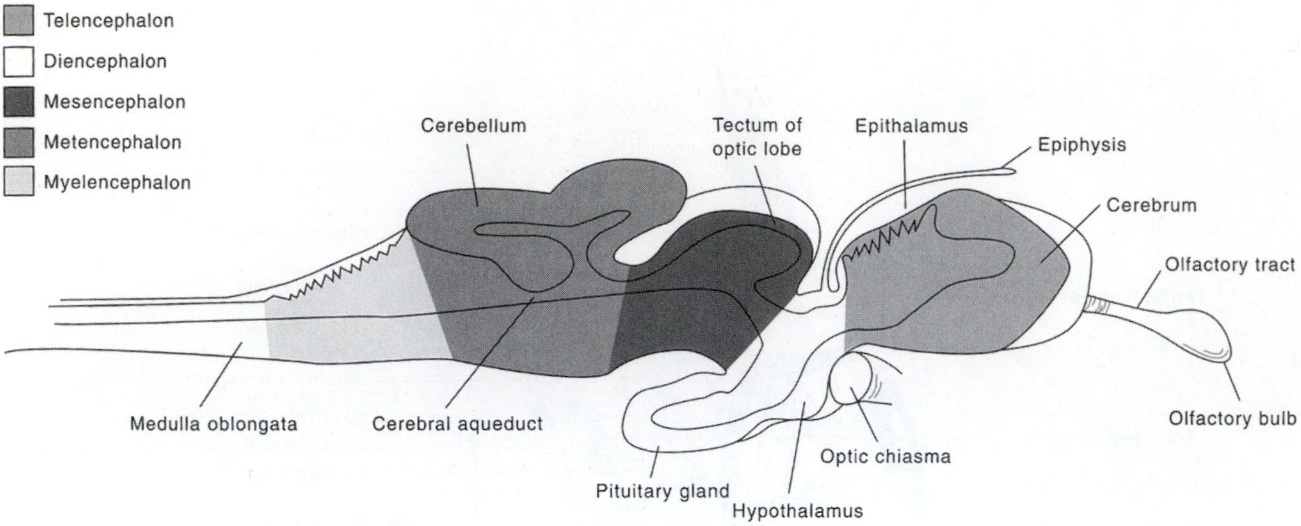
Telencephalon
Diencephalon
Mesencephalon
Metencephalon
Myelencephalon
Cerebellum
Tectum of optic lobe
Epithalamus
Epiphysis
Cerebrum
Olfactory tract
Olfactory bulb
Medulla oblongata
Cerebral aqueduct
Optic chiasma
Pituitary gland
Hypothalamus

Figure 19.8 Sagittal section of a shark brain. Embryological regions are indicated by different shading patterns.

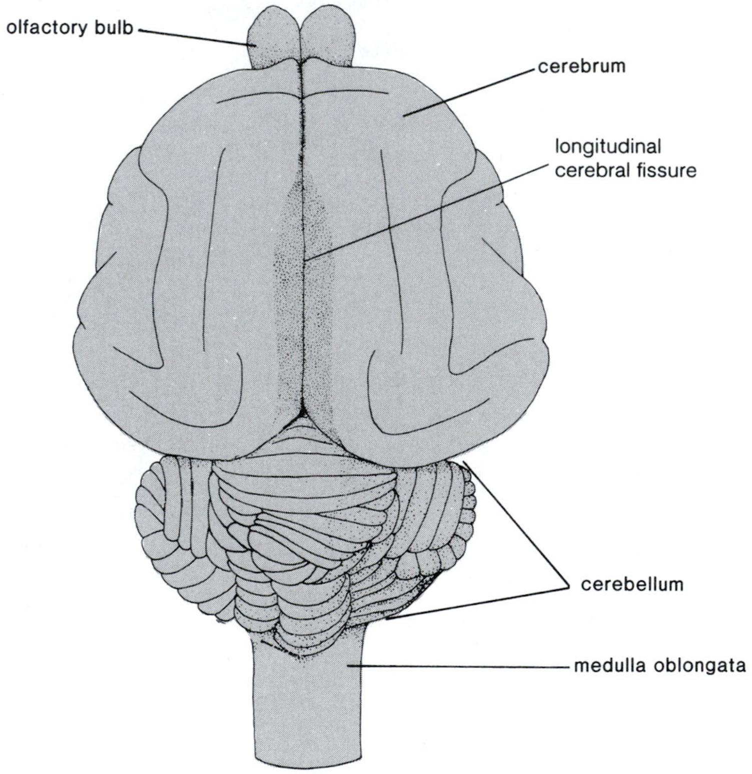
olfactory bulb
cerebrum
longitudinal cerebral fissure
cerebellum
medulla oblongata

Figure 19.9 Dorsal view of the mammalian brain.

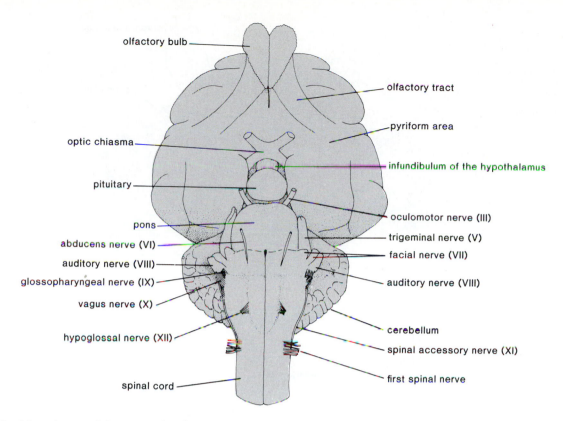

Labels on figure:
- olfactory bulb
- optic chiasma
- pituitary
- pons
- abducens nerve (VI)
- auditory nerve (VIII)
- glossopharyngeal nerve (IX)
- vagus nerve (X)
- hypoglossal nerve (XII)
- spinal cord
- olfactory tract
- pyriform area
- infundibulum of the hypothalamus
- oculomotor nerve (III)
- trigeminal nerve (V)
- facial nerve (VII)
- auditory nerve (VIII)
- cerebellum
- spinal accessory nerve (XI)
- first spinal nerve

Figure 19.10 Ventral view of the mammalian brain.

the **pia mater.** Between these layers **cerebrospinal fluid** circulates and forms a cushion for the brain. All of these layers extend down around the spinal cord as well.

From a dorsal view, the most obvious structure in the sheep brain is the **cerebrum** (fig. 19.9). Note that it is proportionately much larger in the sheep than the shark. This reflects its role in coordination and integration, which it has taken over from the tectum of fishes. Note the highly convoluted **cerebral cortex.** These convolutions greatly increase the surface area. The cerebral cortex consists of masses of gray matter pushed to the outside of the brain due to the increased size and importance of the cerebrum. The two **cerebral hemispheres** are separated by the **longitudinal cerebral fissure.** Posterior to the cerebral hemispheres is the **cerebellum.** The cerebellum of mammals remains a center for motor coordination and equilibrium and balance. Carefully spread the cerebellum from the cerebral hemispheres. Note the four swellings referred to as **corpora quadrigemina.** This is the roof of the midbrain. In the shark this was the tectum and represented the major site for nervous coordination and integration. In mammals the function is limited to optic and auditory reflexes. Extending posteriorly from the cerebellum is the **medulla.** In mammals the medulla functions as a respiratory center, awakening center, cardioregulatory center, and as a pathway for sensory and motor impulses going between the spinal cord and higher brain centers.

In a lateral view (Plate 11), find the **olfactory bulb** on the anterioventral border of the cerebrum. Extending posteriorly is the **olfactory tract.** It ends in the **pyriform area** of the cerebral cortex. Olfactory information is processed here.

Turn the brain to a ventral view (fig. 19.10 and Plate 12). Locate again the olfactory bulb, olfactory tract, and pyriform area. Medial and slightly anterior to the pyriform area is the **optic chiasma,** where **optic nerves** bring sensory information from the eyes to the brain through the diencephalon. Posterior to the optic chiasma is the main visceral control center, the **hypothalamus.** The **pituitary gland** connects to the brain at this point. It has most likely been broken off as the brain was removed from the skull. If available, examine a brain in which the pituitary gland has been left in place. Posterior to the hypothalamus is a broad enlargement that is derived from the metencephalon. This is the **pons.** The pons is a relay center for information going between the cerebrum and cerebellum. Posteriorly, the pons narrows into the medulla. Note the attachment points of one or more cranial nerves. Individual nerves will not be identified.

Finally, examine a midsagittal section of the sheep brain (see fig. 19.11 and Plate 11). Find the structures previously mentioned that can be seen from this view. In addition, find the **thalamus,** which serves as a relay site between upper and lower brain centers. Note also the canal system that is continuous with the central canal of the spinal cord. These canals contain cerebrospinal fluid. ▲

Brain Evolution

We have just examined two extremes in the evolution of the vertebrate brain. You may have noticed that two regions of the brain (the telencephalon and mesencephalon) underwent profound structural changes during vertebrate evolution.

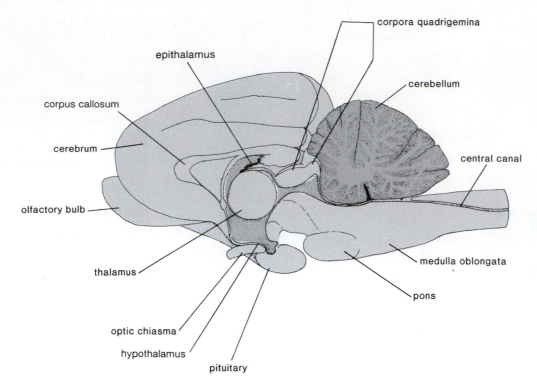

Figure 19.11 Midsagittal section of the mammalian brain.

▼ To appreciate these changes, examine figure 19.12 and any models or preserved brains available for study. Compare the structure and size of each brain region in the vertebrate classes represented. Answer questions 4 through 10 in worksheet 19.1 to guide your study. ▲

THE PERIPHERAL NERVOUS SYSTEM

The final component of the nervous system is the peripheral nervous system. It comprises two subdivisions: the **somatic nervous system,** innervating structures associated with the body wall; and the **autonomic nervous system,** innervating visceral structures (the "fight-or-flight" response). Both consist of nerves (bundles of neurons) leaving the central nervous system and going to the structure being innervated. Some of these nerves are associated with the brain (**cranial nerves).** In the region of the spinal cord, peripheral nerves exit the central nervous system between vertebrae in the cervical, thoracic, lumbar, and sacral regions.

▼ Examine a model of a human torso for these **spinal nerves.** Notice in the cervical, upper thoracic, lumbar, and sacral regions that the nerves fuse with neighboring nerves, forming a **plexus.** A **cervical plexus** represents cervical spinal nerves fusing, then going to the neck and upper thorax. A **brachial plexus** (lower cervicals and upper thoracic spinal nerves) goes to the shoulder and upper arm. A **lumbosacral plexus** (lumbar and sacral spinal nerves) goes to the pelvis and legs. Spinal nerves supplying the lower two-thirds of the thorax come independently to the structures they innervate. We will not distinguish anatomically between the components of the somatic and autonomic nervous systems. ▲

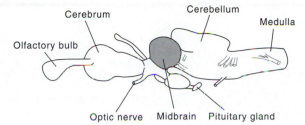

Fish (Shark)

Cerebrum, Cerebellum, Medulla, Olfactory bulb, Optic nerve, Midbrain, Pituitary gland

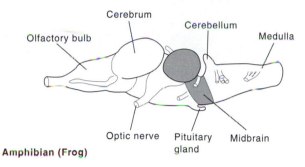

Amphibian (Frog)

Cerebrum, Cerebellum, Medulla, Olfactory bulb, Optic nerve, Pituitary gland, Midbrain

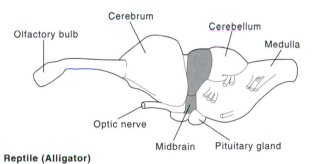

Reptile (Alligator)

Cerebrum, Cerebellum, Medulla, Olfactory bulb, Optic nerve, Midbrain, Pituitary gland

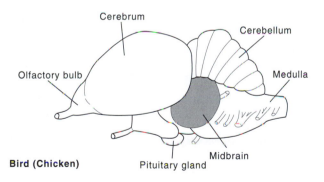

Bird (Chicken)

Cerebrum, Cerebellum, Medulla, Olfactory bulb, Midbrain, Pituitary gland

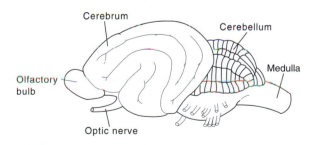

Mammal (Cat)

Cerebrum, Cerebellum, Medulla, Olfactory bulb, Optic nerve

5. Is gray matter inside or outside of white matter in the spinal cord of mammals? _____

6. Is gray matter inside or outside of white matter in the cerebral hemispheres of mammals? _____

   Of fishes?_____

7. Is gray matter inside or outside of white matter in the medulla oblongata of mammals? _____

   Of fishes? _____

8. What kinds of functions occur in gray matter?

   _____

   _____

9. What is the function of the hypothalamus in virtually all vertebrates? _____

   _____

10. What is the function of the cerebellum in virtually all vertebrates? _____

    _____

11. What is a plexus? _____

    _____

    _____

Figure 19.12 Brains of representatives of the vertebrate classes. The midbrain (mesencephalon) is shaded in gray. Note the relative sizes of the brain regions in each class representative. The mesencephalon of the mammalian brain is covered by the expanded and highly convoluted cerebrum.

KEY TERMS

autonomic nervous system   266

axon   259

cerebellum   263

cerebral hemisphere   262

dendrite   259

epithalamus   262

hypothalamus   262

medulla oblongata   263

neurolemmocyte   259

neuron   259

optic lobes   262

plexus   266

somatic nervous system   266

synapse   259

thalamus   262

WORKSHEET 19.1   Vertebrate Nervous Regulation

The Neuron

1. Diagram the structure of a multipolar neuron. Label and describe the function of each part.

2. Describe the association between neurons as they mediate a simple response to a stimulus.

The Central Nervous System

3. What is integration?

4. What is the function of the telencephalon (cerebral hemispheres) in fishes?

5. What is the function of the telencephalon in mammals?

6. How are the functional changes you just described evidenced by structural changes that occurred during vertebrate evolution?

7. What is the function of the mesencephalon (optic lobes) in fishes?

8. What is the function of the mesencephalon in mammals?

9. How are the functional changes you just described evidenced by structural changes that occurred during vertebrate evolution?

10. Regions of the brains of some vertebrates are specialized for sensory functions that have become very important in a particular group. Birds, for example, depend on highly developed visual reflexes. What part of their brain do you think would be enlarged for these functions? _____

Many fishes depend on olfactory reflexes for locating food or home (breeding) waters. For example, fishes such as lampreys and salmon return from the ocean to the stream where they hatched and grew years earlier. Streams have characteristic odors that are used by these fishes for identifying the home stream. What part of their brain would be enlarged for these functions? _____

Exercise 20

Vertebrate Circulation

Learning Objectives

After completing this exercise, you should be able to do the following:

1. Describe the structure and function of vertebrate circulatory systems.

2. Identify and describe the function of each formed element.

3. Describe the role of arterioles and capillaries in vertebrate circulation.

4. Describe circulation through the mammalian heart, identifying and describing the functions of chambers, vessels, and valves.

5. Identify major vessels of the rat and shark and trace circulation through them.

6. Describe the evolutionary relationships among the vertebrate classes as reflected in the evolution of the vertebrate heart.

7. Describe the events of the depolarization and repolarization sequence of the heart as reflected in an electrocardiogram.

8. Describe the interaction between antibodies and antigens.

Prelaboratory Quiz

Study this week's laboratory exercise and then complete the following quiz to assess your preparation for the laboratory.

1. The most abundant kind of white blood cell that functions in phagocytosis of foreign substances is a(n)
   a. neutrophil.
   b. eosinophil.
   c. lymphocyte.
   d. monocyte.
   e. basophil.

2. The white blood cell that functions in immune responses is a(n)
   a. neutrophil.
   b. eosinophil.
   c. lymphocyte.
   d. monocyte.
   e. basophil.

3. All of the following are present in the heart of a fish EXCEPT
   a. a sinus venosus.
   b. a conus arteriosus.
   c. a two-chambered ventricle.
   d. a single-chambered atrium.

4. Exchanges of gases, nutrients, and wastes occur across
   a. venules.
   b. capillaries.
   c. arterioles.
   d. All of the above are correct.

5. Strands of connective tissue that anchor atrioventricular valves to the heart walls of mammals are called
   a. coronary strands.
   b. valve cords.
   c. ventricular strands.
   d. chordae tendineae.

6. True/False   Blood leaving the left ventricle of the mammalian heart enters the pulmonary artery.

7. True/False   The heart of a bird has completely divided atria and ventricles, and the sinus venosus is a large chamber that receives blood from the vena cavae and serves as a pacemaker.

8. True/False   In the electrocardiogram, the QRS complex corresponds to ventricular depolarization.

9. True/False   The dissection of the rat circulatory system in this exercise will begin by observing veins in the abdominal region of the rat.

10. True/False   Blood is pumped from the heart of a shark to the gills and immediately returns to the heart before circulating through the rest of the body.

The structure and function of vertebrate circulatory systems is complex. In this section we will be able to touch on major aspects of both structure and function in only a few representative organisms. The circulatory system is the means of transport for oxygen, carbon dioxide, food, metabolic wastes, and hormones in the vertebrate body. In addition, the circulatory system helps to maintain body temperature and proper pH and to fight disease.

The blood and lymph are functionally the most important parts of the circulatory system because they carry substances to and from all parts of the body. The heart circulates blood through a system of closed vessels. A **closed circulatory system** is one in which blood is confined to vessels throughout the system. On leaving the heart, blood enters **arteries.** Arteries break up into **arterioles** and finally **capillaries.** It is in the very tiny capillaries that exchange with the tissues takes place. Capillaries coalesce into **venules,** and venules into **veins.** Blood pressure in venules and veins is very low. Valves are present here to ensure one-way movement of blood back to the heart. The **lymphatic system** is an arrangement of vessels that provides a means for getting excess tissue fluids and proteins back to the bloodstream, as well as serving in body defenses. At the tissues, tissue fluids and proteins enter the lymphatic system through porous walls of blind-ending lymphatic capillaries. The lymphatic vessels have valves that ensure a one-way movement of fluids back toward the heart. In higher vertebrates, lymph nodes are located along lymphatic vessels. These serve as sites for the storage of white blood cells (lymphocytes) involved with the body's immune system. The spleen is similar to a lymph node, but it also functions with the cardiovascular system in storing red blood cells, destroying worn-out red blood cells, and producing red and white blood cells.

FORMED ELEMENTS

Blood consists of **plasma** and **formed elements.** Plasma (55% by volume) consists of water, dissolved salts, suspended proteins, sugars, gases, and other suspended and dissolved materials. The remaining 45% of blood volume con-

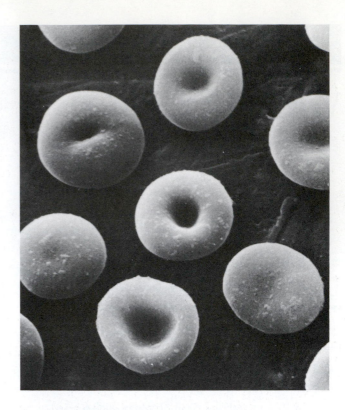

Figure 20.1   Human erythrocytes, scanning electron micrograph (×7,020). From R. G. Kessel and C. Y. Shih. *Scanning Electron Microscopy in Biology.* © Springer-Verlag.

sists of formed elements—blood cells and platelets. Formed elements are classified as follows:

1. **Red blood cells (erythrocytes,** fig. 20.1). Contain hemoglobin that functions in carrying oxygen to and from the tissues.

2. **White blood cells (leukocytes,** fig. 20.2).
   Granular leukocytes—Contain cytoplasmic granules and irregular nuclei.
   a. **Neutrophils**—The most abundant leukocyte. Phagocytizes microbes or other injurious particles. About 60% of WBCs.
   b. **Eosinophils**—Phagocytize foreign proteins and immune complexes. Tend to proliferate in allergic reactions and parasitic infections. About 3% of WBCs.
   c. **Basophils**—Play a role in preventing blood clotting by releasing heparin. Promote inflammatory responses by releasing histamine. About 1% of WBCs.
   Nongranular leukocytes—Lack cytoplasmic granules. Regular nuclei present.
   a. **Lymphocytes**—Function in the body's immune response. Second most abundant leukocyte. About 30% of WBCs.
   b. **Monocytes**—Phagocytize microbes and other injurious particles. About 6% of WBCs.

3. **Platelets.** Cellular fragments containing several compounds that facilitate clotting.

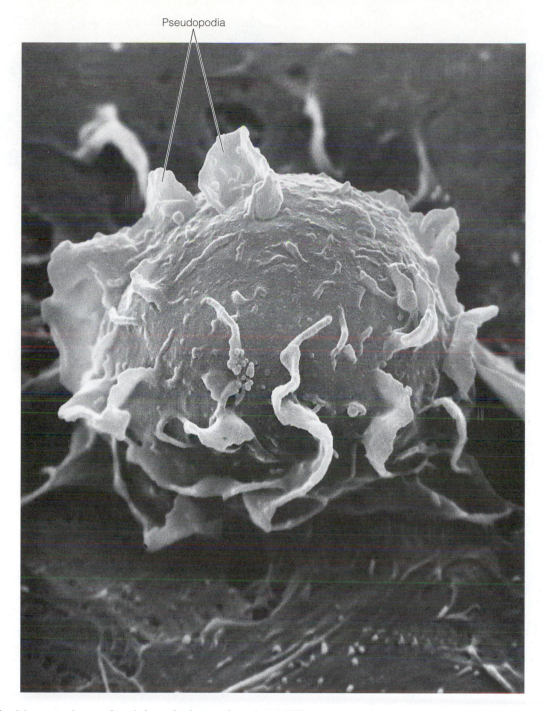

Pseudopodia

Figure 20.2 Monocyte (macrophage) from the human lung (×16,000).

Examining Blood Smears

▼ In this exercise you will examine a slide of human blood. Obtain a prepared slide of human blood. Examine it under low and high power of your compound microscope. Erythrocytes will be stained red, leukocyte nuclei dark blue. Compare what you see to figure 20.3. Find as many kinds of white blood cells as you can. If your instructor directs, examine the slide under oil immersion. Try to appreciate the relative numbers of different cell types. You may also observe small, dark, irregular spindles or disks. These are platelets. ▲

MICROCIRCULATION

The work of the cardiovascular system occurs in capillary beds, where exchanges of gases, nutrients, and wastes occur. Frogs anesthetized with urethane can be used to demonstrate microcirculation. One ml of 10% urethane was injected under the skin of the back. When the frog was anesthetized, it was wrapped in moist cheesecloth, and the web of the foot was spread over an opening of a frog board.

▼ Examine the web under transmitted light with low and high powers of a compound microscope. Can you tell the

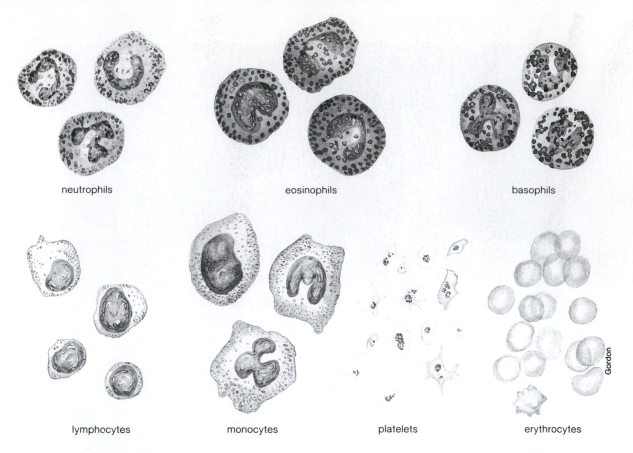

Figure 20.3 Formed elements of human blood.

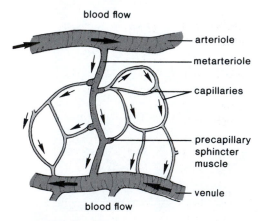

Figure 20.4 Blood flow through a capillary bed. Note that blood flow into capillary beds is regulated by precapillary sphincters (smooth muscle). Blood may be shunted past a capillary bed, passing directly from arteriole to venule through the metarteriole.

difference between arterioles and capillaries? venules and capillaries? Does blood pulsate in some vessels? (See fig. 20.4.) ▲

1. What is a closed circulatory system? _____

_____

2. What is the most common kind of formed element in vertebrate blood? _____

   What is that formed element's function? _____

   _____

3. What is the most common kind of white blood cell in vertebrate blood? _____

   What is that formed element's function? _____

   _____

4. How can you distinguish between an arteriole and a venule as observed in the web of a frog's foot? _____

   _____

   _____

SYSTEMIC CIRCULATION—THE SHARK

In this section you will use demonstration dissections and models to study circulation in fishes.

▼ Examine a dissected *Squalus* specimen (Plate 13). Locate the heart just anterior to the large gray liver. Blood enters the heart at the thin-walled sac at the posterior end; this is the **sinus venosus.** In mammals, the sinus venosus has been reduced to the **sinoatrial node.** In all vertebrates, the sinus venosus initiates the heartbeat. Blood then enters

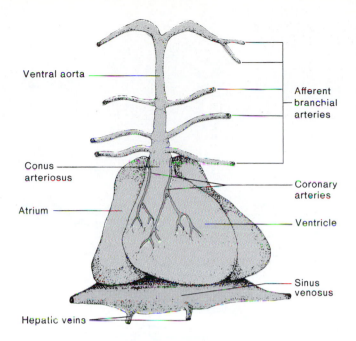

Figure 20.5 Ventral view of the shark heart.

Labels on figure:
Ventral aorta
Afferent branchial arteries
Conus arteriosus
Coronary arteries
Atrium
Ventricle
Sinus venosus
Hepatic veins

the single **atrium** (anterior and dorsal to the sinus venosus). The **ventricle** receives blood from the atrium. (The ventricle is the primary pumping structure of the heart.) The blood then enters the **conus arteriosus** and goes to the **ventral aorta.** The ventral aorta takes blood anteriorly to the gills. The branches to the gills are called **afferent branchial arteries.** Trace one or two of these vessels forward to the gills. In the gills, the vessels break into capillaries for gas exchange. Compare the dissection to figure 20.5 and any models available.

At this point, fold open the lower jaw as shown in figure 20.6. In doing so, notice the vessels on the roof of the mouth that come from the gills (Plate 14). These are **efferent branchial arteries.** They join medially to form the **dorsal aorta.** The dorsal aorta then runs posteriorly, delivering blood to the body. Trace the aorta a short distance. Blood returns to the heart through a series of veins that may not be injected in your specimen (fig. 20.6). Major vessels that return blood to the sinus venosus are the **anterior cardinals** (from the head), **posterior cardinals** (from the body), and the **lateral abdominals.** The latter can be seen along the lateral body wall. The posterior cardinals empty as large sinuses into the sinus venosus and lie along the dorsal body wall. The anterior cardinals enter anteriorly and laterally into the sinus venosus. These cannot be seen easily because they are embedded in muscles of the head. Try to find their point of entrance using a small probe. All three of these veins unite to enter the sinus venosus at the **common cardinal.** The shark also has a **hepatic portal system** that delivers blood from capillaries of the intestine to the liver. **Hepatic veins** deliver blood from the liver to the sinus venosus. Find the hepatic portal vein, but do not try to find the hepatic veins. ▲

THE MAMMALIAN HEART

After studying mammalian circulation, you will study the hearts of other vertebrates to gain an appreciation of the evolutionary changes that led to the mammalian heart. The heart is responsible for pumping blood through arteries and arterioles to the capillaries, where exchanges take place. In studying the mammalian heart, you will use fresh or preserved cow, pig, or sheep hearts.

External Structure

▼ Examine a heart and note the division into two types of chambers (fig. 20.7). The large muscular portion consists of two **ventricles.** The two small earlike lobes on top of the ventricles are the **atria.** The widest part of the heart is the base, and the tapered portion is the apex. Before locating specific structures, one must be able to distinguish left from right on the heart. In doing this, examine the thickness of the ventricular walls. Because the left ventricle must pump blood over the entire body, its wall is much thicker than the wall of the right ventricle, which pumps blood a short distance to the lungs. The left ventricular wall is, therefore, firmer than the right ventricular wall. Feel the ventricular walls to determine right and left. Now orient the heart ventral surface up so the heart's right is on your left.

Before dissecting the heart, identify the pulmonary artery and the aorta. The **pulmonary artery** comes off the right ventricle and arches to the left. The **aorta** comes off the left ventricle and passes dorsal to the base of the pulmonary artery. The aorta has the thicker walls of the two vessels. It arches to the left and then posteriorly. Distend the

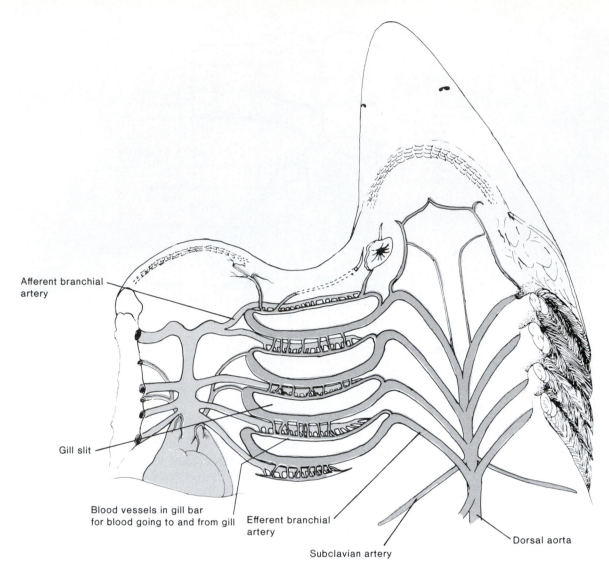

Afferent branchial artery

Gill slit

Blood vessels in gill bar for blood going to and from gill

Efferent branchial artery

Subclavian artery

Dorsal aorta

Figure 20.6 Dorsal view of the heart, ventral aorta, and afferent branchial arteries. Ventral view of the efferent branchial arteries and dorsal aorta.

walls of the aorta and note the elasticity. When the ventricles contract, a pulse of blood enters the aorta, distending its walls. The elastic rebound of the aorta helps propel blood through the vessel.

Internal Structure

Beginning at the base of the pulmonary artery, make an incision through the ventral wall of the right ventricle and extend the incision toward the apex of the heart. Near the apex, continue the incision back toward the base on the dorsal aspect of the heart. You should then be able to lift the outer wall of the ventricle to expose internal structures (fig. 20.8). Do the same for the left ventricle, beginning near the base of the aorta and ending on the dorsal aspect at the base of the heart. Rinse out any clotted blood inside the ventricles.

Blood enters the heart through **vena cavae** that join the right atrium. These vessels are easiest to locate by probing into the right atrium through the right ventricle with a

blunt probe or the handle of a scalpel. Blood passes from the right atrium through the **tricuspid (right atrioventricular) valve** into the right ventricle. Note the three flaps of connective tissue that make up the valve. Strands of connective tissue, called **chordae tendinae,** anchor the valve flaps to prevent the valve from everting into the right atrium during ventricular contraction. This prevents the backflow of blood into the right atrium. Blood is forced into the **pulmonary artery** through the **pulmonary semilunar valve.** Extend your cut far enough into the pulmonary artery to observe the semilunar valve. When the ventricle relaxes, blood in the pulmonary artery falls back against this valve, preventing the reentry of blood into the right ventricle. Blood in the pulmonary artery moves toward the lungs. Note that a vessel is defined as an artery on the basis of the direction of blood flow (away from the heart).

Blood returns to the heart from the lungs through **pulmonary veins** that enter the heart at the left atrium. Probe into the left atrium through the left ventricle to find these

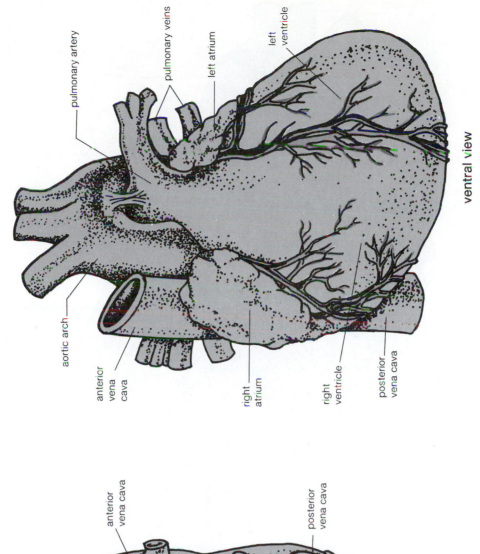

pulmonary artery

pulmonary veins

left atrium

left ventricle

aortic arch

anterior vena cava

right atrium

right ventricle

posterior vena cava

ventral view

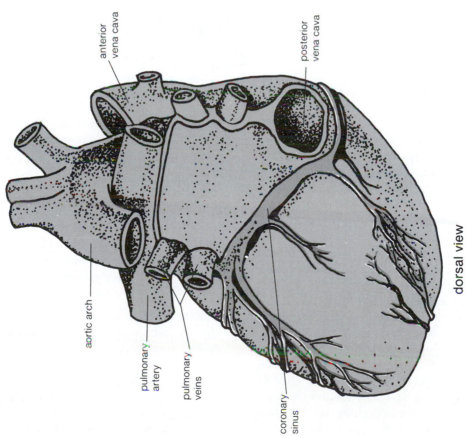

anterior vena cava

posterior vena cava

aortic arch

pulmonary artery

pulmonary veins

coronary sinus

dorsal view

Figure 20.7   The mammal heart, external structure.

277

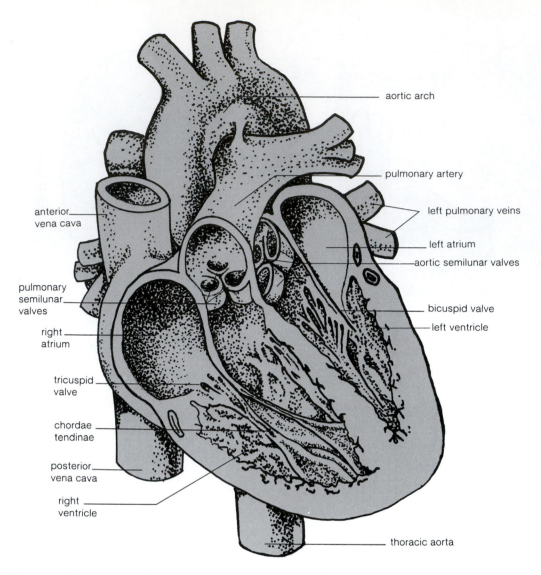

Figure 20.8 The mammal heart, internal structure.

vessels. Blood then passes into the left ventricle through the **bicuspid (left atrioventricular) valve.** It is similar to the tricuspid valve except it is made of two flaps of connective tissue rather than three. From the left ventricle, blood moves to the **aorta** through the **aortic semilunar valve.** Find the aortic semilunar valve by cutting back toward the heart from the distal cut-end of the aorta. Just above the semilunar valve, note two small openings. These are the openings to the **coronary circulation.** During the relaxation period of the ventricles, blood falls back against the aortic semilunar valve and enters the vessels supplying the heart muscle with blood. Blood in the coronary circulation returns to the right atrium through a **coronary sinus,** which can be located by probing through the right atrium along the dorsal surface of the heart near where the atria and ventricles meet. Finally, place your left index finger into the right atrium and your right index finger into the left atrium and feel along the medial walls of the atria. You should detect a thin layer of connective tissue separating the two atria. In the fetus an opening called the **foramen ovale** shunts some blood from the right atrium to the left

atrium, thus bypassing the lungs. Normally, this thin layer of connective tissue closes the foramen ovale after birth. ▲

SYSTEMIC CIRCULATION—THE RAT

We will begin our dissection of the rat circulatory system at the heart. Be very careful in this dissection, because arteries and veins are easily torn. Doubly or triply injected specimens will be used. Arteries are injected with red latex, veins with blue latex, and the hepatic portal system with yellow latex (if triply injected). This scheme is reversed in the pulmonary circulation.

▼ Reread the general instructions for dissection in exercise 12, and remember that you will be using your rat in most of the exercises that follow and while studying for laboratory tests and quizzes. Do not cut your rat unnecessarily or allow it to desiccate.

With your rat positioned ventral side up on your dissecting pan, open the body cavity by making a midventral incision in the abdominal wall (fig. 20.9). Cut through the

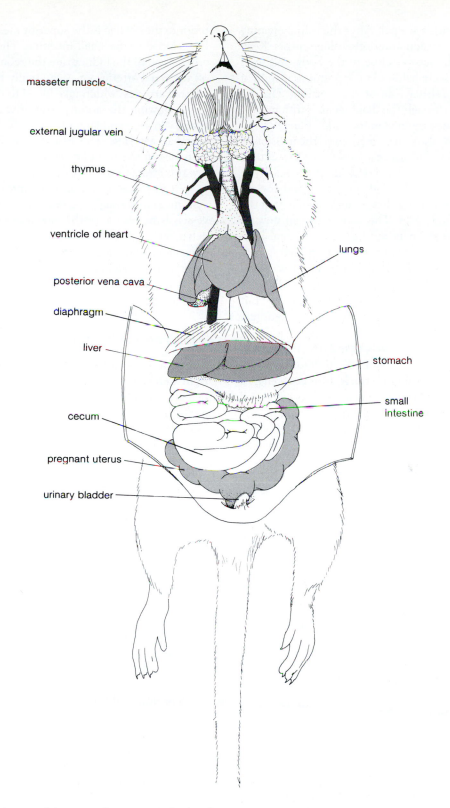

Figure 20.9   Ventral view of the visceral structures of a female rat.

skin and body wall with a scalpel. After the initial opening is made, extend the incision anteriorly using scissors. Keep the lower blade of the scissors close to the body wall so as not to damage the viscera below. At the posterior margin of the rib cage, cut to either side of the sternum. (In the process, you will cut through the diaphragm, a thin muscular sheet that separates the abdominal and thoracic cavities.) Continue anteriorly until your cut joins the incision in the neck that was made during the injection process. Extend your incision in the abdomen posteriorly until you reach the pelvic region, but do not damage genitalia. Find the diaphragm and free it by cutting close to its attachment at the ventral and lateral body wall. The body wall can now be spread and pinned to the bottom of the pan. Lateral incisions through the body wall help keep the body wall spread.

In the following exercise, you will need to follow blood vessels to various internal organs. Use figure 20.9 and Plate 15 to familiarize yourself with the location of major organs. Lift and separate organs gently in the process. Do not cut or tear any tissues.

Covering the heart and vessels in the upper chest and neck will be a large lymphatic gland, the **thymus.** Remove the thymus with forceps. Note that the heart is covered by a sac of connective tissue, the **pericardial sac.** It is composed of two layers. The outer sac is the **parietal pericardium,** and the inner sac that tightly covers the heart and vessels is the **visceral pericardium.** Remove the parietal pericardium. Remove any fat covering the heart and then identify the four chambers and major vessels entering and leaving it (in the rat there are left and right anterior vena cavae as well as the single posterior vena cava).

The Arterial System

Follow the aorta away from the heart, removing connective tissue so you can see the vessels (Plate 16). The **innominate artery** is the first branch of the aorta. Follow the innominate anteriorly. It soon branches into two major arteries (fig. 20.10). The **right subclavian artery** branches off laterally, carrying blood to the foreleg. The medial branch is the **right common carotid artery.** It carries blood to the head. The common carotid branches in the upper region of the neck into an internal carotid (carrying blood to the brain) and an external carotid (carrying blood to the skin and muscles of the head). You need not find these branches.

Return to the aorta at the point of divergence of the innominate. Note that the aorta now curves dorsally and posteriorly. This curve is the **arch of the aorta.** The next two branches of the aorta are the **left common carotid** and the **left subclavian.** In contrast to those on the right side, these arteries come off the aorta individually. Locate and trace these branches a short distance. Trace the aorta posteriorly. In the thoracic region, it gives off branches to the body wall. These are collectively referred to as intercostal arteries.

The next major branch of the aorta is posterior to the diaphragm (Plate 17). It is referred to as the **coeliac artery.** It gives off branches to the spleen, liver, and stomach. Posterior to the coeliac is the **superior mesenteric artery.** It delivers blood to the small intestine. The large arteries delivering blood to the kidneys are the **renal arteries.** Branching off the renal arteries, or occasionally the aorta, are smaller **adrenal arteries,** which take blood to the adrenal glands.

Following the aorta posteriorly, note small branches that go dorsally and laterally to the body wall. These are **iliolumbar arteries.** The **ovarian** or the **spermatic artery** branches off the aorta just posterior to the renal artery. Follow this artery to the ovary or the testis.

At this time, if you have not already done so, continue your incision through the skin and body wall to expose the vessels lying on top of the muscles of one leg. Note in the pelvic region the aorta branches into two main arteries carrying blood to the legs. These are the **common iliacs.** A third branch, the **caudal artery,** goes dorsally and posteriorly into the tail. Numerous smaller branches of these arteries carry blood to all parts of the tail and legs. They will not be located.

The Venous System

Return to the heart (see Plate 16). Again find the **left** and **right anterior vena cavae.** Follow them anteriorly. They branch into **subclavian veins,** bringing blood back from the forelegs, and **internal** and **external jugular veins,** bringing blood back from the head (fig. 20.11). The internal jugulars return blood from the brain. They are small and may not be injected. External jugulars bring blood back from the muscles and skin of the head. Entering into the left anterior vena cava very near the heart is the **azygous vein.** It returns blood to the heart from the chest wall. Follow this vessel to the chest wall. Return to the posterior vena cava and follow it posteriorly. It runs through the diaphragm and the liver. Within the substance of the liver the **hepatic vein** joins the posterior vena cava. We will not attempt to dissect this vessel. Trace the **renal veins** (see Plate 17). Blood from the kidneys enters the vena cava through renal veins. Note that veins from the adrenal gland (**adrenal veins**) empty into the renal vein before reaching the vena cava. The **spermatic** or the **ovarian veins** enter the vena cava directly on the right side of the body (the next posterior vessel), but through the renal vein on the left side.

Continue following the vena cava posteriorly. The next branch following the right spermatic or ovarian vein is the **iliolumbar vein,** delivering blood from the body wall. The vena cava then branches as vessels from the hind legs (**common iliac veins**) unite. The common iliac veins receive numerous branches from all parts of the leg. We will not trace these branches.

Finally, recall that blood entered the gut tract through the coeliac and superior mesenteric arteries. We have not seen how this blood gets back to the vena cava. This is the job of the **hepatic portal system.** A portal system is a vascular system that begins in a capillary bed and ends in a capillary bed. The capillary beds in the hepatic portal system are in the gut tract at one end and the liver at the other. If your specimen is triply injected, the hepatic portal system will be

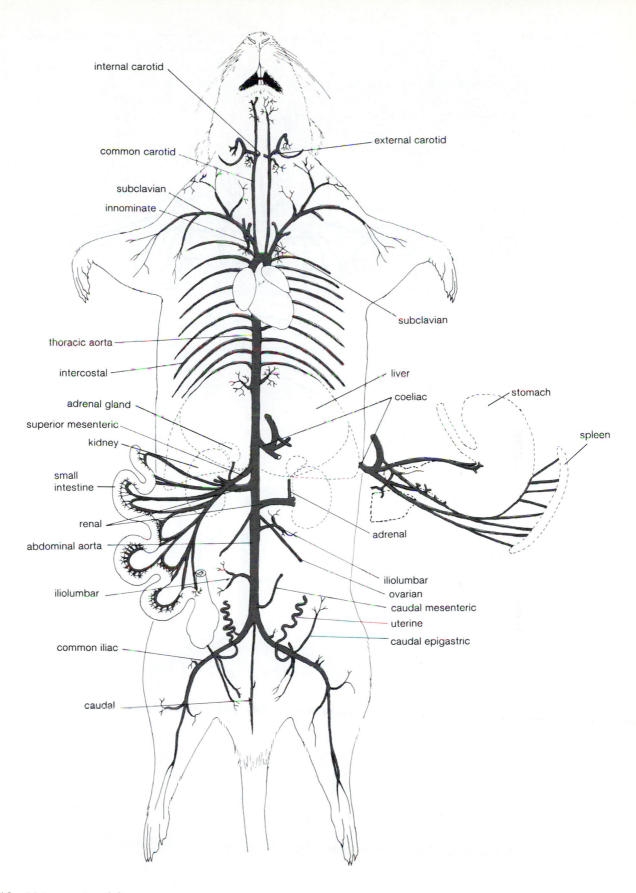

internal carotid

external carotid

common carotid

subclavian

innominate

subclavian

thoracic aorta

intercostal

liver

coeliac

stomach

adrenal gland

superior mesenteric

kidney

spleen

small intestine

renal

adrenal

abdominal aorta

iliolumbar

ovarian

iliolumbar

caudal mesenteric

uterine

caudal epigastric

common iliac

caudal

Figure 20.10 Major arteries of the rat.

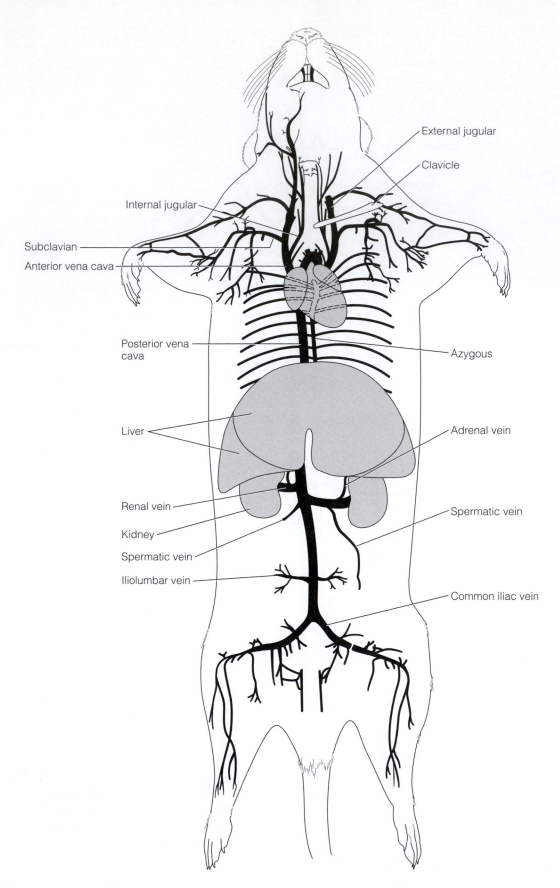

Figure 20.11   Major veins of the rat.

TABLE 20.1   Evolutionary Changes in the Vertebrate Heart

	Sinus Venosus	Atrium	Ventricle	Conus Arteriosus
Shark	Single	Single	Single	Single, leads to ventral aorta
Lungfish	Single	Partially divided, R. and L. to accommodate pulmonary circulation	Partially divided to help separate oxygenated and nonoxygenated blood	Single, with valve to help separate oxygenated and nonoxygenated blood
Reptile				
Crocodile	Smaller	Divided	Completely divided	Incorporated into bases of a single pulmonary artery and two systemic arteries
All others	Smaller	Divided	Partially divided	
Bird	Smaller	Divided	Complete separation of oxygenated and nonoxygenated blood	Same as reptiles, except left systemic is gone
Mammal	Gone as a chamber; represented by sinoatrial node	Divided	Same	Same except systemic arches to the left rather than right as in birds

yellow. If doubly injected, your specimen will have dried blood in the hepatic portal system. Trace the vessels from the intestine to the liver. Blood from the intestine passes to the liver, where it is cleaned of impurities picked up from the intestine. Blood is then passed to the hepatic vein and then to the posterior vena cava. ▲

STOP AND ASK YOURSELF

5. Trace a drop of blood from the ventral aorta to the dorsal aorta of a shark. _____

   _____

6. After passing the tricuspid valve, blood flows

   into the _____
   of the mammalian heart.

7. What is the function of chordae tendinae in the mammalian heart? _____

   _____

8. Trace a drop of blood from the left ventricle of the rat to the kidneys and back to the right atrium of the heart. _____

   _____

   _____

   _____

COMPARATIVE HEART STRUCTURE

You have just observed two extremes in vertebrate circulatory systems. The changes that take place in the heart between fishes and mammals reflect evolutionary relationships among the vertebrate classes. The following exercise is a study of fish, lungfish, reptile, and bird hearts and should help you understand these relationships.

▼ As you work through the following observations, consult any models, preserved hearts, and diagrams available. Also turn frequently to figure 17.5 in worksheet 17.1, which depicts evolutionary relationships among the vertebrates. The evolutionary changes that follow are summarized in table 20.1. ▲

You previously studied the shark heart. List the four chambers of the shark heart in order from the common cardinal vein to the ventral aorta. _____

The following changes in the heart accompanied the adaptation of vertebrates to terrestrial environments by permitting gas exchange with air without excessive water loss.

Internal gas exchange surfaces (lungs) provide large surfaces for gas exchange and at the same time help retard water loss because of the lungs' position within the body cavity. The development of lungs, however, was accompanied by a change in circulatory pathways that brings blood into close contact with air. How this may have been accomplished is reflected in the structure of the lungfish circulatory system (fig. 20.12a, b). The lungfish is used here because it is a living vertebrate that shows a circulatory pattern that may be similar to that present in early amphibians. The lungfish respires primarily through lungs. Lungs allow it to survive both periodic droughts and stagnated water in which it lives. In addition, the lungfish has gills to aid in gas exchange. It has a single sinus venosus but has a partially divided atrium. Blood from the sinus venosus passes into the right atrium, to the right side of a partially divided ventricle, and then through the conus arteriosus to the ventral aorta and gills. In addition, blood from the ventricle can be

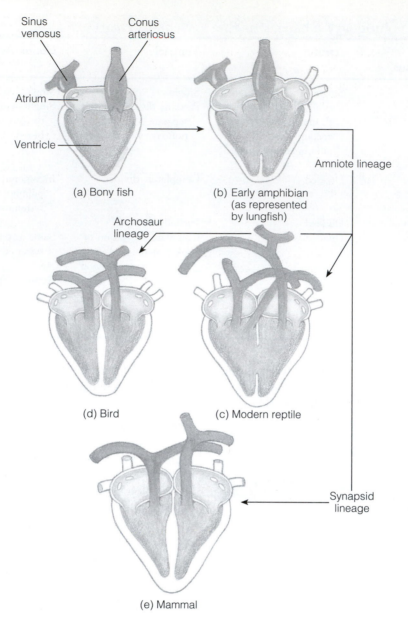

Sinus venosus

Conus arteriosus

Atrium

Ventricle

(a) Bony fish

(b) Early amphibian (as represented by lungfish)

Amniote lineage

Archosaur lineage

(d) Bird

(c) Modern reptile

Synapsid lineage

(e) Mammal

Figure 20.12 A possible sequence in the evolution of the vertebrate heart. *(a)* Fish heart. *(b)* In lungfish, partially divided atria and ventricles separate pulmonary and systemic circuits. The hearts of modern reptiles *(c)* were derived from the form shown in *(b)*. *(d)* The archosaur and *(e)* synapsid lineages resulted in completely separated, four-chambered hearts.

shunted to the lungs through a modified afferent arteriole. After oxygenation at the lungs occurs, blood returns to the left atrium. From the left atrium, blood goes to the left side of the partially divided ventricle and through the ventral aorta to the body. A valve in the conus arteriosus helps keep oxygenated and nonoxygenated blood separate.

From this basic pattern, the lineage leading to modern reptiles and birds diverged from that leading to modern mammals (fig. 20.12c). In the heart of the modern reptile, the sinus venosus is reduced in size (compared to that of the lungfish), the atria are completely divided, the ventricles are partially divided (except in the crocodile where they are completely divided), and the conus arteriosus is incorporated into the bases of the major arteries leaving the heart. These arteries result from the splitting of the ventral

aorta. A single pulmonary artery now carries blood from the right ventricle to the lungs. Right and left systemic arteries (aortae) carry blood from the heart to the body. The separation of oxygenated and nonoxygenated blood is more efficient because a valve directs oxygenated blood to the systemic arteries and nonoxygenated blood to the pulmonary arteries. This valve also diverts blood away from the lungs during periods of apnea (no breathing).

Birds, which evolved from ancient reptiles, have four completely separate heart chambers. However, the left systemic, present in reptiles, is absent, and the sinus venosus is further reduced in size (fig. 20.12d).

In the evolutionary pathway between ancient reptiles and mammals, the conus arteriosus and ventral aorta divided into two vessels rather than three (fig. 20.12e). Complete separation of oxygenated from nonoxygenated blood resulted from the development of a pulmonary artery, aorta (systemic artery), and completely divided atria and ventricles. In mammals, the sinus venosus is represented only by a small patch of tissue in the wall of the right atrium, the sinoatrial node. In all vertebrates, the sinus venosus (sinoatrial node) initiates the heartbeat.

CIRCULATORY PHYSIOLOGY— THE ELECTROCARDIOGRAM

The heart is a self-excitatory pumping structure. The rate at which the heart beats is controlled by the pacemaker, or **sinoatrial (SA) node.** The SA node initiates the heartbeat by sending out an impulse that very rapidly spreads over the heart through specialized conducting fibers. Since the SA node is located in the right atrium just below the point at which the vena cavae enter, the atria are first to receive this impulse. The impulse passes from the atria to the ventricles through the **atrioventricular (AV) node.** The impulse is delayed about 0.1 second at the AV node and then passes down the septum between the ventricles to the apex of the heart. It then proceeds up the sides to the base of the heart. This process is referred to as the **depolarization** of the heart. Immediately after depolarization, the muscle recovers (**repolarization**) and then contracts. What would you guess is the significance of the 0.1 second delay of the impulse at the AV node? _____

_____

The impulses moving over the heart set up electromagnetic fields that circulate through the body. They can be picked up by placing electrodes at various points on the body. What results is the **electrocardiogram (ECG).** A typical ECG is shown in figure 20.13. The various deflections in the ECG are labeled as shown in figure 20.13 and correspond to the following events:

P wave   atrial depolarization
QRS complex   ventricular depolarization
T wave   ventricular repolarization

Atrial repolarization is masked by ventricular depolarization. Your instructor will demonstrate an ECG. Note that these waves are produced from the impulses passing over the surface of heart muscle cells just prior to contraction. They are not a measurement of the contractions of heart muscle.

CIRCULATORY PHYSIOLOGY— ANTIBODY ACTION

Recall from your study of blood cells that lymphocytes are involved with the body's immune response. Lymphocytes are produced in the bone marrow and then go to a lymphoid organ to differentiate (mature). Some lymphocytes differentiate at the thymus and are called **T-cells.** T-cells move from the thymus gland to other lymphoid organs (e.g., lymph

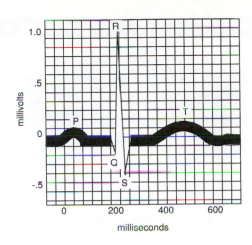

Figure 20.13   The electrocardiogram tracing.

nodes) to serve in the body's cell-mediated immune response. T-cells respond to specific foreign substances (**antigens**) by leaving the lymph node, attaching to the antigen, and activating proteins that destroy the antigen and attract other white blood cells (neutrophils and monocytes) that phagocytize the antigen.

Other lymphocytes, called **B-cells,** differentiate at other lymphoid structures (bursa of Fabricius in birds and probably the spleen and bone marrow of mammals). B-cells then go to lymph nodes to serve in the body's humoral immune response. On contact with a specific antigen, B-cells release **antibodies (immunoglobins)** to the bloodstream. (Most B-cells do not leave the lymph node.) Circulating antibodies attach to the antigen and cause agglutination (sticking together), lysis (breaking apart), or enhancement of phagocytosis by neutrophils and monocytes. In both humoral and cell-mediated immunity, the initial infection results in the formation of **memory cells,** which have an enhanced ability to respond to a second exposure to an antigen. Active immunity differs from passive immunity in that the latter is the result of the transfer of antibodies from one person to another (e.g., the transfer of antibodies from mother to offspring during suckling). Passive immunity is short-lived.

▼ An interesting demonstration of antibody action can be performed using a blood commensal of the rat, *Trypanosoma lewisi.* The same genus in humans causes sleeping sickness, but in rats the infection is harmless. After the initial infection, lasting about two weeks, a rat will build up effective immunity. Examine blood from a rat that was infected with *T. lewisi* about eight days prior to today's laboratory. Under high power of a compound microscope, locate the flagellated protozoans moving in the blood. Obtain a small drop of immune serum from a previously infected rat and add that to your blood sample. Look for agglutination of the trypanosomes. ▲

9. Describe the structure and function of the sinus venosus in the shark. _____

_____

10. Describe the structure and function of the sinus venosus in mammals. _____

_____

11. What major change in the conus arteriosus occurred between fishes and reptiles during evolution?

_____

_____

12. What is the result of the change described in question 11 in terms of the separation of oxygenated and nonoxygenated blood? _____

_____

13. What is occurring on the heart during the P wave of an ECG? _____

_____

14. What part of an ECG corresponds to ventricular repolarization? _____

_____

15. Antibodies are produced by _____.

KEY TERMS

antibodies (immunoglobins)   285
antigens   285
atrioventricular (AV) node   285
atrium   275
B-cells   285

closed circulatory system   272
conus arteriosus   275
electrocardiogram (ECG)   285
formed elements   272
plasma   272

sinoatrial (SA) node   274
sinus venosus   274
T-cells   285
ventricle   275

WORKSHEET 20.1   Vertebrate Circulation

Formed Elements

1. What formed element
   a. may proliferate in allergic reactions and parasitic infections?

   b. functions in the body's immune response?

   c. carries oxygen?

2. Describe how you distinguished visually between white blood cells and red blood cells under the light microscope.

Systemic Circulation

3. What is the function of the hepatic portal system of a vertebrate?

4. Trace a drop of blood from the left ventricle of a rat's heart to the stomach and back to the right atrium of the heart.

5. Trace a drop of blood from the left ventricle to the right arm of the rat and back to the right atrium of the heart.

Comparative Heart Structure

6. Describe the evolutionary changes in the structure of the vertebrate heart that accompanied the movement of vertebrates to land and made gas exchange without excessive water loss possible.

Circulatory Physiology

7. The interval between the P and Q waves of an ECG is normally about 0.1 second. Intervals longer than that would result from impulse conduction problems in what part of the heart?

8. What are memory cells?

Exercise 21

Vertebrate Respiration

Learning Objectives

After completing this exercise, you should be able to do the following:

1. Identify the structures of the shark respiratory system and describe the functions of these structures.
2. Identify the structures of the mammalian respiratory system and describe the functions of these structures.
3. Describe evolutionary trends in vertebrate lung structure.

Prelaboratory Quiz

Study this week's laboratory exercise and then complete the following quiz to assess your preparation for the laboratory.

1. The portion of the alimentary tract shared by the digestive and respiratory systems is the
   a. pharyngeal region.
   b. visceral region.
   c. nasal region.
   d. trachael region.
2. The skeletal framework of the gills of a fish are
   a. gill filaments.
   b. gill arches.
   c. gill lamellae.
   d. gill rakers.
3. Water is in close contact with capillary beds as it passes over
   a. gill filaments.
   b. gill arches.
   c. gill lamellae.
   d. gill rakers.
4. The bony roof of the mouth of a mammal is called the
   a. soft palate.
   b. bony palate.
   c. glottis.
   d. hard palate.
5. The function of goblet cells that line respiratory passageways of mammals is to
   a. secrete mucus that traps foreign material.
   b. engulf foreign material.
   c. promote gas exchange.
   d. warm inspired air.
6. True/False   A countercurrent exchange occurs as blood and water move in opposite directions across gill rakers.
7. True/False   Major respiratory passages of the lungs are lined by simple squamous epithelium.

8. True/False   The spiracle of a shark serves as an alternate route for water entering the pharynx.

9. True/False   Most reptiles carry on cutaneous and buccal respiration.

10. True/False   Bronchioles divide into smaller branches called bronchi.

The respiratory system develops evolutionarily in conjunction with the anterior portion of the digestive tract. The **pharyngeal region** is that portion of the alimentary tract that is shared with the respiratory system. The pharynx is also the location for the evolution of the swim bladder (fishes), openings to the eustachian tubes (remnants of gill pouches), and masses of lymphoidal tissue (palatine and lingual tonsils). With respect to the respiratory system, the pharynx provides a pathway for water moving over the gills of fishes and a pathway for air entering the lung passageways of tetrapods.

In this laboratory, we will again look at the shark and the rat. The two sets of respiratory structures are not homologous; they are simply adaptations to living in different environments.

GILL STRUCTURE AND FUNCTION

Pharyngeal slits primitively served in filter feeding, but in vertebrates they are primarily respiratory.

▼ Observe a demonstration dissection showing the pharyngeal cavity of the shark. In this demonstration, the pharyngeal cavity is opened by cutting into the pharynx at the corner of the jaw and posteriorly through the pharyngeal bars. Just posterior to the pectoral fin, the cut is extended to the opposite side of the body. The mouth cavity can then be folded open (see fig. 20.6).

On the inner surface locate the following:

spiracle   homologous to the first pharyngeal slit, located dorsal to the corners of the jaw. Thought to serve as an alternate route for water entry to the pharynx.

internal pharyngeal slits   five pairs of openings from the pharyngeal cavity into the gill chamber.

gill chamber   the cavity between internal and external pharyngeal slits.

gill rakers   cartilage-supported projections that line internal pharyngeal slits. Prevent food from entering the gill chamber.

On the side of the shark that has been cut, note the following:

gill (visceral) arches   the main skeletal framework of each gill.

gill filaments   membranous branches from the gill arches. Contain blood vessels that carry blood to and from gill lamellae.

gill lamellae   thin, vascular folds of epithelium through which gas exchange takes place.

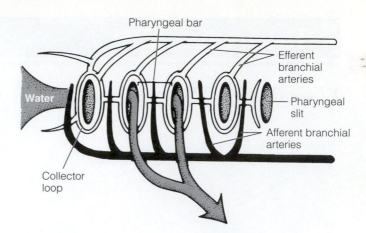

Figure 21.1   Path of blood flow to and from the gills of a fish.

Return briefly to figure 20.6 to recall the arterial pathway: from the heart, through afferent branchials to capillary beds of the gills, to efferent branchials, and then through the dorsal aorta to the rest of the body. If the shark you are studying has the arterial system injected, efferent branchials will be filled with red latex. Efferent branchials arise as collector loops that encircle each pharyngeal slit (fig. 21.1). Collector loops receive blood after blood has passed through gill capillaries. In your specimen, locate sections through collector loop vessels on each side of one of the pharyngeal slits. Envision this collector loop circling the pharyngeal slit. Each pharyngeal bar will have two collector loop vessels: the posterior element of the anterior collector loop, and the anterior element of the posterior collector loop. Between these will be an uninjected afferent branchial that should be located at this time.

The path of blood flow through a gill is diagrammatically illustrated in figure 21.2a. In the pharyngeal bar, each afferent branchial artery gives off branches to the gill filaments. As shown in figure 21.2b, blood then moves across the gill filament through capillary beds in the gill lamellae to efferent vessels that return blood to the collector loop and the efferent branchial. Water enters the pharynx through the mouth and spiracle. It passes through the pharyngeal slits and then between gill lamellae. Note that as water is moving between the lamellae, water and blood are moving in opposite directions. This **countercurrent exchange** provides very efficient gas exchange by maintaining a concentration gradient between the blood and the water over the length of the capillary bed. Compare the diagrams for parallel and countercurrent exchanges in figure 21.2c. ▲

STOP AND ASK YOURSELF

1. What is the function of gill rakers? _____

_____

2. What is the function of gill lamellae? _____

_____

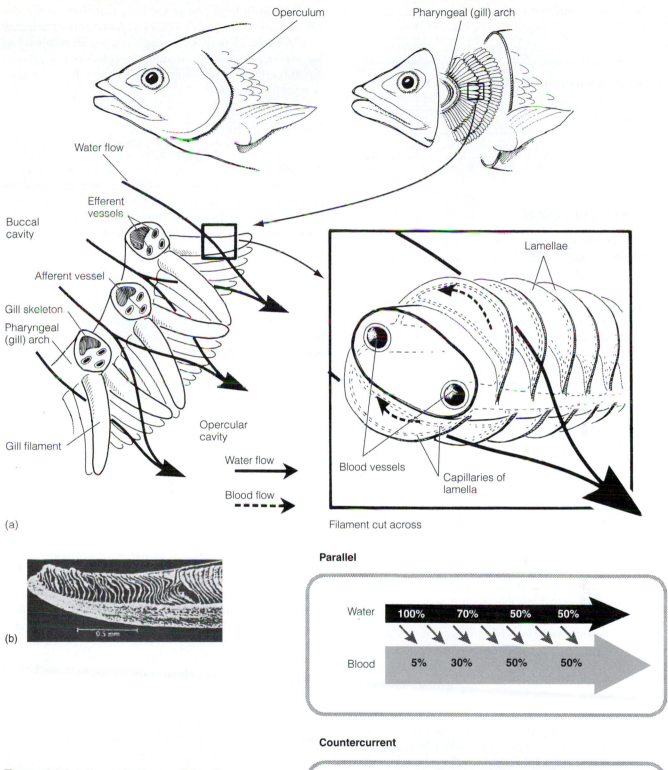

Operculum

Pharyngeal (gill) arch

Water flow

Efferent vessels

Buccal cavity

Afferent vessel

Gill skeleton

Pharyngeal (gill) arch

Gill filament

Opercular cavity

Water flow →

Blood flow ┄►

Lamellae

Blood vessels

Capillaries of lamella

Filament cut across

(a)

(b)

0.5 mm

Parallel

Water | 100% | 70% | 50% | 50%

Blood | 5% | 30% | 50% | 50%

Countercurrent

Water | 100% | 70% | 40% | 15%

90% | 60% | 30% | 5% | Blood

Figure 21.2 Gas exchange at gill lamellae. (*a*) Gill arches support two rows of gill filaments. Arteries break into capillary beds within gill lamellae. Water and blood flow in opposite directions on either side of the lamellae. (*b*) Electron micrograph of the tip of a trout gill showing lamellae. (*c*) A comparison of parallel and countercurrent exchanges. In countercurrent systems, a diffusion gradient is maintained along the length of the lamella.

(c)

3. What is a countercurrent? _____

_____

_____

4. Where is a countercurrent generated in the fish gill?

_____

_____

_____

TETRAPOD LUNGS

Lungs are structurally and evolutionarily distinct from, but functionally similar to, gills. In most tetrapods, the breathing apparatus is similar. It consists of a series of branching tubes originating at the posterior aspect of the pharynx and ending at a moist respiratory epithelium associated with pulmonary capillary beds. Movement of gases between respiratory epithelium and the blood is always by diffusion.

Rat Respiratory Structures

▼ With heavy scissors, cut the lower jaw of the rat at the symphysis of the mandibles. Extend the cut through the tongue and into the mouth cavity and pharynx. In the mouth cavity, note the tongue, teeth, and **hard palate** (the bony roof of the mouth that separates nasal and oral cavities). Follow the hard palate posteriorly. Note that the bone ends, but the palate continues as a fleshy extension. This is the **soft palate.** The opening just dorsal to the posterior margin of the soft palate is the **posterior nares**—the opening of the nasal cavity into the pharynx. In the rat, **eustachian tubes** from the middle ear open into the nasal passageway just above the soft palate. Do not try to locate these openings.

The series of branching passageways leading into the lungs is called the **respiratory tree.** The anterior portion of the respiratory tree is supported by cartilage. The single tube leading to the lungs is the **trachea.** At the anterior end of the trachea is an opening, the **glottis.** A cartilage-supported flap, the **epiglottis,** folds over the glottis during swallowing to prevent food from entering the trachea during swallowing. The **larynx** is the expanded, cartilaginous structure surrounding the glottis. If a vertebrate has vocal chords, they are located here in the larynx. The trachea branches into **bronchi** posteriorly, and bronchi subdivide within the lung—eventually to the level of **bronchioles.** Bronchioles end at **alveolar ducts,** which supply grapelike clusters of **alveoli.** Alveoli are the small air sacs where gas exchange occurs (fig. 21.3).

Examine a fresh cow or pig lung. Cut through the larynx and down the trachea. Follow a branch of the trachea as far as possible. Note that the lung is not just an air-filled bag but a spongelike tissue containing thousands of tiny alveoli. If possible, inflate a fresh lung using a blower. Note the elasticity of the lung. ▲

Lung Histology

Major respiratory passageways are lined by **ciliated, pseudostratified epithelium.** All cells reach the basement membrane, but cells are of different heights, giving a stratified appearance. The lining also contains mucus-secreting **goblet cells.** Mucus traps foreign material that gains entry into the respiratory tree. Cilia beat trapped materials back up to the pharynx (fig. 21.4).

▼ Examine a tracheal section under low power of the compound microscope. Note the epithelial lining and the cartilage that supports the trachea. Observe the epithelial lining under high power. The "fuzzy" border indicates the presence of cilia. Find the translucent goblet cells. Notice also that nuclei are at different levels, giving the tissue a stratified appearance.

Obtain a lung section. Note the very loose appearance. The section should be filled with hundreds of alveoli (see fig. 21.3), as well as sections through larger air passageways and pulmonary blood vessels. Under high power, note that individual alveoli are lined by simple squamous cells. This single-cell boundary is the only barrier between lung gases and the capillary wall. We will not attempt to identify lung passageways further. ▲

Lung Evolution

In evolution, the lung changed from the simple, saclike structures of lungfish and amphibians to the complex construction just observed in mammals. The saclike form seen in lungfish and amphibians presents much less surface area for the diffusion of gases, but it is adequate for the respiratory needs of these groups. It is noteworthy that respiration in amphibians is supplemented by **cutaneous** (via skin) and **buccal** (via mouth epithelium) **respiration.** In reptiles, lung organization becomes more complex, ranging from saclike structures in the primitive lizard, *Sphenodon,* to a more complex spongelike structure in turtles and crocodiles. Birds show a remarkable modification of respiratory structures that allows a continuous, one-way passage of air across respiratory surfaces. This results in greater oxygen availability for metabolically demanding flight.

▼ If demonstrations of these lungs are available, observe them at this time. ▲

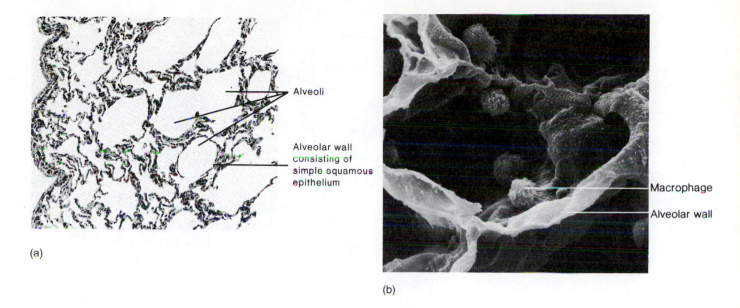

(a)

(b)

Alveoli

Alveolar wall consisting of simple squamous epithelium

Macrophage

Alveolar wall

Figure 21.3 Alveoli of the mammalian lung: (*a*) photomicrograph; (*b*) scanning electron micrograph showing alveolar macrophage.

STOP AND ASK YOURSELF

5. List, in order, the parts of the rat respiratory tree that air moves through before entering lung alveoli.

_____

_____

6. What adaptations of the tracheal epithelium permit the trachea to trap and expel foreign material?

_____

_____

7. What is the function of lung alveoli? _____

_____

8. What is the general trend in the evolution of the vertebrate lung? _____

_____

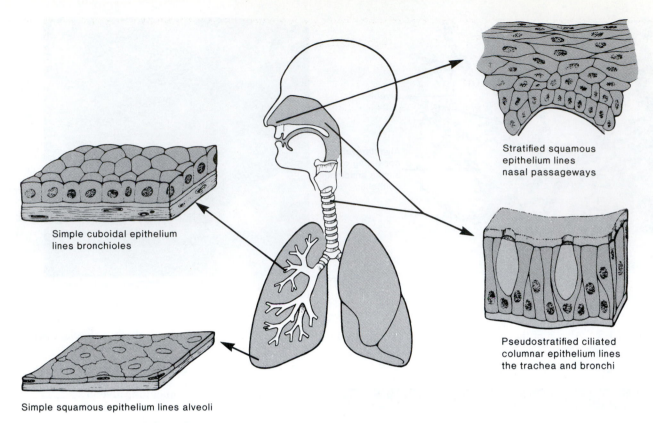

Simple cuboidal epithelium lines bronchioles

Stratified squamous epithelium lines nasal passageways

Pseudostratified ciliated columnar epithelium lines the trachea and bronchi

Simple squamous epithelium lines alveoli

Figure 21.4 Epithelial types found within the mammalian respiratory system.

KEY TERMS

buccal respiration   292

countercurrent exchange   290

cutaneous respiration   292

pharyngeal region   290

respiratory tree   292

WORKSHEET 21.1   Vertebrate Respiration

Gill Structure and Function

1. Describe the path of water through the pharyngeal slits and gill filaments of a fish.

2. Use diagrams to explain why countercurrent exchange systems are more efficient than parallel exchange systems.

Tetrapod Lungs

3. Name the part of the respiratory tree that
   a. opens at the glottis.

   b. branches into bronchioles.

   c. is the location for gas exchange.

   d. leads into alveoli.

4. Why does the epithelial lining of the trachea appear "fuzzy" when viewed under high power of the compound microscope?

5. Why is the pharynx an important body region in vertebrates?

Exercise 22

Vertebrate Digestion

Learning Objectives

After completing this exercise, you should be able to do the following:

1. Identify and describe the functions of components of the mammalian digestive system.

2. Describe specializations of the gut tract in selected vertebrate groups and be able to describe the functional significance of those variations.

3. Describe the histology of the intestinal wall.

4. Describe the conditions in the digestive tract that promote the digestion of starch, protein, and fat.

Prelaboratory Quiz

Study this week's laboratory exercise and then complete the following quiz to assess your preparation for the laboratory.

1. All of the following are features of the stomach EXCEPT the
   a. gastroesophageal sphincter.
   b. pyloric region.
   c. pyloric sphincter.
   d. villi.
   e. rugae.

2. All of the following are organs of the digestive system EXCEPT the
   a. soft palate.
   b. rectum.
   c. spleen.
   d. pancreas.
   e. esophagus.

3. The inner histological layer of the gut tract is the
   a. mucosa.
   b. submucosa.
   c. muscularis.
   d. serosa, or visceral peritoneum.

4. After passing through the ileocecal valve, gut contents enter the
   a. small intestine.
   b. large intestine.
   c. stomach.
   d. rectum.

5. After passing through the pyloric sphincter, gut contents enter the
   a. small intestine.
   b. large intestine.
   c. stomach.
   d. rectum.

6. True/False   The teeth of vertebrates other than mammals are uniformly conical, something known as the homodont condition.

7. True/False   A dental formula depicts the number and kinds of teeth on one side of the upper and lower jaws of a mammal.

8. True/False   The bulk of fat digestion occurs in the stomach of vertebrates.

9. True/False   Protein digestion begins in the stomach of vertebrates.

10. True/False   In this week's exercise, you will study starch digestion using salivary amylase from your saliva.

The digestive system of chordates shows many variations that reflect specializations for various feeding habits. These variations include, but are not limited to, the use of pharyngeal slits in filter feeding (amphioxus), modifications of the small intestine to increase surface area (spiral valve of sharks), modifications of the large intestine to form a fermentation pouch (cecum of rabbits), and modifications of the esophagus for storage and grinding (crop and gizzard of birds). In spite of many variations, the structure of vertebrate digestive systems is consistent in its basic form. In this laboratory, you will study the digestive system of the rat to learn basic structures and then examine a few of the modifications just described.

THE RAT DIGESTIVE SYSTEM

▼ The digestive system of the rat was exposed in exercises 20 and 21, so little cutting will be necessary in the following observations (see fig. 20.9 and Plate 15). As in exercise 21, find the **tongue.** Note the numerous papillae, which contain chemical receptors for the sense of taste. Identify **teeth, hard palate,** and **soft palate.** Just below the jaw, in the neck region, identify **salivary glands.** These secrete amylase into the mouth cavity to promote carbohydrate digestion.

The **esophagus** is a tube that carries food, which has been mixed with saliva, from the mouth to the stomach. The esophagus opens anteriorly just dorsal to the larynx. As you trace the esophagus posteriorly, note that it passes through the diaphragm and liver. The liver is made up of four lobes; however, unlike many vertebrates, the rat has no gallbladder associated with the liver. Bile from the liver empties into the anterior portion of the small intestine through the bile duct and functions in emulsifying fats. (The role of emulsification in fat digestion is demonstrated later in this exercise.)

The esophagus enters the **stomach** at the **cardiac end** (fig. 22.1). A **gastroesophageal sphincter** prevents the backup of stomach contents into the esophagus. Posteriorly, the stomach narrows into the **pyloric region.** A **pyloric sphincter** regulates the passage of stomach contents out of the stomach. Open the stomach and clean out its contents. Note the ridges **(rugae)** on the inner wall of the stomach.

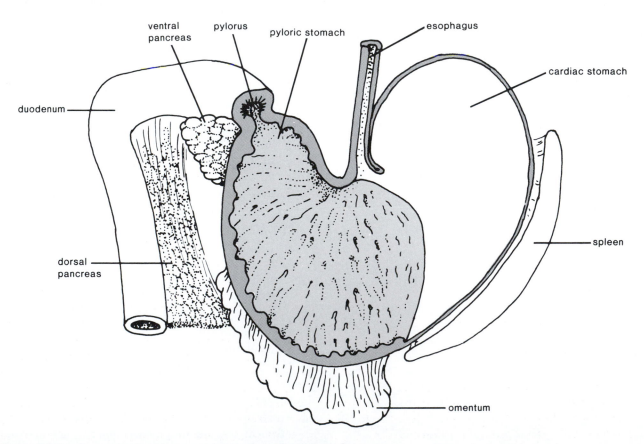

Figure 22.1   The stomach and associated structures. The wall has been cut away to expose the inner lining.

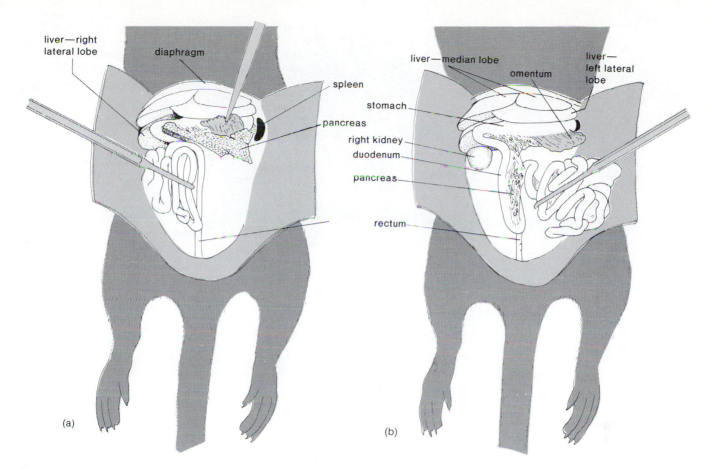

Figure 22.2 Abdominal viscera of the rat (ventral view); *(a)* intestines reflected to the right; *(b)* intestines reflected to the left.

These aid in mixing as well as increasing surface area for the secretion of enzymes and acid. Within the abdominal cavity, note a fat-filled peritoneal membrane (fig. 22.2). This is the **omentum.** Locate the spleen on the left side of the liver. As discussed in exercise 20, the spleen is a possible site for the development of B-cells that mediate a portion of vertebrate immune responses and is the site for the destruction of old red blood cells.

The **small intestine** originates at the distal end of the stomach and is divided into three regions. (There is no clear division between these regions.) The **duodenum** is approximately the anterior one-sixth, the **jejunum** is the middle half, and the **ileum** is the remaining portion. Note that the small intestine, as well as all of the digestive tract, is suspended in the body cavity from the dorsal body wall by **mesentery.** Mesenteries are continuous with the peritoneum, the inner lining of the body wall. In the folds of the duodenum locate the **pancreas.** It is an extremely diffuse gland that empties its digestive enzymes into the small intestine near the bile duct.

At the junction of the large and small intestines is the **ileocecal valve.** It prevents backflow from the large intestine to the small intestine. The **large intestine** consists of the cecum, ascending colon, transverse colon, and descending colon. It is followed by the **rectum,** which descends through the pelvic region and ends at the anus. ▲

SMALL INTESTINE HISTOLOGY

▼ Examine a cross section of the small intestine under low power of the compound microscope (figs. 22.3 and 22.4a). The cavity of the small intestine is called the lumen. Note **villi** that project into the lumen from the intestinal wall. These greatly increase the surface area for digestion and absorption. Villi are part of the **mucosal layer** of the intestine. The mucosa consists of simple columnar epithelium, underlying connective tissue, and some smooth muscle. Under high power note the single layer of columnar cells. Some cells are modified into mucus-secreting **goblet cells.** Mucus is secreted throughout the digestive tract to lubricate and protect the gut wall as well as facilitate the adherence of food and fecal particles. Goblet cells will appear translucent under the compound microscope. Electron micrographs reveal that the outer surface of the simple columnar cells has many **microvilli** that further increase the surface area for absorption and digestion (fig. 22.4b). Villi and the microvilli increase the surface area of the gut 600 times relative to the surface area of a simple cylinder. Next to the mucosal layer is the **submucosa.** Look for connective tissue, blood vessels, and lymphatic vessels associated with this layer. Outside the submucosa are two layers of smooth muscle. In the inner layer, smooth muscle cells are oriented circularly around the gut. These cells were sectioned longitudinally when the

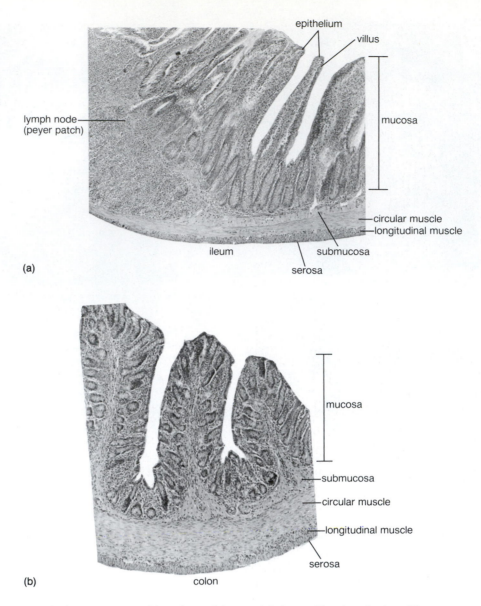

epithelium
villus
lymph node
(peyer patch)
mucosa
circular muscle
longitudinal muscle
ileum
submucosa
serosa

(a)

mucosa
submucosa
circular muscle
longitudinal muscle
serosa

(b)
colon

Figure 22.3 Photomicrograph showing intestinal histology of the rat: *(a)* ileum; *(b)* colon (both ×67).

slide was prepared. When they contract, the gut is constricted. When waves of constriction pass along the gut, gut contents are mixed and moved through the intestine. This series of wavelike contractions is called **peristalsis.** In the outer layer, smooth muscle cells are oriented longitudinally. These cells were cut in cross section when the slide was prepared and individual cells will appear dotlike. When these cells contract, the gut shortens, which aids in mixing gut contents. The outer lining of the intestine is a layer of epithelium called the **serosa,** or **visceral peritoneum.** It is continuous with the mesentery and the peritoneum lining the inner body wall. ▲

THE TEETH OF MAMMALS

The structure and arrangement of teeth are important indicators of mammalian lifestyles. In vertebrates other than mammals, teeth are uniformly conical in shape. This is referred to as the **homodont** condition. In mammals, teeth along the length of the jaw are specialized for different functions. This is referred to as the **heterodont** condition.

There are four kinds of teeth in adult mammals. **Incisors** are anterior on the jaw and are chisel-like for gnawing or nipping. **Canines** are often long, stout, and conical. They are used for catching, killing, and tearing prey. **Premolars** are positioned next to the canines, have one or two

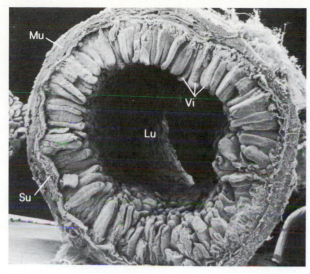

(a)

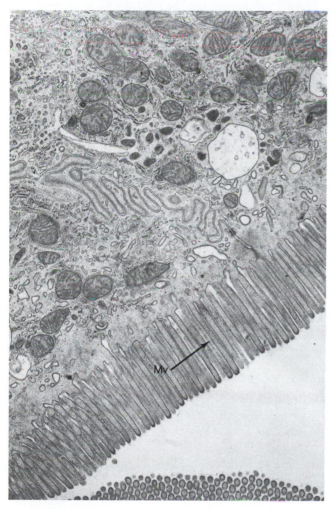

(b)

Figure 22.4 Electron micrographs of the small intestine: (a) cross section—villi (Vi), lumen (Lu), submucosa (Su), muscle layers (Mu)(×45); (b) lumenal surface of simple columnar cells lining each villus—microvilli (Mv). From R. G. Kessel and R. H. Kardon. *Tissues and Organs: A Text-Atlas of Scanning Electron Microscopy.* © 1979 by W. H. Freeman and Company.

roots, and have truncated surfaces for chewing. **Molars** have broad chewing surfaces and two (upper molars) or three (lower molars) roots.

A **dental formula** is an important tool used by zoologists in characterizing particular taxa. It is expressed as the number of teeth of each kind in one side of the upper jaw over the corresponding number in one side of the lower jaw. Teeth are indicated in the following order: incisors, canine, premolars, and molars.

▼ Examine a human skull that has a complete set of teeth. The human dental formula is as follows:

$$\frac{2 \cdot 1 \cdot 2 \cdot 3}{2 \cdot 1 \cdot 2 \cdot 3}$$

Notice the shape and surfaces of each tooth. In the space below, describe the human diet. _____

_____

Describe how, by studying our teeth, one can get clues to what we eat. _____

_____

_____

Examine the skulls of the three other mammals on demonstration. Try to determine the dental formula for each and fill in the information in table 22.1 in worksheet 22.1. ▲

In a dog (or fox, cat) skull, notice the interaction between the fourth upper premolars and the first lower molars when the jaw is closed. This is called the **carnassial apparatus.** In a rodent, notice the large incisors. These incisors grow and wear away throughout the life of the animal. Their chisel-like shape is a result of hard enamel being present only on the outer surfaces. Gnawing causes the posterior margin of the teeth to wear more rapidly than the anterior margin, which keeps the teeth sharp. In a deer (or horse, cattle) notice that the anterior, food-procuring teeth are separated from posterior teeth by a gap, called the **diastema.** The diastema results from elongation of the snout that allows the anterior teeth to reach close to the ground or into narrow openings to procure food. Posterior teeth have high exposed surfaces and continuous growth.

S T O P   A N D   A S K   Y O U R S E L F

1. Which is more ventral in position, the trachea or the esophagus? _____

2. What are the folds of the stomach lining called?

   _____

3. What is the anterior portion of the small intestine called? _____

4. What are folds of the outer plasma membrane of intestinal cells called? _____

What is their function? _____

_____

5. What is a dental formula? How is it useful? _____

_____

_____

_____

VARIATIONS IN VERTEBRATE DIGESTIVE SYSTEMS

▼ Examine demonstrations of the following vertebrates to see variations in digestive systems. Know the taxonomic group in which each occurs and its adaptive significance. ▲

Chordate Group	Modification
Cephalochordata amphioxus	Pharyngeal slits (filter feeding)
Agnatha lamprey	Buccal teeth and rasping tongue (ectoparasitism)
Chondrichthyes shark	Spiral valve (increased digestive and absorptive surface)
Aves	Crop (storage) Gizzard (grinding)
Mammalia rabbit	Enlarged cecum (fermentation)
human	Appendix (reduced cecum)

DIGESTIVE PHYSIOLOGY— ENZYME ACTION

In the following exercises, you will demonstrate the digestion of three classes of food by appropriate enzymes. As you do, think about your observations. Why does digestion occur under one set of conditions and not another? How do the conditions created in this exercise correspond to those actually present in the digestive tract? Where in the digestive tract would the digestion of each class of food occur?

Starch Digestion

Starch digestion begins in the mouth, where starch is mixed with saliva (containing amylase). **Amylase** breaks starch into maltose (a disaccharide containing two glucose subunits).

Procedure

▼

1. Obtain 10 ml of saliva and dilute it with an equal volume of distilled water.
2. Number four test tubes 1 through 4.
3. Add the following to the tubes indicated:
   #1—3 ml distilled water
   #2—3 ml saliva-water mixture
   #3—3 ml saliva-water mixture + 3 drops concentrated HCl
4. Heat the remaining saliva-water mixture to boiling, then allow it to cool. Add 3 ml of boiled saliva-water mixture to tube #4.
5. Add 5 ml of cooked starch to each of the four test tubes.
6. Allow the tubes to incubate for 1 to 1.5 hours at 37° C.
7. After incubation, test the solutions for maltose by adding 5 ml of Benedict's reagent and immersing test tubes in boiling water for 2 minutes. (Benedict's reagent changes from blue to yellow and eventually red with increasing sugar concentrations.)
8. Record and explain your results in table 22.2 in worksheet 22.1. ▲

Protein Digestion

Protein digestion begins in the stomach. **Pepsin** breaks large polypeptides into smaller polypeptides.

Procedure

▼

1. Add the following to separate test tubes.
   #1—5 ml pepsin solution
   #2—5 ml pepsin + 1 drop concentrated HCl
   #3—5 ml pepsin + 1 drop 10N NaOH
   #4—5 ml pepsin + 1 drop concentrated HCl
2. Add a thin slice of egg white from a boiled egg to each tube. (The size of the slices is not critical; however, all four must be very thin and approximately the same size. Slices about 1 cm × 1 cm × 1 mm work well.)
3. Incubate tubes 1–3 at 37° C for 1 hour (shake periodically).
4. Incubate tube #4 at 0° C for 1 hour (shake periodically).
5. Filter each solution and visually examine each egg slice for evidence of digestion.
6. Record and explain your results in table 22.3 in worksheet 22.1. ▲

Fat Digestion

The bulk of fat digestion occurs in the small intestine. The process is dependent on two steps: the **emulsification** of large fat droplets into smaller droplets, and the enzymatic breakdown of fat molecules into fatty acids and glycerol. Why is emulsification a necessary first step? _____

In this exercise, you will monitor the breakdown of fats by observing pH changes. With digestion, an increase in free fatty acids results in a lowered pH.

Procedure

1. Number three test tubes 1 through 3.
2. Add the following to the proper test tubes:*
   #1—1 ml of vegetable oil + 5 ml water + a few grains of bile salts.
   #2—1 ml of vegetable oil + 5 ml of pancreatin solution + a few grains of bile salts.
   #3—1 ml of vegetable oil + 5 ml of pancreatin solution.
3. Incubate the tubes at 37° C for 1 hour. Check the pH of the tubes at the beginning of the experiment and then at 10-minute intervals.
4. Record and explain your results in table 22.4 in worksheet 22.1. ▲

6. What is the function of the spiral valve of a shark?

   _____

7. How does boiling affect enzyme activity? Why?

   _____

8. Which test tube showed evidence of the greatest starch digestion? Why? _____

   _____

9. What is emulsification? _____

   _____

10. Why can pH changes be used as evidence of fat digestion?

    _____

    _____

*Note: It is possible to use a pH meter, litmus paper, or litmus solution in the determination of pH. If the latter is used, add 3 ml of litmus solution to each test tube prior to incubation. If litmus indicators are used, it is sufficient to simply record color changes (blue in alkaline solutions, red in acidic solutions). From Stuart Ira Fox, *A Laboratory Guide to Human Physiology*, 2nd ed. Copyright © 1980 Wm. C. Brown Publishers, Dubuque, Iowa. All Rights Reserved. Reprinted by permission.

KEY TERMS

dental formula   301
emulsification   303
heterodont   300

homodont   300
mucosal layer   299

serosa, or visceral peritoneum   300
submucosa   299

WORKSHEET 22.1   Vertebrate Digestion

TABLE 22.1   Specializations of the Teeth of Selected Mammals

Animal	Dental Formula	Incisors	Canines	Premolars	Molars	Food Habits
Human						
Dog						
Rodent						
Deer						

TABLE 22.2 Starch Digestion	
Tube	Results
1	
2	
3	
4	

Explanation of Results

− No digestion
+ Little to moderate digestion
++ Much digestion

TABLE 22.3 Protein Digestion	
Tube	Results
1	
2	
3	
4	

Explanation of Results

− No digestion
+ Little to moderate digestion
++ Much digestion

TABLE 22.4 Fat Digestion

Time (min)	Tube 1	Tube 2	Tube 3
0			
10			
20			
30			
40			
50			
60			

Explanation of Results

WORKSHEET 22.2 Vertebrate Digestion

The Rat Digestive System

1. After passing each of the following structures, where would gut contents be located next?
   a. gastroesophageal sphincter

   b. pyloric sphincter

   c. ileocecal valve

   d. ascending colon

Small Intestine Histology

2. What aspects of the structure of the gut tract increase surface area for absorption and digestion?

3. What are goblet cells, and where are they found in a cross section of the intestine?

The Teeth of Mammals

4. A mammal has well-developed incisors; reduced canines; a large diastema; and high, continually growing molars. What is the nature of the diet of this animal?

Digestive Physiology—Enzyme Action

5. What experimental conditions were most favorable for protein digestion by pepsin?

6. How do the conditions of your experiment on protein digestion reflect conditions in the stomach of a mammal?

7. Why is emulsification an important step in the digestion of fats?

Exercise 23

Vertebrate Excretion

Learning Objectives

After completing this exercise, you should be able to do the following:

1. Describe the structure of the metanephric and opisthonephric excretory systems.
2. Describe the structure and function of the metanephric kidney.
3. Describe the structure and function of the mammalian nephron and identify its components as viewed in the microscope.
4. Define active transport and describe how active transport was demonstrated in goldfish kidney tubules.

Prelaboratory Quiz

Study this week's laboratory exercise and then complete the following quiz to assess your preparation for the laboratory.

1. The kidneys of a shark are
   a. suspended in the body cavity from the dorsal body wall.
   b. suspended in the body cavity from the ventral body wall.
   c. located adjacent to the large intestine.
   d. located along the dorsal body wall and covered by peritoneum.

2. The structure of the excretory system that runs from the metanephric kidney to the bladder is the
   a. ureter.
   b. urethra.
   c. renal pyramid.
   d. renal cortex.

3. The region of the metanephric kidney where the ureter receives urine from the tubule system of the kidney is called the
   a. medulla.
   b. cortex.
   c. renal pelvis.
   d. glomerulus.

4. The functional unit of the kidney is called the
   a. renal pyramid.
   b. nephron.
   c. calyx.
   d. glomerulus.

5. All of the following occur in the nephron of the metanephric kidney EXCEPT
   a. filtration of the blood.
   b. secretion of ions into the tubule system.
   c. reabsorption of water, ions, glucose, and amino acids from the filtrate.
   d. storage of urine.

6. True/False   In a microscope slide of a metanephric kidney, glomeruli would be found in the cortical region.

7. True/False   In a microscope slide of a metanephric kidney, glomeruli would be found in close association with collecting ducts.

8. True/False   This week's exercise involves a demonstration of active transport across kidney tubules of a goldfish.

9. True/False   The rat has a cloaca.

10. True/False   The loop of the nephron and the collecting duct are located within the medulla of the metanephric kidney.

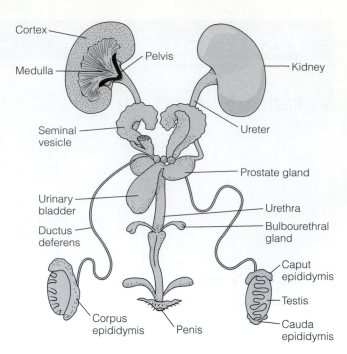

Figure 23.1   Urogenital organs of the male rat.

The excretory and reproductive systems of vertebrates are closely linked embryologically, evolutionarily, and anatomically. It is common, therefore, to study them simultaneously. They are not, however, physiologically linked. In this manual, therefore, they are placed in separate sections to reflect their physiological independence, but they will be covered one after the other to reflect their evolutionary relationship.

Excretory systems of vertebrates function in osmoregulation (water and ion regulation) and ridding the body of nitrogenous wastes. Vertebrates have adapted in various ways to osmoregulatory and excretory demands. Most of these adaptations, however, are not obvious through gross morphological observations. This laboratory, therefore, will focus on describing the anatomical characteristics of two vertebrates, the shark and the rat. These animals possess two of the more common vertebrate kidneys: **opisthonephric** and **metanephric.**

THE OPISTHONEPHRIC KIDNEY

The opisthonephric kidney of the shark is similar to the mesonephric kidney of most other adult fish, as well as adult amphibians, the primary difference being size and position.

▼ Examine a demonstration dissection. The kidneys are located along either side of the middorsal body wall. They are cream-colored, covered by peritoneum, and tightly pressed against the body wall in the mid and posterior region of the body cavity. Do not attempt to find the duct system, but note the **cloaca** (common opening for excretory, reproductive, and digestive waste products). A renal papilla represents the opening of the excretory system in the cloaca (see fig. 24.1). ▲

THE METANEPHRIC KIDNEY AND ASSOCIATED STRUCTURES

▼ Examine the dorsal body wall in the anterior abdominal region of your rat for the **kidneys** (fig. 23.1). They are usually embedded in fat and covered by peritoneum of the body wall rather than being suspended from the body wall by mesentery (as in the case of other abdominal organs).

Working on one side of the rat, remove fat from the kidney and surrounding area. Be careful that you do not tear the ureter in the process. Note the **adrenal gland** located just anterior to the kidney. It also is usually embedded in fat. On the medial side of the kidney is a concavity. This is the point where renal blood vessels enter and leave the kidney and the **ureter** emerges from the kidney. Follow the ureter posteriorly, cleaning fat from it as you proceed. The ureter delivers urine to the **bladder.** The ureter connects to the dorsal/posterior margin of the bladder. The bladder narrows posteriorly to deliver urine to the outside through the **urethra.** Follow the urethra to the pubis. In the male, the urethra receives the duct system from the testes and delivers both excretory and reproductive products to the outside.

Return to the kidneys. Remove the kidney from the body on the side opposite where you were previously working. Using a sharp scalpel, make a frontal section, dividing it into identical halves. Using a hand lens or dissecting microscope, identify the **cortex, medulla, renal pyramids, calices,** and **renal pelvis** (fig. 23.2). The latter three structures are the duct systems that deliver urine to the ureter. See any models or fresh kidneys available. ▲

KIDNEY HISTOLOGY AND THE NEPHRON

▼ The **nephron** is the functional unit of the kidney (fig. 23.3). Examine a slide showing a frontal section of the mammalian kidney. Under low power distinguish between the medulla and the cortex. In the cortex, note the **glomeruli** (figs. 23.4 and 23.5). Examine a glomerulus under

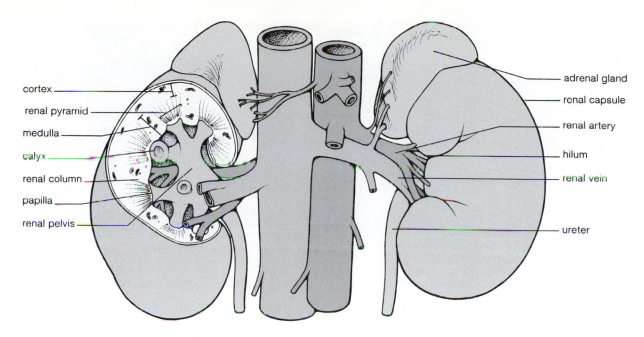

Figure 23.2 Ventral view of the mammalian kidneys. The right kidney is frontally sectioned to show internal structures. The adrenal gland of nonhuman mammals is usually positioned next to, but not attached to, the kidneys.

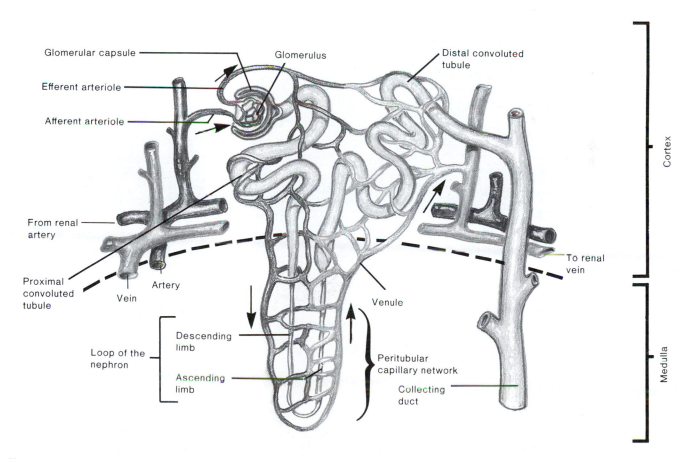

Figure 23.3 The mammalian nephron (diagrammatic).

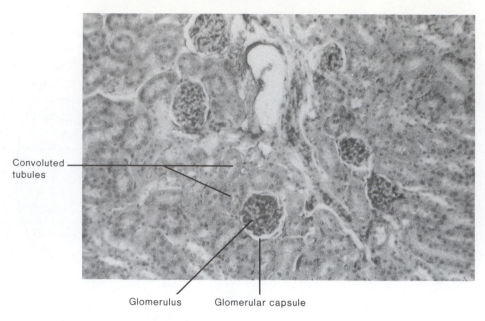

Figure 23.4 Photomicrograph of the cortical region of the mammalian kidney.

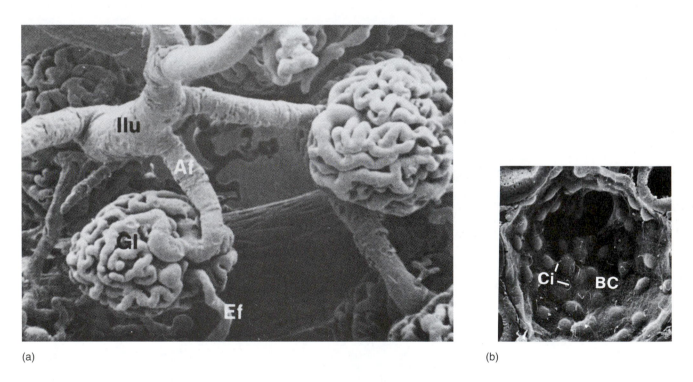

(a)

(b)

Figure 23.5 The mammalian nephron: (*a*) circulatory system of the nephron injected with plastic, kidney tissue digested away—interlobular arteries (Ilu), glomerus (Gl), afferent arteriole (Af), efferent arteriole (Ef)(×260); (*b*) the glomerular (Bowman's) capsule (BC). Note the cilia (Ci) on some cells. Cilia may be involved with moving fluids through the nephron. From R. G. Kessel and R. H. Kardon. *Tissues and Organs: A Text-Atlas of Scanning Electron Microscopy.* © 1979 by W. H. Freeman.

high power. Around it, note the **glomerular capsule.** The glomerulus is a capillary bed modified for filtration. Blood comes to the glomerulus, under pressure, through the afferent arteriole. Openings in the capillary walls of the glomerulus allow water and solutes to move from the bloodstream into the tubule system of the nephron. Blood cells and proteins are too large to pass through the pores; they move from the glomerulus into the efferent arteriole and then to the peritubular capillaries surrounding the tubules of the nephron. The filtrate enters the glomerular capsule and then the tubule system of the nephron. Around the glomerular capsule you should see sections through the tubule systems associated with the cortex. These are sections through the **proximal** and **distal convoluted tubules.** We will not attempt to distinguish between these tubules. They represent areas where water, amino acids, glucose, and salts are reabsorbed from the tubule into the peritubular capillaries and where ions may be secreted into the tubules. Switch back to low power and move to the medulla. There are no glomeruli in the medulla. Instead, there are linear arrays of cells descending toward the renal pelvis. These are the cells that line the **loops of the nephron** and the **collecting ducts.** These tubules allow the kidney to regulate the water content of the urine. The collecting ducts eventually lead to the renal pelvis and ureter. In the process just described, much of the urea filtered from the blood at the glomerulus is excreted, and most of the water, salts, glucose, and amino acids are reabsorbed. ▲

ACTIVE TRANSPORT ACROSS THE KIDNEY TUBULE

▼ Your instructor has dissected out a goldfish kidney and placed it in saline. Cut off a small piece of the kidney, place it in a depression slide, and gently tease the tissue using dissecting pins or teasing needles. Free adhering tissue by sucking the preparation up and down in a dropper. Transfer to fresh saline, cover with a coverslip, and view under low power of a compound microscope. You should be able to see clear tubules protruding from the darker tissue mass. (You may need to repeat the process just described.) Blot saline from the slide and add chlorophenol red (0.05 millimole per liter [mM/L] in goldfish saline). Cover again with a coverslip and observe. Watch a group of tubules (some may have been damaged in the teasing process). You should observe the interior of the tubules gradually becoming pink due to the accumulation of chlorophenol red. After observing active transport, add 2–4 dinitrophenol (0.2 mM/L in goldfish saline) to the preparation. This is a metabolic poison that interrupts ATP generation. What do you observe? _____

_____

_____

_____ ▲

STOP AND ASK YOURSELF

1. What is a cloaca? _____

_____

2. What are the structures urine passes through after leaving the collecting ducts? _____

_____

_____

3. How can one distinguish between the cortex and the medulla of the kidney in a light microscope preparation? _____

_____

4. What is the function of the glomerulus of the nephron? _____

_____

KEY TERMS

cortex   310
medulla   310

metanephric kidney   310
nephron   310

opisthonephric kidney   310

WORKSHEET 23.1 Vertebrate Excretion

1. What portions of the nephron are found in the cortex of the kidney?

2. What portions of the nephron are found in the medulla of the kidney?

3. What is the function of the collecting duct of the nephron?

4. What is the function of the proximal and distal convoluted tubules of the nephron?

5. Mammals have long loops of the nephron, but most other vertebrates do not. Why is the loop of the nephron so important in mammals but not required in other vertebrates?

6. How do you know that what you observed in the goldfish kidney was actually active transport?

Exercise 24

Vertebrate Reproduction

Learning Objectives

After completing this exercise, you should be able to do the following:

1. Identify and describe the functions of structures in the shark reproductive system.

2. Identify and describe the functions of structures in the mammalian reproductive system.

3. Identify and describe the functions of parts of the testis studied in the microscope.

4. Identify and describe the functions of parts of the ovary studied in the microscope.

Prelaboratory Quiz

Study this week's laboratory exercise and then complete the following quiz to assess your preparation for the laboratory.

1. In many sharks, the development of young occurs within a uterus, but most nourishment is derived from yolk present in the egg. This kind of development is
   a. viviparous development.
   b. ovoviviparous development.
   c. oviparous development.

2. Most female mammals and some female sharks nourish young through a placenta during development. This kind of development is
   a. viviparous development.
   b. ovoviviparous development.
   c. oviparous development.

3. The tubule that carries sperm from the testes of a male shark toward the cloaca is the
   a. seminal vesicle.
   b. ductus deferens.
   c. epididymis.
   d. seminiferous tubule.

4. The ovaries of a female shark are located in the
   a. anterior region of the body cavity.
   b. posterior region of the body cavity.
   c. middle region of the body cavity.
   d. cloaca.

5. A mature follicle would contain
   a. a primary oocyte.
   b. a polar body.
   c. a secondary oocyte.
   d. an ootid.

6. True/False   Sperm-producing structures found in testes are called epididymal tubules.

7. True/False   Interstitial cells of the testes are involved in the production of the hormone testosterone.

8. True/False   A corpus luteum remains in the ovary following ovulation and produces the hormones progesterone and estrogen.

9. True/False   One meiotic division in female mammals produces four egg cells.

10. True/False   Seminal vesicles, the bulbourethral glands, and the prostate gland are all accessory glands that create a proper medium for sperm ejaculation and survival.

The evolutionary and developmental relationship between vertebrate excretory and reproductive systems is indicated by their common embryological origin and their common duct systems. This is especially obvious in the males of all vertebrates. The urethra of the male rat, observed in exercise 23, is a common pathway for both urine and sperm. In addition, a cloaca is found in all vertebrates through the monotremes (mammals such as the spiny anteater and the duck-billed platypus). The cloaca represents a common pathway for the exit of excretory and reproductive products as well as undigested food material. The trend in evolution has been to separate the two systems. This separation is essentially complete only in female mammals above the marsupials.

THE SHARK REPRODUCTIVE SYSTEM

The shark will again be used to represent a basic arrangement of reproductive structures. Within the vertebrates, however, there is a considerable variation in arrangements of duct systems. As a result, one must be cautious about generalizing based on what is seen in the shark. This is especially true in regard to the male system.

▼ Examine the external structure of male and female sharks. They can be distinguished on the basis of the structure of the pelvic fin. Unlike most fishes, fertilization in the shark is internal. Sperm is transferred via modified pelvic fins, called **claspers,** that are inserted into the female cloaca. Note the dorsal sperm groove on the clasper. Internally, each clasper has a muscular walled sac, called a siphon, that aids in propelling sperm along the clasper. Sperm must then travel up the female reproductive tract to the cranial end of the oviduct, where fertilization occurs. ▲

The Male

▼ Examine a demonstration dissection of a male shark (fig. 24.1). The **testes** are located near the cranial end of the body cavity and suspended in mesentery (mesorchium). Through this mesentery, small tubules carry sperm to the **ductus deferens.** The ductus deferens is a highly coiled duct lying on the ventral surface of the kidney, which widens cau-

dally to form the **seminal vesicle.** Posteriorly, the seminal vesicle passes dorsal to the **sperm sacs.** The latter two structures function in sperm storage. They enter into the cloaca through a urogenital papilla. A cloaca represents a common opening for excretory, reproductive, and digestive systems ("cloaca" is derived from a Latin word for sewer). The cloaca should be cut open to see the urogenital papilla clearly. ▲

The Female

▼ Examine a demonstration dissection of a female shark (fig. 24.2). The paired **ovaries** are located in the anterior end of the body cavity and suspended by the mesentery (mesovarium). Eggs, if mature, are large, with much yolk. At ovulation, eggs are released into the body cavity and enter the oviducts at a common opening, which lies anterior and ventral to the liver. The **oviduct** runs posteriorly along the dorsal body wall. An enlargement dorsal to the ovary is the **shell gland** (nidamental gland). It secretes a proteinaceous shell around two or three fertilized eggs. A more posterior enlargement of the oviduct (posterior one-third) is referred to as the **uterus.** It is especially large in pregnant females but is almost indistinguishable from the oviduct in young females. Development occurs in the uterus, but the embryo receives its nourishment from the yolk of the egg. This is referred to as **ovoviviparous** development. Animals that have internal development and direct nourishment of the embryo through a placenta are **viviparous** (most mammals and some sharks). Animals that deposit eggs outside the body to develop from stored yolk are **oviparous** (most fishes, some primitive sharks, amphibians, reptiles, and birds). The uteri connect with the cloaca just ventral to the opening of the excretory system, the urinary papilla. The cloaca should be cut open to observe these structures. ▲

THE RAT REPRODUCTIVE SYSTEM

In studying the rat, it should be understood that while the essence of the mammalian reproductive system is constant throughout the class, there are variations in anatomical details that cannot be covered here (e.g., variations in uterine and vaginal structures). In the following dissection, you are responsible for knowing both male and female systems. You should dissect your specimen and then exchange specimens with a classmate who has worked on the opposite sex.

The Male

▼ Turn the rat ventral side up. In living rats, the **testes** are normally retracted into the body cavity, leaving the **scrotum** empty. This is accomplished through the **cremaster muscle,** which pulls the testes into the abdominal cavity through the **inguinal canal.** In forms where the testes permanently descend (e.g., humans), the inguinal canal usually grows almost completely closed.

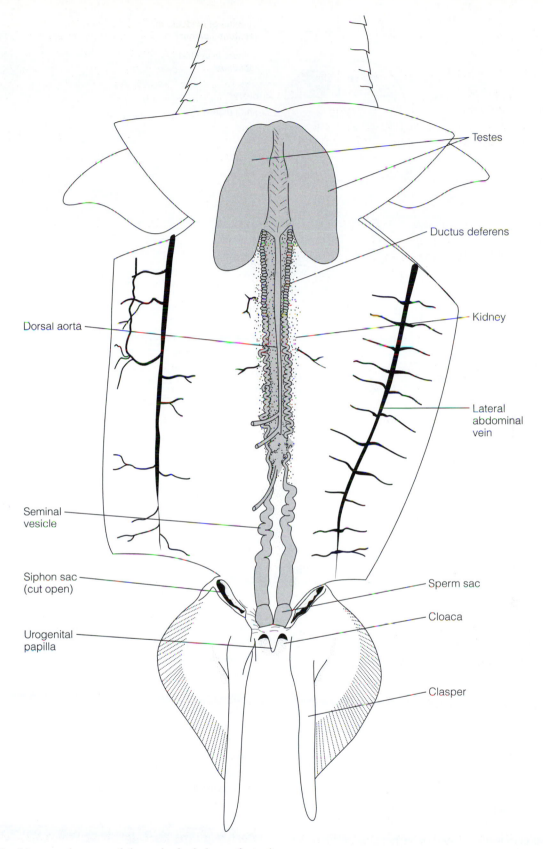

Testes

Ductus deferens

Kidney

Dorsal aorta

Lateral
abdominal
vein

Seminal
vesicle

Siphon sac
(cut open)

Sperm sac

Cloaca

Urogenital
papilla

Clasper

Figure 24.1 Urogenital organs of the male shark (ventral view).

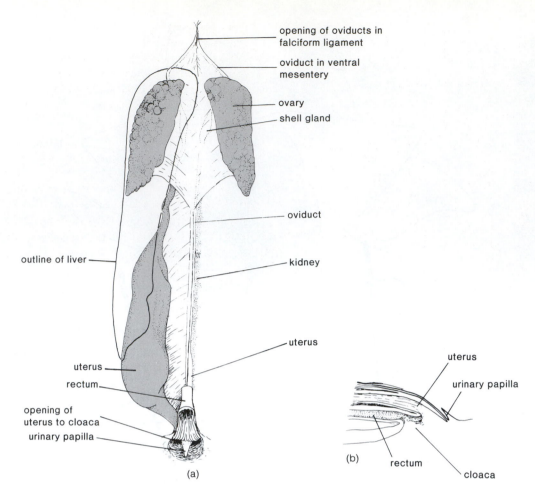

Figure 24.2 Urogenital organs of the female shark. (*a*) A pregnant female is illustrated in the left half of the diagram, and a nonpregnant female is illustrated in the right half of the diagram. (*b*) A diagrammatic sagittal view of the female cloaca.

Open the scrotum and note the paired testes (see Plate 18 and fig. 23.1) Covering the surface of each testis is the **epididymis.** The epididymis consists of a highly coiled tubule that stores and carries sperm from the testis toward the penis. The epididymis covers the anterior, lateral, and posterior borders of the testis. Follow the epididymis to the point that it becomes a distinct, single spermatic cord. The spermatic cord consists of a connective tissue sheath surrounding a spermatic artery and vein, a nerve, and the ductus deferens. The **ductus deferens** carries mature sperm into the penis.

Follow the spermatic cord into the body cavity and toward the penis. The spermatic cord loops behind, then over the ureter before joining the urethra. Using heavy scissors, cut through the pubic bone to follow the urethra to the penis. Associated with the urethra and spermatic cords are **accessory glands** that add secretions to the sperm. These include the large **seminal vesicles,** which unite with the spermatic cord just before the spermatic cord unites with the urethra. The **bulbourethral glands** are located on either side of the proximal end of the penis. The **prostate gland** is located near the junction of the ductus deferens and the urethra. All three function in creating a proper medium for sperm ejaculation and survival, and they contribute to the bulk of the ejaculate.

The Female

Locate the **ovaries** just posterior and slightly lateral to the kidneys (fig. 24.3 and Plate 19). The ovaries of mature females usually show **mature follicles** that represent eggs ready to be discharged. As a result, the surface of the ovary is usually convoluted. The ovary of the typical mammal discharges the egg into the oviduct through a funnel-shaped opening, the **ostium.** In the rat, the ostium completely encloses the ovary. A highly coiled **oviduct** leads to the **uterus.** The uterus of the rat is divided into two "horns," or **cornu,** one associated with each ovary. The cornu unite posteriorly to form the **vagina.** The vagina opens to the exterior through the external vaginal orifice. A **clitoris (glans clitoris)** is associated with the urethral opening. It is the female homolog of the penis. ▲

1. Where in the body cavity of a shark would you find the following:

   a. testes? _____

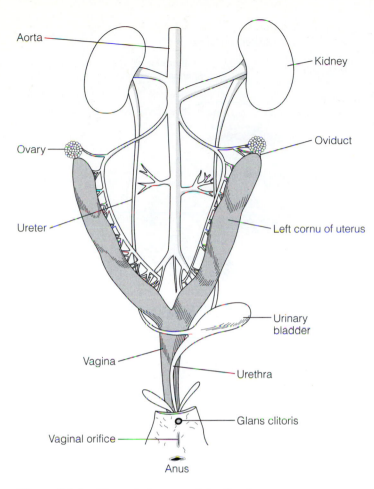

| Aorta |
| Kidney |
| Ovary |
| Oviduct |
| Ureter |
| Left cornu of uterus |
| Urinary bladder |
| Vagina |
| Urethra |
| Glans clitoris |
| Vaginal orifice |
| Anus |

Figure 24.3 Urogenital organs of the female rat.

b. ductus deferens? _____

c. seminal vesicle? _____

2. Sharks are ovoviviparous. What does this mean?

_____

_____

3. What is the function of the cremaster muscle of male rats? _____

_____

4. What is the function of the epididymis of the testes?

_____

_____

GAMETE FORMATION

Gametes are formed by **meiosis** (reduction division). In meiosis, homologous pairs of chromosomes are randomly divided between daughter cells that have half the chromosome number (1N, or haploid) of the original parent cell (2N, or diploid). After a period of differentiation, the re-

sulting gametes may be involved with fertilization. In fertilization, haploid egg and sperm nuclei fuse to restore the diploid chromosome number in the zygote. The zygote then starts its embryological development, as discussed in exercise 5. It is assumed that students have studied meiosis prior to the exercise that follows.

Spermatogenesis

Sperm cells are produced in the thousands of coiled tubules that make up the vertebrate testis. Cells involved with meiosis and sperm differentiation line these **seminiferous tubules.** Between seminiferous tubules are **interstitial cells,** which produce and secrete **testosterone,** one of the major androgens regulating sperm production and secondary sex characteristics.

▼ Obtain a prepared slide of a sectioned mammalian testis. Locate seminiferous tubules and interstitial cells (figs. 24.4 and 24.5). When observing the seminiferous tubules, note the layers of cells surrounding the central cavity of the tubule. These cells are in various stages of sperm development. The outermost cells are **spermatogonia.** These undifferentiated cells divide by mitosis to produce the next layer of cells, **primary spermatocytes.** Primary spermatocytes begin meiosis. The product of the first division of meiosis is represented by those cells within the third cell layer. These are **secondary spermatocytes** and are haploid cells. These cells undergo the second division of meiosis and form **spermatids.** During a period of differentiation, spermatids develop a **flagellum** and a tiny enzyme-filled cap (**acrosome**). These nearly mature sperm cells can be seen lining the lumen of the seminiferous tubule. During its period of differentiation, the spermatid is associated with a **sustentacular cell,** which produces the hormone **estradiol.** Along with testosterone, estradiol helps build up the environment for spermatogenesis. Examine a smear of bull (or other) sperm. Compare your observations to figure 24.6. The flagellum is for locomotion. The caplike acrosome covering the head of a sperm cell is used in egg penetration, and the middle piece, which connects the head to the flagellum, contains mitochondria for energy conversions required for locomotion. ▲

Oogenesis

Oogenesis is similar in all animals. The meiotic processes are the same as in spermatogenesis, but the goal of oogenesis is always to produce an egg cell with a large supply of nutrients to carry the zygote at least through early embryology. Mammals do not require as much food reserve as other vertebrates because of the role of the uterus and placenta in nourishment after the first few days of development. Nevertheless, egg development in mammals is similar to that in other vertebrates. **Oogonia** divide mitotically and differentiate into **primary oocytes** that undergo meiosis in the ovary. The first meiotic division results in the formation of the **secondary oocyte** (which contains the bulk of the cytoplasm) and a small **polar body.** The second meiotic division results

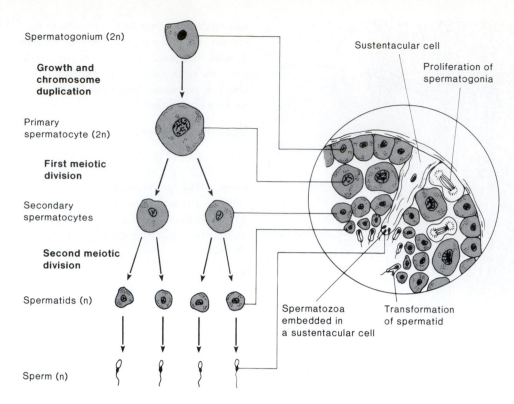

Figure 24.4 Diagrammatic cross section of a seminiferous tubule of the mammal.

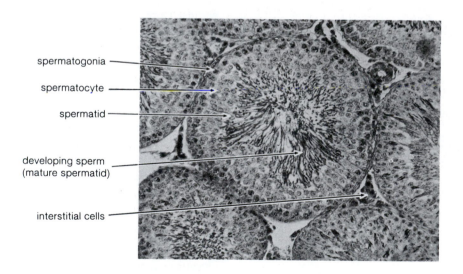

Figure 24.5 Cross section of rat seminiferous tubules (×540).

in the formation of the **ootid** (with the bulk of the cytoplasm) and another polar body. Each meiotic cycle produces one egg with substantial food reserves in the cytoplasm.

Oogenesis begins very early in female mammals. By the fourth month after conception, the human female fetus has all of her primordial germ cells (about three million per ovary) developed by mitosis (oogonia). These oogonia begin meiosis and arrest their development at the first meiotic prophase. Of these, only about four hundred will develop to

mature ova. At puberty, one of these primary oocytes completes its first meiotic division with each menstrual cycle. The second meiotic division occurs after penetration of the ovum by the sperm.

▼ Examine a prepared section of a mammalian ovary (figs. 24.7 and 24.8). Notice the cuboidal epithelium surrounding the ovary. Development of the primary oocyte occurs within ovarian follicles. **Primary follicles** are usually

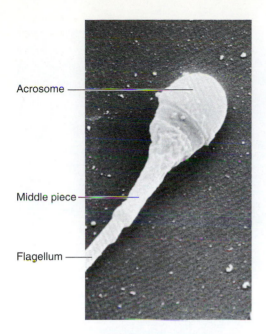

Acrosome

Middle piece

Flagellum

Figure 24.6 Scanning electron micrograph of human sperm (×10,000).

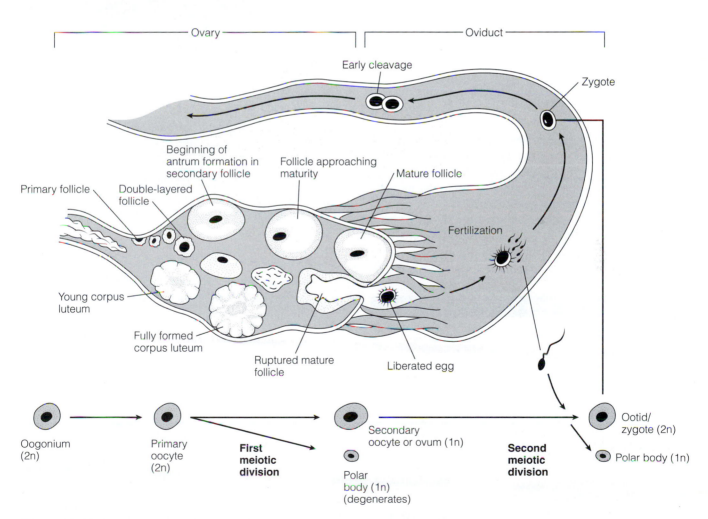

Ovary

Oviduct

Early cleavage

Zygote

Beginning of antrum formation in secondary follicle

Follicle approaching maturity

Mature follicle

Primary follicle

Double-layered follicle

Fertilization

Young corpus luteum

Fully formed corpus luteum

Ruptured mature follicle

Liberated egg

Oogonium (2n)

Primary oocyte (2n)

First meiotic division

Secondary oocyte or ovum (1n)

Polar body (1n) (degenerates)

Second meiotic division

Ootid/ zygote (2n)

Polar body (1n)

Figure 24.7 Diagrammatic section of a mammalian ovary and oviduct. Maturation of the oogonium and the first meiotic division occur during the fetal development of mammals. The development of secondary oocytes and associated follicular cells begins at puberty. When it is mature, the secondary oocyte is released from the ovary as the egg. The second meiotic division occurs, and the ootid exists very briefly, after fertilization but before the fusion of sperm and egg nuclei. Nuclear fusion results in the formation of the zygote. Polar bodies produced during meiosis degenerate.

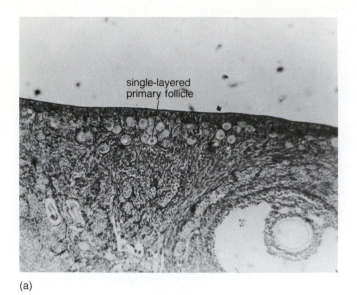

single-layered
primary follicle

(a)

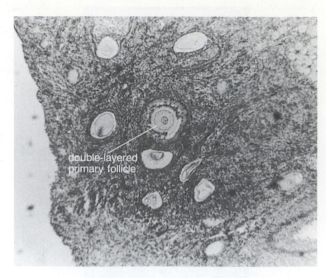

double-layered
primary follicle

(b)

Figure 24.8 Sectioned cat ovary: *(a)* single-layered primary follicles; *(b)* double-layered primary follicle; *(c)* maturing secondary follicle.

located near the ovary wall. They appear as solid spheres with the oocyte inside and are surrounded by one or two layers of cells. A **secondary follicle** has the beginning of a fluid-filled cavity (the **antrum**) surrounding the oocyte. The secondary follicle develops into a **mature follicle** in which the fully developed ovum (secondary oocyte) is held on top of a stalk of cells and is surrounded by a large central cavity that causes the wall of the ovary to bulge. The ovum is released into the coelom and the oviduct when the follicle wall ruptures. What remains in the ovary is transformed into the **corpus luteum,** which produces hormones (estrogen and progesterone) that maintain the uterine wall during early pregnancy. If implantation of the embryo at the uterus does not occur, the corpus luteum degenerates into a patch of scar tissue (corpus albicans). ▲

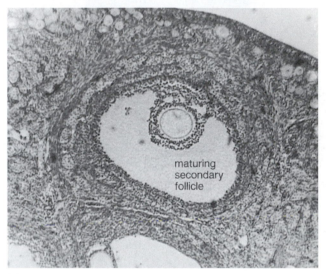

maturing
secondary
follicle

(c)

STOP AND ASK YOURSELF

5. What structures within the testes are the sites of spermatogenesis? _____

6. What are the immediate products of the second division of meiosis in spermatogenesis? _____

7. How can one distinguish a secondary follicle from a primary follicle when viewing an ovary section in the microscope? _____

   _____

8. What is a corpus luteum? _____

   _____

KEY TERMS

accessory glands   320
mature follicle   320
ovaries   318
oviparous   318

ovoviviparous   318
primary follicle   322
secondary follicle   323
seminiferous tubules   321

testes   318
uterus   318
viviparous   318

WORKSHEET 24.1 Vertebrate Reproduction

The Shark Reproductive System

1. Trace the path of sperm from the testes to the cloaca of a male shark.

2. How can one distinguish between male and female sharks when viewing them externally?

The Rat Reproductive System

3. Name three accessory glands present in male rats. What is their function?

4. Trace the path of sperm from the vagina of a rat to an egg waiting to be fertilized.

Gamete Formation

5. Where are interstitial cells found, and what is their function?

6. Where would one find a primary follicle? What is its function?

7. In what ways are egg production and sperm production different?

8. Would a secondary follicle contain a secondary oocyte? Why, or why not?

Glossary and Pronunciation Guide

In this pronunciation system, long vowels that stand alone or occur at the end of a syllable are unmarked. When a long vowel occurs in the middle of a syllable, it is marked with a macron (¯).

Short vowels that occur in the middle of a syllable are unmarked. When a short vowel occurs at the end of a syllable, or when it stands alone, it is marked with a breve (˘). The following are examples: cape (cāp), cell (sel), vacation (va-ka'shun), and vegetal (vej"ĕ-tal').

The vowel heard in *foot* and *book* is indicated as "fŏŏt" and "bŏŏk." The vowel heard in *loose* and *too* is unmarked and is indicated as "loos" and "too."

A

Acanthocephala (*ah-kan'thah-sef'ah-lah*)
Spiny-headed worms. A group of pseudocoelomates that are parasites of the digestive tract of fishes, turtles, swine, and other vertebrates.

acoelomate (*a-se'lŏ-māt*)
Without a body cavity.

actin (*ak'tin*)
One of the proteins involved with muscle contraction.

active transport (*ak'tiv trans'pōrt*)
The transport of substances across a cell membrane using protein carriers and energy.

Agnathan (*ag-nāth-an*)
A nontaxonomic grouping of fishlike vertebrates that lack jaws and paired appendages.

allele (*al-ēl'*)
Alternative forms of genes that occur at the same locus. Genes for white eyes and wild-type eyes of the fruit fly are alleles for the eye color trait.

alveolus (*al-ve'o-lus*)
A small cavity. In the lungs, alveoli are the small sacs where gas exchange occurs.

amoebocyte (*ah-me'bo-sīt*)
One of the cell types in the Porifera. Amoebocytes function in spicule formation, food distribution, and reproduction.

Amphibia (*am-fib'e-ah*)
A vertebrate class whose members are characterized by skin with mucoid secretions, moist skin functioning as a respiratory organ, and, usually, metamorphosis. Frogs, toads, and salamanders.

anatomy (*ah-nat'ah-me*)
The study of the structure of living organisms.

Animalia (*an"ah-māl'ēah*)
The animal kingdom. It includes organisms that are eukaryotic and multicellular, lack cell walls, and are nourished by ingestion rather than by photosynthesis or absorption.

Annelida (*ah'nel-ĭ-dah*)
The phylum containing the segmented worms. Its members are characterized by metamerism (the serial repetition of body parts).

Anthozoa (*an'thŏ-zo-ah*)
The class of cnidarians that includes the sea anemones, corals, sea fans, and sea pens. Anthozoans lack the medusoid body form.

antibody (*an'tĭ-bod"e*)
A specific protein (immunoglobin) that is produced by B-lymphocytes to combat an antigen.

antigen (*an'tĭ-jen*)
A foreign substance that can cause an immune response in the body.

Apicomplexa (*ap'ĭ-kom-pleks'ah*)
A protistan phylum. Spores or oocysts present, cilia and flagella lacking (except in gametes), all parasites.

Aplacophora (*a-plah-kŏ'fŏr-ah*)
Solenogastres. A class of wormlike molluscs whose members lack a shell, mantle, and foot.

appendicular skeleton
(*ă-pen'dĭ-ku-lar skel'-ĕ-ton*)
The bones or cartilages of the paired appendages.

Arachnida (*ah-rak'nĭ-dah*)
A class of arthropods whose members are characterized as having four pairs of walking legs, one pair of chelicerae, and one pair of pedipalps. Spiders, mites, ticks, scorpions, etc.

arteriole (*ar-te're-ōl*)
A branch of the arterial system. Smooth muscle in its wall allows an arteriole to constrict or dilate to regulate blood flow to a tissue.

artery (*ar'ter-re*)
One of the vessels of the arterial system. Arteries carry blood away from the heart. Elastic connective tissue in their walls allows arteries to distend on receiving a pulse of blood and then rebound to help propel blood.

Arthropoda (*ar-throp'ah-dah*)
The phylum whose members are characterized as having a jointed exoskeleton, metamerism, and a high degree of tagmatization.

ascon (*as'kon*)
The simplest of the three sponge body forms. Asconoid sponges are vaselike, with choanocytes directly lining the spongocoel.

association neuron (*ah-so"se-a'shun nur"on*)
See interneuron.

aster (*as'ter*)
The star-shaped structure seen in a cell during the prophase of mitosis; composed of a system of microtubules arranged in astral rays around the centriole; may emanate from a centriole or from a pole of a mitotic spindle.

Asteroidea (*as"tar-oid'ēah*)
A class of echinoderms that includes the sea stars. Its members are characterized as being star shaped, with rays not distinctly set off from the central disk, having ambulacral grooves with tube feet, having suction disks on tube feet, and having pedicellariae.

atrium (*a'tre-um*)
1. In the Cephalochordata, the atrium is the chamber that water passes into after passing through the pharyngeal slits. 2. In the Vertebrata, the atrium is the chamber(s) of the heart that receives blood from major veins or the sinus venosus and delivers blood to the ventricle(s) on atrial contraction.

Aves (*a'vēs*)
A class of vertebrates whose members are characterized by feathers and other flight modifications, endothermy, and amniotic eggs. Birds.

axial skeleton (*ak'se-al skel-ĕ-ton*)
Bones on the longitudinal axis of a vertebrate. Skull, vertebral column, ribs, and sternum.

axon (*ak'son*)
The region of a neuron that propagates an all-or-none action potential (nerve impulse).

B

basal body (*ba'sel bod'e*)
A centriole that has given rise to the microtubular system of a cilium or flagellum, and is located just beneath the plasma membrane. Serves as a nucleation site for the growth of the axomene.

basement membrane (*bas'ment mem'brān*)
A noncellular layer upon which an epithelium rests. It is produced by the epithelium and the underlying connective tissue.

binomial nomenclature (*bi-no"me-al no'men-kla"cher*)
A system for naming organisms in which each kind of organism (a species) has a name of two parts, the genus and the species epithet.

biramous (*bi-ra'mus*)
Divided into two parts. The biramous appendages of crustaceans consist of a single proximal element, the protopodite, and two distal elements, the exopodite and the endopodite.

Bivalvia (*bi'valv"e-ah*)
The class of molluscs having a shell of two hinged valves. Clams, mussels, etc.

blastula (*blas'tu-lah*)
An early stage in the development of an embryo. It consists of a sphere of cells enclosing a fluid-filled cavity (blastocoel).

Brownian movement (*brow'ne-an moov'ment*)
The movement of particles in water that results from water molecules colliding with the particles in question.

brush border (*brush bor'der*)
Microvilli lining the luminal surface of some epithelial tissues that increase the surface area for cellular transport processes.

bulk transport (*bulk trans-pōrt'*)
The transport of large compounds across a cell membrane by invaginating or evaginating large areas of the cell membrane.

C

cancellous bone (*kan'sel-us bōn*)
Spongy bone. Bone made up of a network of loosely organized bony fibers called trabeculae. As in the epiphyses of long bones.

capillary (*kap'ĭ-ler"e*)
Vessel within the circulatory system having the smallest diameter. Capillaries have walls consisting of a single cell layer, thus allowing exchanges of gases and nutrients between blood and surrounding tissues.

cardiac muscle (*kar'de-ak mus"el*)
Muscle of the heart. Muscle characterized by branching fibers, striations, and cells separated at their ends by intercalary disks.

cartilage (*kar'tĭ-lij*)
A connective tissue that gives support to structures such as the nose and earlobes, and often the intervening material between bones. Cartilage consists of cells (chondrocytes) and fibers embedded in a semisolid matrix.

cell cycle (*sel si'kel*)
The regular sequence of events during which a cell grows, prepares for division, duplicates its contents, and divides to form two daughter cells.

cell membrane (*sel mem'brān*)
See plasma membrane.

cell theory (*sel ther'e*)
The concept that cells (1) are the basic units of life, (2) possess all characteristics of life, and (3) arise only from preexisting cells.

cellular respiration (*sel'u-lar res"pĭ-ra"shun*)
The series of cellular metabolic reactions by which energy in the chemical bonds of glucose is converted into a form of energy directly usable by the cell (adenosine triphosphate).

central nervous system (*sen'tral ner'vus sis'tem*)
The brain and the spinal cord. The region of the nervous system where integration takes place.

centriole (*sen'tre-ōl'*)
A cellular organelle located just outside the nucleus of the nondividing animal cell. Centrioles may play a role in the organization of spindle fibers during mitosis and meiosis.

centromere (*sen'tro-mer*)
The constricted region of a mitotic chromosome that holds sister chromatids together; also the site on the DNA where the kinetochore forms and then captures microtubules from the mitotic spindle.

centrosome (*sen'tro-sōm*)
The two centrioles found outside the nucleus of a nondividing cell.

cephalization (*sef"al-īz-a'shun*)
The development of a head, with nervous tissue accumulating to form a brain.

Cephalochordata (*sef"al-o-kor-da'tă*)
A chordate subphylum whose members are characterized as having a laterally compressed, transparent, fishlike body. All four chordate characteristics persist throughout life.

Cephalopoda (*sef"ah-lah-po'dah*)
A class of Mollusca whose members are characterized as having the foot modified into a circle of tentacles and a siphon, the shell reduced or absent, and the head in line with the visceral mass.

Cestoidea (*ses-toid'e-ah*)
A class of the phylum Platyhelminthes. Tapeworms.

Chilopoda (*ki-lah'pōd-ah*)
A class of Arthropoda. Chilopods have one pair of appendages per segment and a body that is oval in cross section. Centipedes.

chordamesoderm (*kōr"dah-mez'ŏ-derm*)
Tissue in the amphibian gastrula that forms between ectoderm and endoderm in the region of the dorsal lip of the blastopore. It develops into mesoderm and the notochord.

Chordata (*kōr-da'tah*)
The phylum characterized by a dorsal tubular nerve cord, pharyngeal pouches, a notochord, and a postanal tail.

chromatid (*kro'mah-tid*)
One of the two (normally identical) coils of DNA that make up a chromosome.

chromatin (*kro'mah-tin*)
Nuclear material that gives rise to chromosomes during mitosis; complex of DNA, histones, and nonhistone proteins.

chromosome (*kro"mŏ-sōm*)
DNA and protein (chromatin) condensed or wound into discrete threadlike structures seen during cell division.

cilia (*sil'e-ah*)
Microscopic, hairlike processes on the exposed surfaces of certain eukaryotic cells. Cilia contain a core bundle of microtubules and are capable of performing repeated beating movements. They are also responsible for the swimming of many single-celled organisms.

Ciliophora (*sil-e-of'ōr-ah*)
The phylum of Protista whose members are characterized as having cilia. A macronucleus and a micronucleus are usually present.

cisternae (*sis'tern-a*)
1. Membranous vesicles that stack to form the Golgi apparatus. 2. The vesicular portion of sarcoplasmic reticulum (endoplasmic reticulum) of skeletal muscle.

cladogram (*klad'o-gram*)
Cladograms are a result of one approach to systematics called phylogenetic systematics (cladistics). Cladograms depict a sequence in the origin of taxonomic characteristics. The entire cladogram can be considered a hypothesis regarding a monophyletic lineage.

Cnidaria (*ni-dār'ēah*)
The phylum whose members are characterized by nematocysts, polymorphism, and diploblastic tissue-level organization. Jellyfish, sea anemones, corals, sea fans, etc. Also Coelenterata.

coelom (*se'lōm*)
The fluid-filled body cavity of many animals. The term is also used to designate the eucoelom, a body cavity lined by mesoderm.

compact bone (*kom'pakt bōn*)
The hard, dense bone making up the shaft of a long bone such as the femur.

concentration gradient (*kon-sen"tra'shen gra'de-ent*)
A difference in the concentration of a substance between two points of reference. A substance diffuses along its concentration gradient from an area of higher concentration to an area of lower concentration.

Concentricycloidea (*kon-sen'trĭ-si-kloi'de-ah*)
A class of echinoderms whose members have two concentric water-vascular rings encircling a disklike body. No digestive system. Sea daisies.

connective tissue (*kŏ-nek'tiv tish'u*)
A tissue that binds and/or supports body parts. It consists of scattered cells and fibers embedded in a gel-like matrix.

control group (*kon'trol groop*)
In a scientific experiment, a control group is treated the same as the experimental group except that the variable being tested is omitted. A control group serves as a basis for comparing the data derived in an experiment.

controlled variables (*kon-trol'ĕd var'e-ă-bels*)
Controlled variables are factors or conditions of an experiment that are held constant so that they do not affect the outcome.

cortex (*kŏr'teks*)
The outer region of an organ. For example, the cortex of the kidney, adrenal gland, or cerebrum.

Crinoidea (*krīn-oid'ēah*)
A class of Echinoderms that includes sea lilies and feather stars. Crinoids are frequently attached to the substrate through an aboral stalk of ossicles.

cristae (*kris'tae*)
Shelf-like folds of the inner membrane of a mitochondrion. Enzymes located on cristae catalyze cellular energy conversions.

Crustacea (*krus-tās'ēah*)
A class of Arthropoda including crabs, crayfish, lobsters, etc. Crustaceans possess biramous appendages and two pairs of antennae.

Ctenophora (*tēn-of'er-ah*)
A phylum of radial animals whose members are characterized by eight meridional rows of fused cilia and retractable tentacles with colloblasts.

cytokinesis (*si'ta-kin-e'sis*)
The division of the cytoplasm of a cell into two parts, as distinct from the division of the nucleus (which is mitosis).

cytoplasm (*si'ta-plazm*)
The contents of a cell surrounding the nucleus; consists of a semifluid medium and organelles.

D

dendrite (*den'drīt*)
The part of a neuron that initiates a graded action potential.

dependent variable (*dĭ-pen'dent var'e-ă-bel*)
The dependent variable is the parameter that the scientist is measuring. It is what will be affected during an experiment. In many experiments, scientists must monitor more than one dependent variable.

diencephalon (*di"en-sef'ah-lon*)
The portion of the embryonic vertebrate brain that develops to form the thalamus, epithalamus, and hypothalamus.

diffusion (*di-fu'zhun*)
The movement of a substance from an area of higher concentration to an area of lower concentration.

dioecious (*di-ēs'ēus*)
Having separate sexes. Derived from Greek meaning "two houses."

diploblastic (*dip"lo-blast-ik*)
Having a body wall derived from two embryonic layers, ectoderm and endoderm (e.g., the Cnidaria).

diploid (*dip'loid*)
A cell that has two members of each set of homologous chromosomes. Somatic cells are usually diploid. 2N.

Diplopoda (*dip"lo-pōd'ah*)
The class of arthropods that contains the millipedes. Its members are characterized by two appendages per apparent segment and a body that is round in cross section.

domain (*do-mān'*)
The broadest taxonomic grouping. Recent evidence from molecular biology indicates that there are three domains: Archaea, Eubacteria, and Eukarya.

dominant (*dom'ah-nent*)
A gene that masks one or more of its alleles. In order for a dominant trait to be expressed, at least one member of the gene pair must be the dominant allele. *See* recessive.

E

Echinodermata (*ek"īn-o-derm'ah-tah*)
The phylum that includes the sea stars, sea urchins, sea cucumbers, and sea lilies. Its members are characterized as having a water-vascular system, pentaradial symmetry, and an internal calcareous skeleton.

Echinoidea (*ek"i-noi'de-ah*)
The class of echinoderms that includes the sea urchins and sand dollars. Members are characterized by a globular or disk-shaped test, no rays, movable spines, and a skeleton (test) of closely fitting plates.

ecosystem (*ek"o-sis'tem*)
All of the populations of organisms living in a certain place plus their physical environment.

ectoderm (*ek"to-derm'*)
The outer embryological tissue layer. It gives rise to skin epidermis, nervous tissue, glands, hair, nails, and other structures.

effector organ (*ĕ-fek'tōr ōr'gan*)
A muscle or gland that responds to an impulse from the central nervous system.

Elasmobranchiomorphi
(*el-as"mo-brank'e-o-mōr"fi*)
A class of the phylum Chordata that includes the sharks, skates, rays, and fossil placoderms. Its members are characterized by jaws, paired appendages, a cartilaginous skeleton, no swim bladder, and no operculum.

embryology (*em'bre-ol'-ŏ-je*)
The study of development from the egg to the point at which all major organ systems have formed.

endocytosis (*en"do-si-to'sis*)
Physiological process by which substances may move through a plasma membrane into the cell via membranous vesicles or vacuoles.

endoderm (*en'do-derm"*)
The innermost embryological tissue layer. It gives rise to the inner lining of the gut tract.

endoplasmic reticulum (ER)
(*en"do-plaz'mik rĭ-tik"ku-lum*)
A membrane system spreading through the cytoplasm of a cell. Its function is transporting proteins within the cell. Rough ER (granular ER) has ribosomes associated with it. Smooth ER (agranular ER) lacks ribosomes.

epiboly (*ep-ib'ŏl-e*)
A spreading and thinning of ectoderm from the animal pole of the amphibian gastrula toward the vegetal pole.

epithelium (*ep"ĭ-the'le-um*)
A sheet of tissue that covers a surface or lines a cavity of an organism. Epithelium can be protective or provide surfaces for diffusion and other forms of transport.

erythrocyte (*ĕ-rith'ro-sīt*)
A mature red blood cell. It contains the respiratory pigment hemoglobin and carries oxygen and some carbon dioxide in the vertebrate circulatory system.

eukaryotic (*u"kār-e-ot'ik*)
Cells that have a membrane-bound nucleus as well as other internal membranous organelles.

evolution (*ev"ol-u'shun*)
Biological evolution is a series of changes in the genetic composition of a population over time.

evolutionary adaptation (*ev"ah-loo'she-nere ad"ap-ta'shen*)
Structures or processes that increase an organism's potential to successfully reproduce in a specified environment.

exocytosis (*eks'o-si-to"sis*)
The process by which substances are moved out of a cell; the substances are transported in the cytoplasmic vesicles, the surrounding membrane of which merges with the plasma membrane in such a way that the substances are dumped outside.

F

flagellum (*flah-jel'um*)
A long, hairlike process. Beating flagella move some protistans and sperm cells through their media, and they move fluids and particles across the surface of tissues. Flagellar organization is based on a 9+2 microtubule arrangement. Flagella are similar to cilia, though longer and fewer in number.

fluid mosaic model (*floo'id mo-za'ik mod'el*)
A model that describes the plasma membrane as a bimolecular lipid layer in which hydrophilic ends of lipid molecules are oriented to the outside of the membrane, while hydrophobic ends are oriented to the inside of the membrane. Proteins are associated with the surface of the membrane or embedded within the membrane.

food web (*food web*)
A sequence of organisms through which energy is transferred in an ecosystem. Rather than being a linear series, a food web has highly branched energy pathways.

formed element (*fōrmd el'ĕ-ment*)
Blood cells or fragments of cells. White blood cells (leukocytes), red blood cells (erythrocytes), and platelets.

Fungi (*fun'ji*)
One of the five kingdoms. Its members are characterized as being eukaryotic, multicellular, sedentary, and saprophytic.

furrowing (*fur'o-ing*)
The process by which division of the cytoplasm (cytokinesis) occurs during cell division. Microfilaments at the equator of a cell contract to constrict the plasma membrane of the cell.

G

gametes (*gam'ets*)
Mature haploid cells (sperm or eggs) that function in sexual reproduction.

gap 1 (*gap 1*)
A portion of interphase of the cell cycle characterized by rapid RNA and protein synthesis. It immediately follows mitosis.

gap 2 (*gap 2*)
A portion of interphase of the cell cycle characterized by protein synthesis prior to mitosis.

Gastropoda (*gas"tro-po'dah*)
The class of molluscs that includes the snails, slugs, and limpets. Its members are characterized by a coiled shell (when a shell is present) and distorted symmetry. Both monoecious and dioecious taxa are present in this class.

Gastrotricha (*gas'trŏ-trik-ah*)
A phylum of microscopic, freshwater, pseudocoelomate animals that have a ventrally flattened and bristly or scaly body.

gastrovascular cavity (*gas'tro-vas'ku-lar kav'ĭ-te*)
The large, central cavity of cnidarians that serves to receive and digest food.

gastrula (*gast'ru-lah*)
An early stage in the development of an embryo. The gastrula is the first developmental stage to show ectoderm and endoderm. The archenteron or primitive gut is formed, but nervous elements are not developed.

genetics (*jĕ-net'iks*)
The study of the mechanisms of transmission of genes from parents to offspring.

Gnathostome (*na'tho-stōm*)
A group of vertebrates whose members possess hinged jaws and paired appendages.

glomerulus (*glo-mer'u-lus*)
The bundle of capillarylike vessels within the glomerular capsule of the nephron. It is the place where filtration of blood occurs.

goblet cells (*gob'let selz*)
Simple columnar cells in the digestive tract modified for the production and secretion of mucus.

Golgi apparatus (*gōl"je ap"ah-ra'tus*)
A cellular organelle that packages secretory products.

gonad (*go'nad*)
An organ that produces gametes by meiotic cell division (the ovary or testis).

H

haploid (*hap'loid*)
A cell having one member of each pair of homologous chromosomes. Haploid cells are usually the product of meiosis. Gametes are haploid.

Haversian system (*ha-ver'shan sis'tem*)
The organizational unit of compact bone. It consists of concentric rings of bone called lamellae, organized around a central Haversian canal. Bone cells are located in lacunae between adjacent lamellae.

head-foot (*hed-fŏŏt*)
The body region of a mollusc that contains the head and is responsible for locomotion as well as retracting the visceral mass into the shell.

heterozygous (*het"er-o-zīg'us*)
The condition of an individual or a cell having different alleles at the same locus on a pair of homologous chromosomes.

Hexapoda (*hex'să-pōd'ah*)
See Insecta.

Hirudinea (*hīr"oo-din'ēah*)
The leeches. The class of Annelida whose members are characterized as having 34 segments with many annuli and anterior and posterior suckers. Most leeches are predators on other invertebrates. A few leeches feed on vertebrate blood.

histology (*his"tol'o-je*)
The study of tissue structure.

holoblastic cleavage (*hōl'o-blas'tik kle"vij*)
Division of the egg or zygote that results in separate blastomeres. Equal holoblastic cleavage results in similarly sized blastomeres (e.g., the sea urchin). Unequal holoblastic cleavage results in dissimilarly sized blastomeres (e.g., amphibians).

Holothuroidea (*hōl"o-thur-oid'ēah*)
The class of echinoderms whose members are characterized as being elongate along the oral-aboral axis, having microscopic ossicles embedded in a muscular body wall, having circumoral tentacles, and lacking rays. The sea cucumbers.

homologous (*ho-mol'ŏ-gus*)
1. Structures that have a common evolutionary history. The wing of a bat and the arm of a human are homologous; each can be traced back to a common ancestral appendage. 2. Chromosomes with genes that code for the same traits, but not necessarily the same expression of those traits.

homozygous (ho"mo-zīg'us)
The condition of having identical alleles for a particular gene pair.

Hydrozoa (hi"drŏ-zo'ah)
A class of cnidarians whose members are characterized as usually having well-developed polypoid and medusoid stages. Medusa with velum.

hypertonic (hi"per-ton'ik)
A solution having a greater number of solute particles than another solution to which it is compared.

hypothesis (hi-poth'es-is)
A tentative explanation of a question; an explanation based on careful observations.

hypotonic (hi"po-ton'ik)
A solution having a lesser number of solute particles than another solution to which it is compared.

I

independent variable (in-dī-pen'dent var'e-ă-bel)
The parameter being tested is the independent variable. It is what the scientist varies. In any experiment, there must be only one independent variable. When there is only one independent variable, any changes in the dependent variable(s) must be the result of the scientist's manipulation of the independent variable.

Insecta (in-sekt'ah)
The class of arthropods whose members are characterized as having three pairs of legs, one pair of antennae, and often wings (usually two pairs). The insects. Also Hexapoda.

integration (in"tě-gra'shun)
Combining parts into an integral whole. The role of the central nervous system is to combine sensory input from a variety of sources and initiate a coordinated response to that input.

interneuron (in'ter-nu"ron)
A nerve cell entirely confined to the central nervous system. Interneurons carry information from one part of the central nervous system to another. Also association neuron.

interphase (in'ter-faz)
Period between two cell divisions when a cell is carrying on its normal functions. Long period of the cell cycle between one mitosis and the next. Includes G_1 phase, S phase, and G_2 phase. The replication of DNA occurs during interphase.

isotonic (i"so-ton'ik)
In a comparison of two solutions, both have equal concentrations of solutes.

K

kingdom (king'dom)
The level of classification above phylum; the traditional classification system includes five kingdoms: Monera, Protista, Fungi, Plantae, and Animalia. Recent evidence from molecular biology indicates that these five kingdoms may not be monophyletic lineages.

Kinorhyncha (kin'o-ring'kah)
A phylum of pseudocoelomate marine worms. Its members are characterized by a body of 13 segments, spines, and a retractile head and proboscis.

L

leucon (lu'kon)
The sponge body form that has an extensively branched canal system. The canals lead to flagellated chambers lined by choanocytes. Flagellated chambers discharge water into excurrent canals that eventually lead to an osculum.

leukocyte (lu'ko-sīt")
A white blood cell.

lungs (lungz)
1. The highly vascular tissues at the distal end of the respiratory tree that allow gas exchange between air and the blood of lungfish and tetrapods. 2. The highly vascular modification of the mantle of some gastropod molluscs that allows gas exchange between the air and blood.

lymphatic system (lim-fat'ik sis'tem)
The system of vessels, glands, and nodes that is responsible for returning excess tissue fluids and proteins to the circulatory system and for developing and maintaining acquired immunity.

lysosome (li'so-sōm)
Vesicular intracellular structures that contain enzymes capable of digesting all biologically important macromolecules. They digest ingested particles as well as old, nonfunctional organelles.

M

Mammalia (mam-āl'e-ah)
The vertebrate class whose members are characterized by hair, mammary glands, amniotic eggs, and endothermy (Chordata).

mantle (man'tel)
The outer, fleshy tissue of molluscs that secretes the shell.

Mastigophora (mas-tig'of-er-ah)
The protistan subphylum whose members are characterized as having one or more flagella; being autotrophic, heterotrophic, or both; and usually reproducing through fission (Sarcomastigophora).

matrix (ma'trix)
The gelatinous interior of a mitochondrion containing DNA, ribosomes, and enzymes. Enzymes located in the matrix catalyze cellular energy conversions.

meiosis (mi-o'sis)
The type of cell division in which homologous chromosomes are separated into different cells, thus reducing chromosome number by one half. Meiosis usually results in the formation of haploid gametes.

meninges (mě-nin'jēs)
The three connective tissue membranes that cover and protect the brain and spinal cord of vertebrates.

Merostomata (mer'o-sto'mah-tah)
The class of arthropods whose members are characterized as having the first pair of appendages modified to form chelicerae, having one pair of pedipalps and four pairs of legs, having a body divided into a cephalothorax and abdomen, and lacking antennae. Eurypterids (extinct) and horseshoe crabs.

mesencephalon (mez"en-sef'el-on)
The region of the embryonic brain that gives rise to the adult midbrain.

mesentery (mez'en-ter-e)
A sheet of peritoneum (mesodermal in origin) that suspends most abdominal viscera from the body wall. It is continuous with the parietal and visceral peritoneum.

mesoderm (mez"o-durm)
The middle of the three embryological germ layers. Mesoderm gives rise to muscles, blood, bone, cartilage, etc. Mesoderm is found only in triploblastic animals.

metamerism (met-am'er-izm)
The serial repetition of body parts. The two major metameric phyla are Annelida and Arthropoda.

metamorphosis (met-ah-morf'ě-sis)
A transformation from one form to another. As used here, metamorphosis generally refers to a transformation from the larval to the adult body form.

metencephalon (met"en-sef'al-on)
One of the five embryonic regions of the brain of vertebrates. It develops into the pons and cerebellum of the adult.

microfilament (mi'kro-fil'ah-ment)
Intracellular proteins arranged in linear arrays or networks. Microfilaments are involved with movement within cells and with maintaining or changing cell shape.

microtubule (*mi"kro-tu'būl*)
Cylindrical arrangements of proteins that, like microfilaments, are the basis of cellular locomotion and structure.

microvillus (*mi"kro-vil'us*)
An extension of the cell membrane of some epithelial cells that increases surface area for secretion, digestion, and/or absorption.

mitochondrion (*mi"to-kon'dre-on*)
The organelle that is responsible for aerobic energy conversions within a cell.

mitosis (*mi-to'sis*)
Cell division resulting in two cells that are genetically identical to a parent cell.

mitotic spindle (*mi-to'tic spin'del*)
In a cell undergoing mitosis, the centrioles and the microtubules radiating between them form the mitotic spindle.

Mollusca (*mol-usk'ah*)
A phylum of triploblastic coelomate animals whose members are characterized by a body of three regions: the visceral mass, the head-foot, and the mantle. Molluscs include the snails, bivalves, chitons, squid, octopuses, and others.

Monera (*mon'er-ah*)
One of the five kingdoms into which living organisms can be classified. Its members are characterized as having cells that lack a membrane-enclosed nucleus, as well as other internal membrane-bound organelles (they are prokaryotic). The Monera include bacteria and blue-green algae.

monoecious (*mon-ēs'e-es*)
A species in which both male and female sex organs occur in the same individual. Derived from Greek meaning "one house."

Monoplacophora (*mon"o-plah-kof'o-rah*)
The molluscan class whose members are characterized by a single arched shell, a broad and flat foot, and serial repetition of certain structures.

morula (*mor'u-lah*)
A stage in animal embryology consisting of a solid ball of blastomeres.

motor unit (*mo'tor unit*)
A nerve cell plus the muscle cells associated with it.

mucosa (*mu-ko'sah*)
The inner lining of the digestive tract or other tube or cavity that opens to the outside of the body. Mucous membrane.

muscular tissue (*mus'kŭ-lar tish'u*)
A tissue characterized by the ability to shorten and conduct action potentials.

myelencephalon (*mi"ĕ-len-sef'al-lon*)
One of the five regions of the embryonic vertebrate brain. It develops into the medulla oblongata of the adult.

myosin (*mi'o-sin*)
One of the proteins involved with muscle contraction.

N

nematocyst (*nĭ-mat'ah-sist*)
An organelle characteristic of the Cnidaria that is used in defense, food gathering, and as a holdfast structure. Nematocysts develop in and are discharged from cnidoblasts.

Nematoda (*nem'ah-tōd"ah*)
The pseudocoelomate phylum whose members are characterized as being wormlike and round in cross section; having a flexible, nonliving cuticle; lacking motile cilia and flagella; and having only longitudinal muscles. The roundworms.

Nematomorpha (*ne"mat-o-morf'ah*)
A phylum of pseudocoelomate worms commonly called horsehair worms. Adults are elongate and free-living in streams and ponds. Larvae are parasites of arthropods.

neoteny (*ne-ot'en-e*)
The retention of the larval body form in a reproductive individual.

nephron (*nef'ron*)
The functional unit of the kidney of vertebrates. It filters excretory products from the blood, reabsorbs water and nutrients from the filtrate, and actively transports ions and other substances from the blood into the filtrate.

nerve (*nerv*)
A bundle of nerve fibers (axons or dendrites) outside of the central nervous system.

nervous tissue (*ner'vus tish'u*)
The type of tissue composed of individual cells called neurons and supporting neurological cells.

neuron (*nur'on*)
A nerve cell. The basic cellular unit of the nervous system.

neurotransmitter (*nu"ro-trans-mit'er*)
A chemical released from synaptic vesicles at the presynaptic membrane. It binds to receptors at the postsynaptic membrane to either cause or inhibit depolarization of the postsynaptic membrane.

notochord (*no'to-kord*)
A connective tissue sheath that is filled with vacuolated cells and that runs along the dorsal longitudinal axis of chordates. This supportive, yet flexible, structure is present in all embryonic and many adult chordates.

nuclear envelope (*nu'kle-ar en-vel'op*)
The double membrane forming the surface boundary of a eukaryotic cell nucleus.

nucleolus (*nu-kle'o-lus*)
A darkly stained region within the nucleus where ribosomal RNA is synthesized.

nucleus (*nu'kle-us*)
The genetic control center of the cell. It contains DNA and protein in a highly dispersed state called chromatin.

O

objective lens system (*ob-jek'tiv lenz sis'tem*)
The lens system of a compound microscope closest to the specimen.

ocular lens system (*ok'ya-ler lenz sis'tem*)
The lens system of a compound microscope closest to the eye of the observer.

ocular micrometer (*ok'ya-ler mi'kro-me-ter*)
A thin circle of glass or plastic etched with a nonunit scale. When placed in the ocular lens of a compound microscope the scale image is superimposed over the specimen and is used to measure the size of microscopic objects.

Oligochaeta (*ol"ig-o-ke'tah*)
A class of annelids whose members are characterized by few setae, no parapodia, no distinct head, direct development, and monoecious individuals. The earthworms and related organisms.

Ophiuroidea (*ōf"e-ūr-oid'ēah*)
The class of echinoderms whose members are characterized by arms that are sharply set off from the central disk, a body cavity limited to the central disk, and tube feet without suckers. Brittle stars.

organ (*or'gan*)
An assemblage of tissues cooperating to perform a specialized function.

organ system (*or'gan sis'tem*)
An assemblage of organs cooperating to perform specialized functions.

osmosis (*oz-mō'sis*)
The movement of water from an area of high concentration to an area of low concentration through a selectively permeable membrane.

osmotic pressure (*oz-mot'ik presh'er*)
A measure of the pressure that would be needed to prevent the net diffusion of water from one solution to another.

Osteichthyes (*os"te-ik'the-ēz"*)
The class of vertebrates whose members are characterized by a bony skeleton, an operculum, and a swim bladder (Chordata).

ovary (*o'var-e*)
The primary female reproductive organ. An egg-producing organ.

oviduct (*o'vĭ-dukt*)
A tube that carries eggs from an ovary to an area where eggs develop or are stored. In animals with internal fertilization, the oviduct is frequently where eggs are fertilized.

oviparous (*ōv-ip'er-us*)
Organisms that lay eggs that develop outside of the body of the female.

ovovivparous (*ōv"o-vĭv-ip'er-us*)
Organisms with eggs that develop within the reproductive tract of the female. Embryos are nourished by yolk stored within the egg.

P

parfocal (*par-fo'kel*)
The property of a compound light microscope that allows one to switch from one objective lens to another without refocusing. An object that is in focus using one objective lens will remain in focus when another objective lens is moved into place.

parthenogenesis (*par"then-o-jen'es-is*)
The development of an egg without its first being fertilized.

passive transport (*pas'iv trans'pōrt*)
The movement of substances into and out of cells without the use of energy.

peripheral nervous system (*pĕ-rif'er-al ner'vus sis'tem*)
The portion of the nervous system that includes the nerves going to somatic and visceral structures outside of the brain and spinal cord.

phagocytosis (*fag'o-si-to'sis*)
Process by which a cell engulfs bacteria, foreign proteins, macromolecules, and other cells and digests their substances; cellular eating.

pharyngeal slits (pouches) (*far-in'je-al slitz [pouches]*)
A series of openings between the digestive tract, in the pharyngeal region, and the outside of the body. One of the four unique characteristics of the chordates.

pharynx (*far'ingks*)
The region of the alimentary tract shared by digestive and respiratory systems.

physiology (*fiz"e-ol'o-je*)
The study of the function of cells, tissues, organs, or organ systems.

pinocytosis (*pin'o-si-to"sis*)
Cell-drinking; the engulfment into the cell of liquid and of dissolved solutes by way of small membranous vesicles.

Plantae (*plant'a*)
One of the five kingdoms of life. Its members are characterized as being eukaryotic and multicellular and as having rigid cell walls and chloroplasts.

plasma membrane (*plas'ma mem'brān*)
The selectively permeable outer boundary of a cell. It is composed of proteins and lipids. Also cell membrane and plasmalemma.

Platyhelminthes (*plat"e-hel-min'thēz*)
The animal phylum whose members are characterized as being triploblastic and acoelomate and as having dorsoventrally flattened bodies, tissues organized into organs, and bilateral symmetry with cephalization.

Polychaeta (*pol"e-kēt'ah*)
The class of annelids whose members are characterized by parapodia bearing numerous setae, and a head with eyes and tentacles. Mostly marine.

polymorphism (*pol"e-morf'izm*)
The condition in which a species has more than one body form (as in the Cnidaria).

Polyplacophora (*pol"e-plak-of'er-ah*)
The class of molluscs whose members are characterized by an elongate, dorsoventrally flattened body; a reduced head; and a shell of eight dorsal plates. The chitons.

Porifera (*por-if'er-ah*)
The animal phylum whose members consist of loosely organized cells, including choanocytes, amoebocytes, and pinacocytes.

postanal tail (*post-a'nal tāl*)
A projection that extends posterior to the anus of chordates and that is supported by either the notochord or the vertebral column. It is one of the four unique chordate characteristics.

predictions (*pri-dik'shens*)
A forecast of the outcome of a scientific experiment that is based upon the hypothesis—usually in the form of an if/then statement.

prokaryotic (*pro"kar-e-ot'ik*)
A cell without a membrane-bound nucleus or other internal membranous compartments. As in the Monera.

Protista (*pro-tist'ah*)
The kingdom whose members are characterized as being eukaryotic and unicellular or colonial.

pseudocoelom (*sūd"o-se'lōm*)
A body cavity that is not lined by a mesodermal, epidermal sheet.

Pycnogonida (*pik"no-gon'id-ah*)
A class of marine arthropods called sea spiders.

R

radula (*raj'oo-lah*)
The rasping tonguelike structure of most molluscs. It is composed of minute chitinous teeth that move over the cartilaginous odontophore. The radula is used in scraping food.

receptor (*rĭ-sep'tor*)
Something that receives. A sensory receptor receives an environmental stimulus and converts the stimulus into a nerve impulse.

recessive (*ri-ses'iv*)
A gene whose expression is masked when it is present in combination with a dominant gene. In order for a recessive gene to be expressed, both members of the gene pair must be the recessive form of the gene.

replicate (*rep'le-kit*)
The duplication of an experiment to ensure that the results are consistent and the experiment is well conceived.

Reptilia (*rep-til'e-ah*)
A class of vertebrates whose members are ectothermic, lay amniotic eggs, have dry skin, and have epidermal scales.

respiration (*res"pah-ra'shun*)
1. The energy-converting reactions of the cell. The energy in organic compounds is converted to that in adenosine triphosphate. 2. The act of breathing.

ribosomes (*ri-bo-soms*)
Cytoplasmic organelles that consist of protein and RNA, and function in protein synthesis.

Rotifera (*ro'tĭ-fer-ah*)
A pseudocoelomate phylum whose members have a body consisting of head, trunk, and foot and who have a ciliated corona that functions in gathering food. Rotifers.

S

Sarcodina (*sar"ko-dīn'ah*)
A subphylum of Sarcomastigophora that includes the amoebas.

Sarcomastigophora (*sar'ko-mas-tig'of-er-ah*)
The protistan phylum whose members are characterized by flagella and/or pseudopodia.

sarcomere (*sar'ko-mēr"*)
A contractile unit of skeletal muscle.

Scaphopoda (*skaf"o-po'dah*)
A class of marine molluscs whose members have a tubular shell, tentacles, and no head. Tooth shells or elephant-tusk shells.

science (*si'ens*)
Knowledge of the natural world gained by observation.

scientific method (*si'en-tif"ik meth'od*)
A series of steps that reflects a logical approach to solving problems in our physical world. It usually involves an experimental approach involving questions based upon careful observation, the formulation of a hypothesis, a test of the hypothesis, and a conclusion based upon the results of the test.

scientific theories (*si″en-tif′ik ther′es*)
Generalized statements of predictive value that are supported by many individual experiments, each testing one small aspect of a larger concept.

Scyphozoa (*sīf″ŏ-zo′ah*)
The cnidarian class in which the polyp is reduced or absent and the medusa is without a velum.

sensory neuron (*sen′so-re nu′ron*)
A neuron that carries nerve impulses from a sensory receptor toward the central nervous system.

serosa (*si-ro′sah*)
The peritoneal lining on the outside of a visceral organ. The serosa is a serous membrane that is sometimes called the visceral peritoneum.

sinus venosus (*si′nus ven-o′sus*)
A chamber of the vertebrate heart (exception, the mammals). It receives blood from the venous system and initiates the heartbeat. Evolution of the vertebrates has resulted in a gradual reduction in the size of the sinus venosus. The pacemaker function, however, remains unchanged.

skeletal muscle (*skel′ĕ-tal mus′el*)
Muscle in which a regular arrangement of contractile filaments gives a banded appearance when viewed through the compound microscope. Skeletal muscle is voluntary and usually associated with the skeleton.

smooth muscle (*smooth mus′el*)
Muscle that lacks a regular arrangement of contractile filaments, that is involuntary, and that usually is associated with visceral structures.

species (*spe′shez*)
The fundamental unit of classification of living organisms.

spindle fibers (*spin′del fi′berz*)
Microtubles that originate at the centriole and attach at the centromere of individual chromosomes.

submucosa (*sub″mu-ko′sah*)
A layer of connective tissue that lies below the mucosa of an organ.

sycon (*si′kon*)
A sponge body form characterized as having choanocytes lining radial canals (Porifera).

Symphyla (*sim-fil′ah*)
A class of centipedelike arthropods whose members occupy soil and leaf mold.

symplesiomorphy (*sim-ples′e-o-mor′fe*)
Characters common to all members of the lineage. Symplesiomorphies indicate that all members of a study group are related. Because they are present in ancestors of all members of the study group, they cannot be used to distinguish between members of the study group.

synapomorphy (*sin-ap′o-mor′fe*)
A character that has arisen within a group since it arose from a common ancestor. It is also called a derived character. A synapomorphy can be used to indicate relatedness within a group.

synapse (*sin′aps*)
The region where an impulse is conducted from an axon of one neuron to a dendrite of a second neuron, or from an axon of a neuron to a muscle cell.

synthesis (*sin′the-sis*)
The portion of interphase of the cell cycle during which DNA is replicated so that each chromosome will consist of two identical chromatids.

systematics (*sis-tem-at′iks*)
The study of the classification and phylogeny of animals.

T

tagmatization (*tag″mah-ti-za′shun*)
The specialization of body regions of a metameric animal for specific functions, usually, feeding and sensory functions (head tagma); locomotion (thoracic tagma); and visceral functions (abdominal tagma).

telencephalon (*tēl″en-sef′al-on*)
The portion of the embryonic vertebrate brain that develops into cerebral hemispheres.

testis (*tes′tis*)
The primary male reproductive organ. A sperm-producing organ.

tissue (*tish′u*)
A group of similar cells that cooperate in a common function.

Trematoda (*trem′ah-tōd-″ah*)
The class of Platyhelminthes whose members are characterized as being leaflike or cylindrical, having oral and ventral suckers, and having a nonciliated, syncytial tegument. Adults are parasites of vertebrates. The flukes.

Trilobitamorpha (*tri″lo-bit′a-mor′fah*)
A subphylum of extinct arthropods that is believed to be ancestral to many (if not all) arthropods. Trilobites had a body divided into three longitudinal lobes as well as head, thoracic, and abdominal tagmata; compound eyes; and biramous appendages.

triploblastic (*trip′lo-blast″ik*)
Having a body derived from three embryonic layers: ectoderm, mesoderm, and endoderm. All animals above the Cnidaria are triploblastic.

trochophore (*trok′o-for″*)
A name given to the ciliated, free-swimming larval stages of some annelids and some molluscs.

Turbellaria (*tur″bel-ar′e-ah*)
The class of Platyhelminthes whose members are free-living, have a ciliated epidermis containing rhabdites, and have a ventral mouth opening.

U

Urochordata (*ūr″o-kord-a′tah*)
A subphylum of chordates in which the adults are sessile or occasionally planktonic and are enclosed in a cellulose tunic. A notochord, a nerve cord, and a postanal tail are present only in free-swimming larvae. The tunicates.

V

vacuole (*vak′u-ōl*)
A membrane-bound vesicle within the cytoplasm of a cell. It is used for storage and transport.

vein (*vān*)
A blood vessel that carries blood toward the heart.

veliger (*vēl′ĭ-jer*)
The second free-swimming larval stage of many molluscs. It develops from the trochophore and develops rudiments of the shell, visceral mass, and head-foot before settling to the substrate and undergoing metamorphosis.

ventricle (*ven′tri-kl*)
A fluid-filled cavity. The ventricles of the brain are filled with cerebrospinal fluid. The ventricles of the heart are filled with blood. This term is also used to refer to the muscle that pumps blood from the heart to the systemic circulation of an animal.

venule (*ven′ūl*)
A blood vessel that carries blood from a capillary bed to a vein.

Vertebrata (*ver″te-brah′tah*)
The chordate subphylum whose members have vertebrae that provide primary axial support and protect the nerve cord. The skeleton is modified anteriorly into a skull for protection of the brain. Also Craniata, fishes, amphibians, reptiles, birds, and mammals.

visceral mass (*vis′er-al mas*)
The region of a mollusc's body that contains visceral organs.

viviparous (*viv-ip′er-us*)
An animal in which young develop within the reproductive tract of the female and are nourished directly by the female.

Z

zygote (*zi′gōt*)
The cell that results from the fusion of gametes of opposite mating types.

Credits

Photographs

Chapter 1
Figure 1.1b: Courtesy of Carl Zeiss, Inc., Thornwood, NY; **1.2a, b:** Courtesy of Reichert Scientific Instruments; **1.5:** Courtesy of Bausch and Lomb, Inc.

Chapter 2
Figure 2.2b: © Gordon Leedale/Biophoto Associates; **2.3a, b:** © D. Fawcett/Visuals Unlimited; **2.4:** © Gordon Leedale/Biophoto Associates; **2.5:** Courtesy of Dr. Paul Heidgar; **2.6:** Courtesy of T. Tanaka; **2.7a:** Courtesy of Dr. Keith Porter; **2.8:** Courtesy of James Burbach, University of South Dakota, School of Medicine; **2.9a, b:** © L.E. Roth, Y. Shigenaka, and D.J. Pihlaja/Biological Photo Service; **2.10a:** Courtesy of Jan Lofberg, Institute of Zoology, Upsala University, Sweden; **2.10b:** Courtesy of J. Andre and E. Favret-Fremiet, Institute of Zoology, Upsala University, Sweden; **2.11a:** © Gordon Leedale/Biophoto Associates; **2.12:** Courtesy of Joseph Viles, Iowa State University; **2.13a:** © Dwight Kuhn; **2.13b:** © Ed Reschke; **2.13c:** © B.F. King/ Biological Photo Service; **2.14b, 2.15b, 2.16b, 2.17b:** © Ed Reschke.

Chapter 3
Figure 3.1a-c: © Kunkel/Phototake; **3.7a, b:** © Carolina Biological Supply Company/ Phototake.

Chapter 4
Figure 4.2a, b: © Steve Miller.

Chapter 5
Figure 5.1a: Courtesy of William Byrd, Louisana State University; **5.1b:** Courtesy of Gerald Schatten, Florida State University; **5.1c, d:** Reprinted with permission from M. Tegner, "Sea Urchin Sperm-Egg Interaction Studied with the Scanning Electron Microscope" in *Science*, 179:687, February 1973. Copyright © 1973 by American Association for the Advancement of Science, **5.2a-m:** © Carolina Biological Supply/ Photolake **5.2n:** Courtesy of Leland Johnson; **5.4a-d:** © Carolina Biological Supply/Photolake.

Chapter 6
Figure 6.2: © Bob Pool/Tom Stack & Associates; **6.3a:** © R. Calentine/Visuals Unlimited; **6.3b:** © David M. Dennis/Tom Stack & Associates; **6.4:** © D. Wilder/Tom Stack & Associates; **6.5:** © G. Meszaros/Visuals Unlimited; **6.6b, 6.7:** © William H. Amos; **6.8:** © Stephen Dalton/Photo Researchers, Inc; **6.9:** Courtesy of Douglas A. Craig, Department of Biological Sciences, University of Alberta, Edmonton; **6.10:** © Lawrence Pringle/Photo Researchers, Inc.

Chapter 8
Figure 8.1b: © Paul W. Johnson/Biological Photo Service; **8.2a, 8.3a, 8.4a:** © Carolina Biological Supply/Photolake **8.5b:** © M.I. Walker/Photo Researchers, Inc.; **8.8:** Courtesy Barbara Grimes; **8.9:** © Bruce Russell/BioMedia Associates; **8.10:** © Carolina Biological Supply/Photolake.

Chapter 9
Figure 9.4, 9.5: Courtesy of Dr. Louis De Vos; **9.6:** Courtesy Leland Johnson.

Chapter 11
Figure 11.7: Courtesy of Fred Whittaker; **11.8:** From J.E. Ubelaker, V.F. Allison and R.D. Specian, *Journal of Parasitology* 59:667-671, © American Society of Parasitologists.

Chapter 12
Figure 12.6: © Carolina Biological Supply/ Photolake **12.8:** (photo) © Steve Miller.

Chapter 13
Figure 13.2a: © Rick Poley Photography, Inc.; **13.7c:** © Ed Reschke; **13.8:** © Carolina Biological Supply/Photolake.

Chapter 14
Figure 14.1, 14.2: Courtesy of Betsey Brown; **14.5a:** © John D. Cunningham/Visuals Unlimited; **14.6, 14.8a:** © Carolina Biological Supply/Photolake.

Chapter 15
Figure 15.1: © Michael DiSpezio.

Chapter 16
Figure 16.1: © Carolina Biological Supply/Photolake **16.5:** © John Cunningham/ Visuals Unlimited; **16.6:** © Carolina Biological Supply/Photolake **16.7:** Courtesy of the Museum of New Zealand Te Papa Tongarewa.

Chapter 17
Figure 17.3a: © Runk/Schoenberger/Grant Heilman; **17.4a, b:** © Carolina Biological Supply/Photolake.

Chapter 18
Figure 18.1b: From Kessel, R.G. und Kardon, R.H. *Tissues and Organs: A Text-Atlas of Scanning Electron Microscopy.* © 1979 by W.H. Freeman and Company/Photolake **18.2, 18.3, 18.6b:** © Carolina Biological Supply/Photolake **18.7:** Courtesy of Michael Reedy; **18.9b:** © Carolina Biological Supply/Photolake **18.10b, 18.11a, b:** Courtesy of Stuart Fox.

Chapter 19
Figure 19.3: © Carolina Biological Supply/ Photolake **19.4b:** Courtesy of J.D. Robertson; **19.5:** © John Cunningham/Visuals Unlimited; **19.6:** Courtesy of R.B. Szamier.

Chapter 20
Figure 20.1: From R.G. Kessel and C.Y. Shih, *Scanning Electron Microscopy in Biology*, © Springer-Verlag; **20.2:** © D.M. Phillips/Visuals Unlimited.

Chapter 21
Figure 21.2b: Courtesy of Dr. G.M. Hughes; **21.3a, b:** © Carolina Biological Supply/ Photolake **22.3a, b:** From R.B. Chiasson, *Laboratory Anatomy of the White Rat*, 4th edition, 1980, Wm. C. Brown Communications.

Chapter 22
Figure 22.4a, b: From R.G. Kessel and R.H. Kardon, *Tissues and Organs; A Text-Atlas of Scanning Electron Microscopy*, © 1979 by W.H. Freeman.

Chapter 23
Figure 23.4: © Carolina Biological Supply/ Photolake **23.5a, b:** From R.G. Kessel and R.H. Kardon, *Tissues and Organs; A Text-Atlas of Scanning Electron Microscopy*, © 1979 by W.H. Freeman.

Chapter 24
Figure 24.5: From R.B. Chiasson, *Laboratory Anatomy of the White Rat*, 4th edition, 1980, Wm. C. Brown Communications; **24.6:** © Manfred Kage/Peter Arnold, Inc.; **24.8a-c:** © Steve Miller.

Plates
Plates 1-19: © Steve Miller.

Line Art

Preface

Figure 001a, b: From Charles F. Lytle and J. E. Wodsedalek, *General Zoology Laboratory Guide, Complete Version*, 11th ed. Copyright © 1991 Wm. C. Brown Communications, Inc., Dubuque, Iowa. All rights reserved. Reprinted by permission. **Figure 001c:** From Charles F. Lytle and J. E. Wodsedalek, *General Zoology Laboratory Guide, Complete Version*, 11th ed. Copyright © 1991 Wm. C. Brown Communications, Inc., Dubuque, Iowa. All rights reserved. Reprinted by permission.

Chapter 1

Figure1.3a-d: From Warren D. Dolphin, *Biology Laboratory Manual*, 3d ed. Copyright © 1992 Wm. C. Brown Communications, Inc., Dubuque, Iowa. All rights reserved. Reprinted by permission. **Figure 1.4a-c:** From Warren D. Dolphin, *Biology Laboratory Manual*, 3d ed. Copyright © 1992 Wm. C. Brown Communications, Inc., Dubuque, Iowa. All rights reserved. Reprinted by permission. **Figure 1.4d:** From Warren D. Dolphin, *Biology Laboratory Manual*, 2d ed. Copyright © 1987 Wm. C. Brown Communications, Inc., Dubuque, Iowa. All rights reserved. Reprinted by permission.

Chapter 2

Figure 2.1: From Stephen A. Miller and John P. Harley, *Zoology*, 3d ed. Copyright © 1994 Wm. C. Brown Communications, Inc., Dubuque, Iowa. All rights reserved. Reprinted by permission. **Figure 2.2a:** From S. J. Singer and G. L. Nicholson, "The Fluid Mosaic Model of the Structure of Cell Membrane" in *Science*, 175:720–731, 1972. Copyright © 1972 American Association for the Advancement of Science, Washington, DC. Reprinted by permission of the publisher and authors. **Figure 2.7b:** From Eldon D. Enger, et al., *Concepts in Biology*, 6th ed. Copyright © 1991 Wm. C. Brown Communications, Inc., Dubuque, Iowa. All rights reserved. Reprinted by permission. **Figure 2.11b:** From Leland G. Johnson, *Biology*, Copyright © 1986 Wm. C. Brown Communications, Inc., Dubuque, Iowa. All rights reserved. Reprinted by permission. **Figure 2.13a,b:** From John W. Hole, Jr., *Human Anatomy and Physiology*, 5th ed. Copyright © 1990 Wm. C. Brown Communications, Inc., Dubuque, Iowa All rights reserved. Reprinted by permission. **Figure 2.13c:** From Leland G. Johnson, *Biology*, 2d ed. Copyright © 1986 Wm. C. Brown Communications, Inc., Dubuque, Iowa. All rights reserved. Reprinted by permission. **Figure 2.14:** From John W. Hole, Jr., *Human Anatomy and Physiology*, 5th ed. Copyright © 1990 Wm. C. Brown Communications, Inc., Dubuque, Iowa. All rights reserved. Reprinted by permission. **Figure 2.15a:** From John W. Hole, Jr., *Human Anatomy and Physiology*, 3d ed. Copyright © 1984 Wm. C. Brown Communications, Inc., Dubuque, Iowa. All rights reserved. Reprinted by permission. **Figure 2.16a:** From John W. Hole, Jr., *Human Anatomy and Physiology*, 3d ed. Copyright © 1984 Wm. C. Brown Communications, Inc., Dubuque, Iowa. All rights reserved. Reprinted by permission. **Figure 2.17a:** From John W. Hole, Jr., *Human Anatomy and Physiology* 5th ed. Copyright ©

1990 Wm. C. Brown Communications, Inc., Dubuque, Iowa. All rights reserved. Reprinted by permission.

Chapter 3

Figure 3.5: From Stuart Ira Fox, *Human Physiology*, 4th ed. Copyright © 1993 Wm. C. Brown Communications, Inc., Dubuque, Iowa. All rights reserved. Reprinted by permission. **Figure 3.6:** From John W. Hole, Jr., *Human Anatomy and Physiology*, 6th ed. Copyright © 1993 Wm. C. Brown Communications, Inc., Dubuque, Iowa. All rights reserved. Reprinted by permission. **Figure 3.8:** From Leland G. Johnson, *Biology*, 2d ed. Copyright © 1987 Wm. C. Brown Communications, Inc., Dubuque, Iowa. All rights reserved. Reprinted by permission.

Chapter 4

Figure 4.1: Reprinted with the permission of Carolina Biological Supply Company.

Chapter 8

Figure 8.4b: From Charles F. Lytle and J. E. Wodsedalek, *General Zoology Laboratory Guide, Complete Version*, 11th ed. Copyright © 1991 Wm. C. Brown Communications, Inc., Dubuque, Iowa. All rights reserved. Reprinted by permission. **Figure 8.5a:** From Charles F. Lytle and J. E. Wodsedalek, *General Zoology Laboratory Guide, Complete Version*, 11th ed. Copyright © 1991 Wm. C. Brown Communications, Inc., Dubuque, Iowa. All rights reserved. Reprinted by permission. **Figure 8.7:** From Charles F. Lytle and J. E. Wodsedalek, *General Zoology Laboratory Guide, Complete Version*, 11th ed. Copyright © 1991 Wm. C. Brown Communications, Inc., Dubuque, Iowa. All rights reserved. Reprinted by permission.

Chapter 9

Figure 9.1a, b: From Stephen A. Miller and John P. Harley, *Zoology*, 3d ed. Copyright © 1994 Wm. C. Brown Communications, Inc., Dubuque, Iowa. All rights reserved. Reprinted by permission. **Figure 9.2a-c:** From Charles F. Lytle and J. E. Wodsedalek, *General Zoology Laboratory Guide, Complete Version*, 11th ed. Copyright © 1991 Wm. C. Brown Communications, Inc., Dubuque, Iowa. All rights reserved. Reprinted by permission. **Figure 9.3a-c:** From J. E. Wodsedalek and Charles F. Lytle, *General Zoology Laboratory Guide, Complete Version*, 8th ed. Copyright © 1981 Wm. C. Brown Communications, Inc., Dubuque, Iowa. All rights reserved. Reprinted by permission.

Chapter 10

Figure 10.1: From Leland G. Johnson, *Biology*, 2d ed. Copyright © 1986 Wm. C. Brown Communications, Inc., Dubuque, Iowa. All rights reserved. Reprinted by permission. **Figure 10.2:** From Stephen A. Miller and John P. Harley, *Zoology*, 3d ed. Copyright © 1994 Wm. C. Brown Communications, Inc., Dubuque, Iowa. All rights reserved. Reprinted by permission. **Figure 10.3:** From Charles F. Lytle and J. E. Wodsedalek, *General Zoology Laboratory Guide, Complete Version*, 11th ed. Copyright © 1991 Wm. C. Brown Communications, Inc., Dubuque, Iowa. All rights reserved. Reprinted by permission.

Figure 10.4: From Charles F. Lytle and J. E. Wodsedalek, *General Zoology Laboratory Guide, Complete Version*, 11th ed. Copyright © 1991 Wm. C. Brown Communications, Inc., Dubuque, Iowa. All rights reserved. Reprinted by permission. **Figure 10.5:** From J. E. Wodsedalek and Charles F. Lytle, *General Zoology Laboratory Guide, Complete Version*, 8th ed. Copyright © 1981 Wm. C. Brown Communications, Inc., Dubuque, Iowa. All rights reserved. Reprinted by permission. **Figure 10.6a, b:** From Charles F. Lytle and J. E. Wodsedalek, *General Zoology Laboratory Guide, Complete Version*, 11th ed. Copyright © 1991 Wm. C. Brown Communications, Inc., Dubuque, Iowa. All rights reserved. Reprinted by permission. **Figure 10.7:** From Charles F. Lytle and J. E. Wodsedalek, *General Zoology Laboratory Guide, Complete Version*, 11th ed. Copyright © 1991 Wm. C. Brown Communications, Inc., Dubuque, Iowa. All rights reserved. Reprinted by permission. **Figure 10.8:** From Charles F. Lytle and J. E. Wodsedalek, *General Zoology Laboratory Guide, Complete Version*, 11th ed. Copyright © 1991 Wm. C. Brown Communications, Inc., Dubuque, Iowa. All rights reserved. Reprinted by permission.

Chapter 11

Figure 11.3: From J. E. Wodsedalek and Charles F. Lytle, *General Zoology Laboratory Guide, Complete Version*, 8th ed. Copyright © 1981 Wm. C. Brown Communications, Inc., Dubuque, Iowa. All rights reserved. Reprinted by permission. **Figure 11.5:** From Leland G. Johnson, *Biology*, 2d ed. Copyright © 1986 Wm. C. Brown Communications, Inc., Dubuque, Iowa. All rights reserved. Reprinted by permission.

Chapter 12

Figure 12.1: From Charles F. Lytle and J. E. Wodsedalek, *General Zoology Laboratory Guide, Complete Version*, 11th ed. Copyright © 1991 Wm. C. Brown Communications, Inc., Dubuque, Iowa. All rights reserved. Reprinted by permission. **Figure 12.2a-d:** From Leland G. Johnson, *Biology*, 2d ed. Copyright © 1986 Wm. C. Brown Communications, Inc., Dubuque, Iowa. All rights reserved. Reprinted by permission. **Figure 12.3:** From Charles F. Lytle and J. E. Wodsedalek, *General Zoology Laboratory Guide, Complete Version*, 11th ed. Copyright © 1991 Wm. C. Brown Communications, Inc., Dubuque, Iowa. All rights reserved. Reprinted by permission. **Figure 12.4a-c:** From Charles F. Lytle and J. E. Wodsedalek, *General Zoology Laboratory Guide, Complete Version*, 11th ed. Copyright © 1991 Wm. C. Brown Communications, Inc., Dubuque, Iowa. All rights reserved. Reprinted by permission. **Figure 12.9a, b:** From Charles F. Lytle and J. E. Wodsedalek, *General Zoology Laboratory Guide, Complete Version*, 11th ed. Copyright © 1991 Wm. C. Brown Communications, Inc., Dubuque, Iowa. All rights reserved. Reprinted by permission.

Chapter 13

Figure 13.1: From Leland G. Johnson, *Biology*, 2d ed. Copyright © 1986 Wm. C. Brown Communications, Inc., Dubuque, Iowa. All Rights Reserved. Reprinted by permission.

Chapter 14

Figure 14.3a-c: From *Life of Invertebrates*, © 1979, W. D. Russell-Hunter.

Chapter 15

Figure 15.2: From Charles F. Lytle and J. E. Wodsedalek, *General Zoology Laboratory Guide, Complete Version*, 11th ed. Copyright © 1991 Wm. C. Brown Communications, Inc., Dubuque, Iowa. All rights reserved. Reprinted by permission. **Figure 15.3:** From Charles F. Lytle and J. E. Wodsedalek, *General Zoology Laboratory Guide, Complete Version*, 11th ed. Copyright © 1991 Wm. C. Brown Communications, Inc., Dubuque, Iowa. All rights reserved. Reprinted by permission. **Figure 15.4:** From Charles F. Lytle and J. E. Wodsedalek, *General Zoology Laboratory Guide, Complete Version*, 11th ed. Copyright © 1991 Wm. C. Brown Communications, Inc., Dubuque, Iowa. All rights reserved. Reprinted by permission. **Figure 15.8:** From Charles F. Lytle and J. E. Wodsedalek, *General Zoology Laboratory Guide, Complete Version*, 11th ed. Copyright © 1991 Wm. C. Brown Communications, Inc., Dubuque, Iowa. All rights reserved. Reprinted by permission. **Figure 15.9:** From Hickman/Roberts/Larson, *Integrated Principles of Zoology*, 11th ed. Copyright © 2001 The McGraw-Hill Companies. **Figure 15.10:** From Raven/Johnson, *Biology*, 3d/e. Copyright © 1999 The McGraw-Hill Companies.

Chapter 16

Figure 16.2: From Stephen A. Miller and John P. Harley, *Zoology*, 3d ed. Copyright © 1994 Wm. C. Brown Communications, Inc., Dubuque, Iowa. All rights reserved. Reprinted by permission.

Chapter 17

Figure 17.2: From Charles F. Lytle and J. E. Wodsedalek, *General Zoology Laboratory Guide, Complete Version*, 11th ed. Copyright © 1991 Wm. C. Brown Communications, Inc., Dubuque, Iowa. All rights reserved. Reprinted by permission. **Figure 17.3b:** From Stephen A. Miller and John P. Harley, *Zoology*, 3d ed. Copyright © 1994 Wm. C. Brown Communication, Inc., Dubuque, Iowa. All rights reserved. Reprinted by permission.

Chapter 18

Figure 18.1a: From John W. Hole, Jr., *Human Anatomy and Physiology*, 5th ed. Copyright © 1990 Wm. C. Brown Communication, Inc., Dubuque, Iowa. All rights reserved. Reprinted by permission. **Figure 18.4:** From Robert B. Chiasson and William J. Radke, *Laboratory Anatomy of the Perch*, 4th ed. Copyright © 1991 Wm. C. Brown Communications, Inc., Dubuque, Iowa. All rights reserved. Reprinted by permission. **Figure 18.5:** From Robert B. Chiasson, *Laboratory Anatomy of the White Rat*, 5th ed. Copyright © 1988 Wm. C. Brown Communications, Inc., Dubuque, Iowa. All rights reserved. Reprinted by permission. **Figure 18.6a:** From Stephen A. Miller and John P. Harley, *Zoology*, 3d ed. Copyright © 1994 Wm. C. Brown Communications, Inc., Dubuque, Iowa. All rights reserved. Reprinted by permission. **Figure 18.9a:** From John W. Hole, Jr., *Human Anatomy and Physiology*, 5th ed. Copyright © 1990 Wm. C. Brown Communications, Inc., Dubuque, Iowa. All rights reserved. Reprinted by permission. **Figure 18.10a:** From Stephen A. Miller and John P. Harley, *Zoology*, 3d ed. Copyright © 1994 Wm. C. Brown Communications, Inc., Dubuque, Iowa. All rights reserved. Reprinted by permission.

Chapter 19

Figure 19.2a, b: From John W. Hole, Jr., *Human Anatomy and Physiology*, 3d ed. Copyright © 1984 Wm. C. Brown Communication, Inc., Dubuque, Iowa. All rights reserved. Reprinted by permission. **Figure 19.4a:** From Warren D. Dolphin, *Biology Laboratory Manual*, 2d ed. Copyright © 1987 Wm. C. Brown Communications, Inc., Dubuque, Iowa. All rights reserved. Reprinted by permission. **Figure 19.7:** From Charles F. Lytle and J. E. Wodsedalek, *General Zoology Laboratory Guide, Complete Version*, 11th ed. Copyright © 1991 Wm. C. Brown Communications, Inc., Dubuque, Iowa. All rights reserved. Reprinted by permission. **Figure 19.9:** From Robert B. Chiasson and Ernest S. Booth, *Laboratory Anatomy of the Cat*, 8th ed. Copyright © 1989 Wm. C. Brown Communications, Inc., Dubuque, Iowa. All rights reserved. Reprinted by permission. **Figure 19.10:** From Robert B. Chiasson and Ernest S. Booth, *Laboratory Anatomy of the Cat*, 8th ed. Copyright © 1989 Wm. C. Brown Communications, Inc., Dubuque, Iowa. All rights reserved. Reprinted by permission. **Figure 19.11:** From Robert B. Chiasson and Ernest S. Booth, *Laboratory Anatomy of the Cat*, 8th ed. Copyright © 1989 Wm. C. Brown Communications, Inc., Dubuque, Iowa. All Rights Reserved. Reprinted by permission. **Figure 19. 12:** From Kent M. Van De Graaff, *Human Anatomy*, 3d ed. Copyright © 1992 Wm. C. Brown Communications, Inc., Dubuque, Iowa. All rights reserved. Reprinted by permission.

Chapter 20

Figure 20.4: From John W. Hole, Jr., *Human Anatomy and Physiology*, 3d ed. Copyright © 1984 Wm. C. Brown Communications, Inc., Dubuque, Iowa. All Rights Reserved. Reprinted by permission. **Figure 20.5:** From Charles F. Lytle and J. E. Wodsedalek, *General Zoology Laboratory Guide, Complete Version*, 11th ed. Copyright © 1991 Wm. C. Brown Communications, Inc., Dubuque, Iowa. All Rights Reserved. Reprinted by permission. **Figure 20.6:** From Laurence M. Ashley and Robert B. Chiasson, *Laboratory Anatomy of the Shark*, 5th ed. Copyright © 1988 Wm. C. Brown Communications, Inc., Dubuque, Iowa. All rights reserved. Reprinted by permission. **Figure 20.7:** From Kent M. Van De Graaff, *Laboratory Manual for Human Anatomy*, 3d ed. Copyright © 1992 Wm. C. Brown Communications, Inc., Dubuque, Iowa. All rights reserved. Reprinted by permission. **Figure 20.8:** From Kent M. Van De Graaff, *Laboratory Manual for Human Anatomy*, 3d ed. Copyright © 1992 Wm. C. Brown Communications, Inc., Dubuque, Iowa. All rights reserved. Reprinted by permission. **Figure 20.9:** From Robert B. Chiasson, *Laboratory Anatomy of the White Rat*, 5th ed. Copyright © 1988 Wm. C. Brown Communications, Inc., Dubuque, Iowa. All rights reserved. Reprinted by permission. **Figure 20.10:** From Robert B. Chiasson, *Laboratory Anatomy of the White Rat*, 5th ed. Copyright © 1988 Wm. C. Brown Communications, Inc., Dubuque, Iowa. All rights reserved. Reprinted by permission. **Figure 20.11:** From Robert B. Chiasson, *Laboratory Anatomy of the White Rat*, 5th ed. Copyright © 1988 Wm. C. Brown Communications, Inc., Dubuque, Iowa. All rights reserved. Reprinted by permission. **Figure 20.12:** From Stephen A. Miller and John P. Harley, *Zoology*, 3d ed. Copyright © 1994 Wm. C. Brown Communications, Inc., Dubuque, Iowa. All rights reserved. Reprinted by permission.

Chapter 21

Figure 21.4: From Kent M. Van De Graaff, *Human Anatomy*, 3d ed. Copyright © 1992 Wm. C. Brown Communications, Inc., Dubuque, Iowa. All rights reserved. Reprinted by permission.

Chapter 22

Figure 22.1: From Robert B. Chiasson, *Laboratory Anatomy of the White Rat*, 5th ed. Copyright © 1988 Wm. C. Brown Communications, Inc., Dubuque, Iowa. All rights reserved. Reprinted by permission. **Figure 22.2:** From Robert B. Chiasson, *Laboratory Anatomy of the White Rat*, 5th ed. Copyright © 1988 Wm. C. Brown Communications, Inc., Dubuque, Iowa. All rights reserved. Reprinted by permission.

Chapter 23

Figure 23.1: From Robert B. Chiasson, *Laboratory Anatomy of the White Rat*, 5th ed. Copyright © 1988 Wm. C. Brown Communications, Inc., Dubuque, Iowa. All rights reserved. Reprinted by permission. **Figure 23.2:** From Kent M. Van De Graaff, *Human Anatomy*, 2d ed. Copyright © 1989 Wm. C. Brown Communications, Inc., Dubuque, Iowa. All rights reserved. Reprinted by permission. **Figure 23.3:** From John W. Hole, Jr., *Human Anatomy and Physiology*, 6th ed. Copyright © 1993 Wm. C. Brown Communications, Inc., Dubuque, Iowa. All rights reserved. Reprinted by permission.

Chapter 24

Figure 24.1: From Laurence M. Ashley and Robert B. Chiasson, *Laboratory Anatomy of the Shark*, 5th ed. Copyright © 1988 Wm. C. Brown Communications, Inc., Dubuque, Iowa. All rights reserved. Reprinted by permission. **Figure 24.2:** From Laurence M. Ashley and Robert B. Chiasson, *Laboratory Anatomy of the Shark*, 5th ed. Copyright © 1988 Wm. C. Brown Communications, Inc., Dubuque, Iowa. All rights reserved. Reprinted by permission. **Figure 24.3:** From Robert B. Chiasson, *Laboratory Anatomy of the White Rat*, 5th ed. Copyright © 1988 Wm. C. Brown Communications, Inc., Dubuque, Iowa. All rights reserved. Reprinted by permission. **Figure 24.4:** From Warren D. Dolphin, *Biology Laboratory Manual*, 2d ed. Copyright © 1987 Wm. C. Brown Communications, Inc., Dubuque, Iowa. All rights reserved. Reprinted by permission.